NATO ASI Series

Advanced Science Institutes Series

A series presenting the results of activities sponsored by the NATO Science Committee, which aims at the dissemination of advanced scientific and technological knowledge, with a view to strengthening links between scientific communities.

The Series is published by an international board of publishers in conjunction with the NATO Scientific Affairs Division.

A Life Sciences B Physics	Plenum Publishing Corporation London and New York
C Mathematical and Physical Sciences D Behavioural and Social Sciences E Applied Sciences	Kluwer Academic Publishers Dordrecht, Boston and London
F Computer and Systems Sciences G Ecological Sciences H Cell Biology I Global Environmental Change	Springer-Verlag Berlin Heidelberg New York Barcelona Budapest Hong Kong London Milan Paris Santa Clara Singapore Tokyo

Partnership Sub-Series

1. Disarmament Technologies	Kluwer Academic Publishers
2. Environment	Springer-Verlag / Kluwer Academic Publishers
3. High Technology	Kluwer Academic Publishers
4. Science and Technology Policy	Kluwer Academic Publishers
5. Computer Networking	Kluwer Academic Publishers

The Partnership Sub-Series incorporates activities undertaken in collaboration with NATO's Cooperation Partners, the countries of the CIS and Central and Eastern Europe, in Priority Areas of concern to those countries.

NATO-PCO Database

The electronic index to the NATO ASI Series provides full bibliographical references (with keywords and/or abstracts) to about 50 000 contributions from international scientists published in all sections of the NATO ASI Series. Access to the NATO-PCO Database is possible via the CD-ROM "NATO Science & Technology Disk" with user-friendly retrieval software in English, French and German (© WTV GmbH and DATAWARE Technologies Inc. 1992).

The CD-ROM can be ordered through any member of the Board of Publishers or through NATO-PCO, B-3090 Overijse, Belgium.

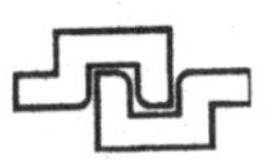

Series F: Computer and Systems Sciences, Vol. 166

Springer

Berlin
Heidelberg
New York
Barcelona
Hong Kong
London
Milan
Paris
Singapore
Tokyo

Operations Research and Decision Aid Methodologies in Traffic and Transportation Management

Edited by

Martine Labbé
Université Libre de Bruxelles
ISRO and SMG
Boulevard du Triomphe
B-1050 Brussels, Belgium

Gilbert Laporte
Université de Montréal
Centre de Recherche sur les Transports
CP 6128
Montréal H3C 3JT, Canada

Katalin Tanczos
Technical University of Budapest
Department of Transport Economics
Bertallan L. u. 2
H-1111 Budapest, Hungary

Philippe Toint
Université Notre Dame de la Paix
Département de Mathématiques
Rempart de la Vierge, 8
B-5000 Namur, Belgium

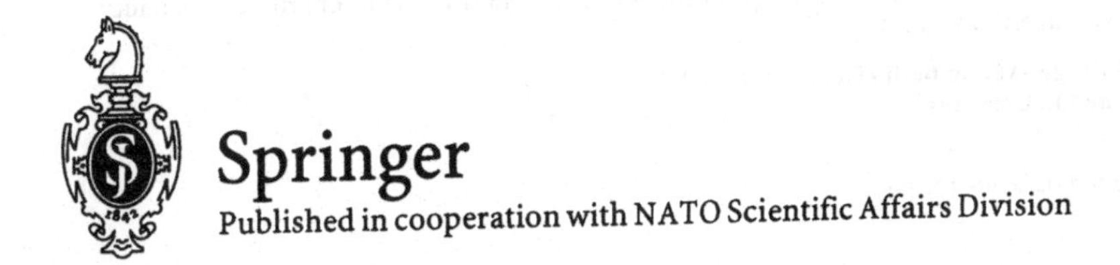

Proceedings of the NATO Advanced Study Institute on Operations Research and Decision Aid Methodologies in Traffic and Transportation Management held in Balatonfüred, Hungary, March 10–21, 1997

Library of Congress Cataloging-in-Publication Data applied for

Die Deutsche Bibliothek - CIP-Einheitsaufnahme

Operations research and decision aid methodologies in traffic and transportation management : [proceedings of the NATO Advanced Study Institute on Operations Research and Decision Aid Methodologies in Traffic and Transportation Management, held in Baltonfüred, Hungary, March 10 - 21, 1997] / ed. by Martine Labbé ... Publ. in cooperation with NATO Scientific Affairs Division. - Berlin ; Heidelberg ; New York ; Barcelona ; Budapest ; Hong Kong ; London ; Milan ; Paris ; Santa Clara ; Singapore ; Tokyo : Springer, 1998
(NATO ASI series : Ser. F, Computer and systems sciences ; Vol. 166)

ISBN 978-3-642-08428-7

ACM Computing Classification (1998): G.1.8, G.2.1, I.6.3

Printed in Germany

Printed on acid-free paper

Preface

Every one relies on some kind of transportation system nearly every day. Going to work, shopping, dropping children at school and many other cultural or social activities imply leaving home, and using some form of transportation, which we expect to be efficient and reliable. Of course, efficiency and reliability do not occur by chance, but require careful and often relatively complex planning by transportation system managers, both in the public and private sectors.

It has long been recognized that mathematics, and, more specifically, operations research is an important tool of this planning process. However, the range of skills required to cover both fields, even partially, is very large, and the opportunities to gather people with this very diverse expertise are too few. The organization of the NATO Advanced Studies Institute on "Operations Research and Decision Aid Methodologies in Traffic and Transportation Management" in March 1997 in Balatonfüred, Hungary, was therefore more than welcome and the group of people that gathered for a very studious two weeks on the shores of the beautiful lake Balaton did really enjoy the truly multidisciplinary and high scientific level of the meeting. The purpose of the present volume is to report, in a chronological order, the various questions that were considered by the lecturers and the students at the institute.

After a general introduction to the topic, the first week focused on issues related to traffic modeling, mostly in an urban context. The contributions by Cornélis and Toint and by Mahmassani explore the inner workings of a new generation of traffic models, where dynamism, advanced user information systems and route guidance are prominent, making the analysis of today's technologies possible and rigorous. The contribution by Papageorgiou considers similar issues, and provides methodological guidance for problems such as signal and parking control, ramp metering and variable speed limits. Ferris and Dirkse introduce the latest developments in tools for specifying and solving the traffic assignment problem, one of the most pervasive mathematical optimization problems arising in transportation modeling. Finally, Duchateau indicates how the use of tools like those discussed by the previous authors can be applied in the context of general transportation strategies in a large city, directly leading to the question of evaluating transportation scenarios, a subject covered by Tanczos.

The second week of the meeting was first concerned with issues of a combinatorial nature. Laporte provides an in-depth review of routing techniques, an important ingredient in many logistics applications, while facility location, another aspect of the same domain, is surveyed by Labbé. An interesting application of combinatorial techniques is presented by Caprara, Fischetti, Guida, Toth and Vigo, who describe solution techniques for the problem of establishing crew schedules and rosters for a large-scale railway operation. Uncertainty and other dimensions of the choice process were also discussed. Uncertainty is the main subject of Louveaux's contribution, while Bierlaire reviews classical and recent discrete choice models.

Several very interesting presentations by ASI students were also included in the program. We have selected three of these for the present volume. The first, by Cirillo, Daly and Lindveld, provides deeper insight into the modeling of discrete choice using random utility models. The paper by Bulavsky and Kalashnikov looks at the economics of transportation from the point of view of spatial competition, while Nonato discusses a combined crew and vehicle scheduling technique.

In summary, a great variety of topics were presented at an advanced level to an enthusiastic and sophisticated audience.

Of course, such a meeting could never have taken place without the help and dedication of organizers. It is my pleasure to acknowledge here the constant enthusiastic support of C. Nys and C. Houyoux from the Institute for International Education in Transport (IFIT, Belgium), and also of M. Jetzin and Zs. Somlyody, from the TRIVENT Conference Office (Hungary), without whom local organization would have been impossible. Special thanks are also due to NATO for funding the event. Last but not least, the help of G. Laporte, M. Labbé and K. Tanczos as co-editors, and that of Ch. Ippersiel as editorial assistant and A. Langley as corrector, made producing this volume not only possible but enjoyable.

Namur, July 1998

Ph. Toint,
ASI Director

Table of Contents

List of Contributors

M. Bierlaire
Intelligent Transportation Systems Program
Massachusetts Institute of Technology
Cambridge, USA

V. A. Bulavsky
Central Economics & Mathematics Institute
Nakhimovsky pr. 47, Moscow 117418
Russia

C. Cirillo
Politecnico di Torino
DITIC Area Transporti
C.SO Duca degli Abruzzi, 24
I-10129 Torino, Italy

E. Cornelis
Transportation Research Group
Facultés Universitaires
Notre-Dame de la Paix
Rue de Bruxelles, 61
B-5000 Namur, Belgium

A. Daly
Hague Consulting Group
Surinamestraat, 4
2585 GS Den Haag
The Netherlands

S. P. Dirkse
GAMS Development Corporation
1217 Potomac Street NW
Washington, D.C. 20007, USA

H. Duchateau
STRATEC
Boulevard A. Reyers, 156
B-1030 Brussels, Belgium

M. C. Ferris
University of Wisconsin – Madison
1210 West Dayton Street,
Madison, Wisconsin 53706, USA

V. V. Kalashnikov
Sumy University
Rimsky-Korsakov st. 2,
Sumy 244007, Ukraine

M. Labbé
Université Libre de Bruxelles
ISRO and SMG
Boulevard du Triomphe
B-1050 Brussels, Belgium

G. Laporte
Centre de Recherche sur les Transports
Université de Montréal,
Montréal, Canada

K. Lindveld
Hague Consulting Group
Surinamestraat, 4
2585 GS Den Haag
The Netherlands

F. Louveaux
Département de Méthodes
Quantitatives
Facultés Universitaires
Notre-Dame de la Paix
Rempart de la Vierge, 8
B-5000 Namur, Belgium

H. Mahmassani
Dept of Civil Engineering
ECJ 6.2
The University of Texas at Austin
Texas 78712, USA

M. Nonato
Istituto di Elettronica
Facolta di Ingegneria
Universita di Perugia
Via G. Duranti 1 S. Lucia Canetola
I-06131 Perugia, Italy

M. Papageorgiou
Dynamic Systems and
Simulation Laboratory
Department of Production
Engineering & Management
Technical University of Crete
Chania, Greece

K. Tanczos
Technical University of Budapest
Department of Transport Economics
Bertalan Lajos u. 2.
H-1111 Budapest, Hungary

Ph. L. Toint
Transportation Research Group
Facultés Universitaires
Notre-Dame de la Paix
Rue de Bruxelles, 61
B-5000 Namur, Belgium

P. Toth
Dipartimento di Elettronica
Informatica E Sistemistica
University of Bologna
Viale Risorgimento 2
I-40136 Bologna, Italy

Transportation Modelling and Operations Research: A Fruitful Connection

Philippe L. Toint

Transportation Research Group
Department of Mathematics
Facultés Universitaires ND de la Paix,
B-5000 Namur, Belgium
email: pht@math.fundp.ac.be

Summary. The purpose of this paper is twofold. It first aims at introducing the subject of transportation modelling and therefore at setting the stage for the other contributions of this volume. At the same time, it also aims at pointing out the many connections between transportation modelling and operations research, and the rich and insightful nature of these connections.

1. Introduction

Although traffic modelling and analysis do correspond to the basic curiosity about the behaviour of fellow human beings, the true utility of these techniques really shows up when one realizes that they are absolutely necessary for defining transportation strategies at the level of today's society as a whole. And the necessity of such strategies can no longer be discussed today, given the staggering societal cost of the current transportation system.

- Around 55,000 people are killed every year on the roads on the countries of the European Communities only. A further 1.7 million are injured and 150,000 more are handicaped for life. The human and societal cost of these accidents just cannot be measured. Their purely financial cost exceeds 50 millions ECU per year.
- The financial cost of traffic itself is estimated to be around 500 billions ECU per year for the same countries, most of which is caused by congestion and inefficient route choice.
- Pollution caused by transportation (such as private vehicles, trains, busses, planes,...) is a serious concern. Its annual cost in Europe only is over 5 billions ECU.

These numbers, extracted from official UE documents [1] reveal the urgent need for a better planning of transportation. This urgence is further reinforced by the commonly accepted forecast that transportation demand is likely to grow at a sustained rate. For instance, the level of motorisation of peri-urbain dwellers in France has grown from around 4 to 6 cars for 10 adults between 1975 and 1990 (see [48]), and this growth is not likely to slow down in the coming years...

Understanding the transportation phenomenon in depth is therefore important. This understanding must ideally lead to tools (the transportation models) allowing traffic forecasts when the transportation network and/or demand and/or behaviour is modified. They first allow a better understanding of the process itself, but their principal use is for transportation *management*, both in its daily operation (traffic regulation), and for long term planning (strategic planning and investment policies). The many questions raised in these contexts are of course of extremely diverse nature. The group in which the author works has, for instance, studied the global impact of the displacement of a high school and an hospital on the traffic in a small city, the long term effects of ring roads, the efficiency of relief routes, the feasibility of automated urban transportation and the impact on congestion of variable message signs for incident warning, just to cite a few cases.

Quite naturally, transportation models are adapted to this diversity of application fields, but they all share a common purpose: that of making possible the establishment of strategies for improving comfort and safety of users, in view of political, economical and environmental constraints.

The acutest transportation problems arise of course in big cities. Most of these suffer from the negative effects of an extremely dense traffic and are indeed faced (to a varying degree) with all the nuisances caused by daily commuting. The basic traffic modelling methodologies naturally appeared in this context, but are the subject of a necessary and permanent re-evaluation. Indeed, the introduction of new information technologies in both urban and regional traffic has already modified the behaviour of many transportation network users in many cities. Frequent radio bulletins indicating congested areas and incidents are now part of the daily commuter routine, and have substantial impact on his/her transportation strategy. Further and more radical changes are afoot: route guidance techniques (autonomous or interactive), city map updating via radio beacons, variable message signs for traffic control and parking information, road pricing systems are but a few of the techniques studied and applied in the contemporary management of transportation networks. In Europe, international programs, as EUREKA PROMETHEUS and the DGVII and DGXIII efforts, have been set up to analyze and optimize the use of these new information and control technologies in a quickly evolving environment and society, but this line of research and development is also very active in the United States and Japan, where similar programs are proceeding. More generally, it is safe to say that the design and management of transportation systems today require to exploit every possible advanced and innovative technique with a potential role in this domain.

Clearly, operations research is a field which provides such techniques. It has been and still is one of the most active areas of development in mathematics over the past thirty years. Moreover, its object shares much of the preoccupations of transportation system analysts: the main idea is to provide a rigorous and constructive (algorithmic) analysis of a large number of management problems, of varying degree of abstraction, from the very practical application to the more theoretical aspects. The progress in this domain have been truly remarkable, especially in sectors covering both continuous and discrete optimization problems (see the collection of Handbooks published by North-Holland, and more particularly [56], for a global view of the field).

It is thus natural to attempt to explore the interface between transportation modelling and operations research and it is the purpose of this paper to point out some of the fruitful connections between these domains. This is not the first such enterprise (see [34], for instance) and the ambition is not of to be exhaustive, but rather to provide a broad framework in which the other papers can be situated and motivated, albeit in a relatively broad manner. The methodology chosen is to develop the main ideas of transportation modelling and to indicate, when suitable, the many connections with operations research. It would also be possible to proceed in the other direction, that is to provide links to transportation modelling from a development of the main ideas in operations research. We have chosen the first approach for its more practical nature and its more immediate motivation.

Section 2 considers, in a first stage, the context in which transportation modelling is used. Section 3 then covers some basic concepts in this domain, already providing some bridges with operations research. These bridges are even more apparent in Section 4, which is concerned with the description of the main methodological approaches. Some conclusions and perspectives are finally mentioned in Section 5.

2. The context of traffic modelling

2.1 Political evaluation

The first major arena where transportation modelling is widely used is that of the political evaluation of transportation policies at city, regional, national or even European level.

As alluded to in the introduction, transportation problems arise in the everyday life of most citizens. Any policy with a strong impact on the daily "way of life" of so many voters clearly warrants a careful evaluation from the political point of view. Furthermore, as the decision issues become more and more urgent and complex, decision makers are less and less willing to take strong options on the basis of their understanding alone and rely on (hide behind?) expert advice in a steadily larger degree.

Very importantly, the dialog between transportation modelling, as a technical field, and both the public and the decision makers requires a real professional skill. The paper by Duchateau in this volume will illustrate these points from the point of view of a practitioner.

2.2 Technological assessment

Transportation models are also used in technology assessment. Indeed, the technological development in transportation is often costly, which makes a sound economical evaluation of its effects and market crucial. The preoccupation is therefore different here from that in the previous paragraph: instead of considering the effects of transportation policies at a society-wide level, the impact of transportation related techniques is now considered from the point of view of specific actors (productors, implementation organizers) within the society.

The areas of transportation technique concerned by these considerations broadly fall in the following categories:

- *Infrastructural design.* Probably the oldest field of application for transportation modelling is that of improvement and updating of transport infrastructures. Measuring the impact on traffic of network modifications, both in private and public transport, is indeed a very common application. The diversity of situations is nevertheless very large: motorways and urban roads, trains, busses, cars, cyclists, pedestrian routing, general infrastructure maintenance and all combinations thereof (multimodal transportation systems) provide a wide domain of investigation.
- *Technological advances in vehicles.* A number of applications have been made in relation with today's technological advances in vehicles. Started many years ago with the "cruising" features (mostly in american cars), the innovations giving the vehicle some intelligence and more acute perception of its environment have flourished. Gap detection and warning, automatic speed adjustment, synchronized driving, all have their effect on the general traffic, not to mention the simple improvement in car performances.
- *Traffic management and control.* Besides the problems posed by the availability to users of more sophisticate technology in their vehicle, the problems of traffic management and control also constitute a large class of applications where transportation modelling and associated methodologies play an important role: traffic lights settings, traffic control measures as speed enforcement or ramp metering on motorway access are just a few of the possible interesting topics in this category.
- *Technological advances in communications.* Possibly the greatest challenge in today's modelling arises from the advent of advanced information systems for transportation users. These developments are based on the premice that informing transportation users modifies their behaviour. The behavioural modification, at the user's level, is then reckoned to cause significant change in the global transportation scene at city-wide level.
  Of course, the problem is then to infer these global changes from the sometimes rather detailed knowledge of the information distributed. Questions of interest are then related to the broadness of information distribution, its accuracy, reliability, timeliness and method of transmission. The attitude of users with respect to this information is also a much studied subject.

- *Fleet management.* The daily or strategic management of a private or public fleet of vehicles is also an important source of practical transportation modelling problems of true economic importance. They typically arise in freight distribution systems, routing of community services as garbage collection, school bussing, fire brigade or urgent medical care. They also arise in modern public transportation systems designed for handling variable demand by adapting their route definition, either on a daily basis on even in real time. Finally, such problems also arise in intelligent guidance systems, in particular those designed to bring the car users to free parking areas or to park-and-ride centres.
- *Regulations in the transport arena.* Finally, the establishment of new regulations dealing with transportation also raises the question of their impact on traffic. One immediately thinks of parking policies, scheduling of goods delivery in city centers, road pricing or speed limits enforcement on motorways.

In many of these sectors, strong emphasis is put by the user on the reliability aspects of the involved models.

2.3 Land use strategies

We also stress here the intimate relationship between transportation planning and land-use regulations. Most transportation experts are very well aware of this relationship, but it remains a difficult task to integrate both approaches in coherent, practical and politically viable policies. The problems involved are the effect of transportation infrastructure on the localization of human activities and also that of human activities on travel demand and thus on transportation networks. This area provides a first connection between operations research and transportation modelling: the problem of optimizing the location of various activities, ranging from goods' distribution centres to hospitals, from manpower intensive industries to intermodal platforms, has been indeed widely studied in both fields (see [28] or [47], for instance). Optimal location problems are the subject of the paper by Labbé in this volume.

Of particular interest are the interactions of the land use strategies and transportation systems in time. The dynamics of their interrelated evalutions is a fascinating subject, in which progress has already been obtained (see [69], [59], [60], for example), but where very active research is still ongoing (see [17] or [42]). The scope for applying operations research techniques in this direction is obviously very important.

2.4 Environmental analysis

A last domain of application for transportation models is that of environmental analysis and the complex question of designing transportation systems that are environmentally sustainable. The idea is that a number of nuisances generated by traffic are somehow proportional to traffic itself, and thus that traffic forecasts might help in predicting the evolution of these nuisances. Noise, gas emission and the related atmospheric modifications, vibrations and chemical effects of winter road salting are part of a set of preoccupations of growing importance in western societies.

This type of application is the subject of intensive study. It is complicated by the fact that the relation between pollution and traffic is often complex and not always fully understood: this leaves room for considerable debate and substantial methodological improvements. For some recent work, see [2] and other papers in the same volume.

An important question for a sound environmentally conscious transportation policy is that of generating cooperation between various agencies, regions or even countries on the effect of pollution generated by international traffic. This remains a very open subject, although possible approaches have been considered in [31] or [32] and the references therein.

3. The basics of transportation modelling

3.1 The spatial dimension

3.1.1 Zones and centroids. Like all models, transportation models considerably simplify reality. For instance, it is nearly always impossible to handle a level of detail where trips can be specified from building to building or from address to address in a city. It is thus customary to define *zones* that cover the considered area, the population and activity within these zones then being concentrated in a point, *the zone centroid.* Trips are then considered from an *origin zone* to a *destination zone.*

It is important to note at this stage that the definition of these zones is intrinsically tied with the availability of data. Indeed, if one wishes to estimate the transportation demand from or to one of the zones, one must clearly know some of their characteristics: population, sociological structure, attractivity for various purposes and accessibility, for example. It is therefore frequent that the very definition of the zones explicitly takes the availability of these informations into account, or, at least, the cost and effort necessary for obtaining them.

Besides the problems linked to data collection, other factors also influence the definition of zones. The purpose of the study being made may also suggest some natural aggregation. For instance, if one considers the traffic perturbations caused by the opening of a new shopping mall, it is clear that the immediate surroundings of the mail should be modelled with a certain amount of detail, while a more and more aggregate description might be adequate when the distance to the mall increases. Depending on the application, a city as Brussels could count from twenty zones to more than a thousand...

3.1.2 The transportation network. Once the definition of zones established, it is then possible to specify the transportation network itself, as the set of links between these zones. There will generally be more than one subnetwork, each corresponding to a different *transportation mode*: roads for cars, bus lines, cycle lanes etc. Such a complex network is then called a *multimodal* transportation network.

As for the zone's definition, different considerations influence the network description to be retained in the model and its level of detail. Availability of data about the network links, study purpose and integration of the network with the zone typically come into play. It is, in fact, rather fallacious to consider the definition of zones and the network specification as two successive and distinct phases: they are intimately linked to each other and should be integrated in a global and simultaneous approach.

The modelling of an urban area using these techniques is usually represented by a mathematical object called an *oriented graph*: *nodes* (points in the plane, for example) are linked together by "arrows" indicating the possible direction of travel (the *links*). Zones are then associated with nodes and network links to arcs.

3.1.3 Local or specific models. The above paragraphs focussed on modelling transportation at the level of a city or a region. But this is not the only type of modelling and much more local models are also of importance. The typical situation is when one desires to analyze, forecast or control a limited set of connected intersections. In this case, the definition of the zone under study often has nearly immediate implications in the definition of centroids and links: links are given by the actual roads and possible turns within the zone, while centroids and nodes are associated to the entry/exit points of the zones and with the inner intersections.

Modelling at other levels is also common when dealing with another part of the transportation system. For instance, techniques for modelling the transportation demand do not necessarily proceed by building an explicit representation of the transportation network, although there is still some concept of zones in this case.

3.1.4 New technologies and spatial modelling. The impact of new advanced technologies in transportation has an important consequence on the spatial description of models. The main observation is that the object of study is no longer restricted to the movements of users in the transportation network, but also covers the exchange of information across its various parts: sophisticate route guidance or incident detection as well as relatively classical traffic light synchronization involve that exchange very clearly. Furthermore, the efficiency of advanced transportation technologies is often said to be dependent on the quality, reliability, accuracy of the exchanged information, and on its distribution method.

Typical examples of relevant factors are listed below, where we attempt to indicate, in each case, the possible connections with operations research.

- A first factor is the location of traffic monitoring equipment (the actual locations of detection and measurement equipment on the network of course conditions the final information in that it specifies the parts of the network about which information could be available). The optimal localisation of this equipment (in order to cover the largest or most interesting part of the transportation system) may be viewed as a location problem on a network.
- The second factor is the location of information distribution mechanisms (conditioning in turn the locations at which re-routing can occur). Again, this can be considered as a location problem, either of continuous nature (a radio transmitter with limited range may be localized nearly anywhere) or discrete (variable message signs must be placed at specific locations on the network).
- Finally, one can consider the actual interconnection pattern (topology) of monitoring devices, information processing centres and distribution mechanisms. This is a network design problem, a very widely studied class of combinatorial optimization applications.

Besides the classical transportation network, we therefore discover another network, which describes the information flow (the feedback loop) between the transportation network and its users. We call this network the *information network*. It is the author's view that explicit modelling of this information network is an important part of the present research challenges in this area. Of course, these information network models should be flexible, in order to adapt to the many different cases of potential interest. In particular, they should include the relevant factors mentioned above, but also parameters that influence their efficiency (communication links' or detectors' reliabilities, for instance) and thus also permit the modelling of information system breakdowns.

3.2 The time dimension

A fundamental distinction appears at this stage between two distinct model classes: *static* and *dynamic* models. As could be inferred from these terms, both approaches differ by the way in which they consider the traffic evolution in time.

Historically the first, *static models* describe traffic on a time period that is long enough to ensure that all travellers starting a trip in this period also reach their destination within the same period.

One therefore obtains a considerable conceptual simplification, because it is no longer necessary to analyze traffic on the network links at successive moments. It is indeed sufficient to consider the cumulative value on the considered time period. This simplification is of course made at the expense of some realism, but the results obtained with this approach can however be adequate for some kinds of studies. (As the wear of road surface is roughly proportional to the total volume of traffic on the road, but is independent of its repartition in time, it is thus enough to know an approximation to this total volume for establishing a maintenance program for the city roads...).

Dynamic models explicitly introduce the passing of time within their mechanism. These models originate from the deficiency of static models in correctly describing route choice in peak period, where the traffic on the network links can significantly differ from its average value on the day, say. Moreover, not only the routes but also transportation modes and departure times can vary: dynamic models are thus *necessary* to describe transient traffic phenomena, as congestion experienced by commuters on the big cities' approaches.

Practically, these models are much harder to build because the quantity of needed parameters and results is usually much larger than for static ones. It is indeed necessary to make every quantity fully time dependent. The requested data collection as well as the a posteriori validation are therefore considerably more complex.

As most models for time dependent phenomena, dynamic traffic models can be classified into three broad categories, depending on the method they use for representing the passing of time:

- Continuous models, first, consider traffic parameters as continuous functions of time, and they aim at determining some of these functions (as traffic density, for instance) as a consequence of other such functions. It is in this context that Lighthill and Whitnam developed in 1955 [45] their famous model for a phenomenon well known by many road users: the "moving traffic shock waves" delimited by zones of slowing and zones of accelerating traffic.
- The discretized models, on the other hand, describe the time dependence of traffic parameters by representing the time in successive "slices" during which the transportation network has a constant behaviour. The studied effects (as congestion) therefore vary from one timeslice to the next. They are often calculated from laws that link their value at a given timeslice with other values from preceding timeslices.
  This technique has the further advantage to allow a progressive transition between static and dynamic models. A static model can indeed be considered as a dynamic model with only one (long) timeslice. The model is then made "dynamic" by increasing the number of timeslices and reducing their length. As a result, one obtains mixed models whose behaviour is similar to that of static ones for long timeslices (from 0 to 6am, for instance) but still allow for a finer time analysis at peak periods, when timeslices are shorter.
- A third way to handle the passing of time in models is the so-called "event driven" technique. Time is considered as a sequence of events, each of them implying some modification in the state of the model, with the possibility for an event to define one or more other events in its future. Such events might be traffic lights changes, arrival of vehicles at intersections or bus-

stops, a traveler's departure or arrival, network incidents or information transmission. This technique is thus especially adapted in the treatment of transient phenomena, since these are mostly caused by such "events" in the network, and to models involving some type of incident detection and warning.
Another advantage of this methodology is that it is most suited to exploiting parallel computer architecture, whenever available. Different events in different geographical areas can indeed be handled in a completely asynchronous fashion by different processors. This feature is far from being anecdotic, because the computational requirements of dynamic models is in general substantial, and techniques to speed processing up are very worthwhile in the framework of real time applications.

The distinction between continuous and discretized models is further blurred in practice because continuous models often use numerical techniques that rely on an internal time discretization: timeslicing is thus implicitly present even in purely continuous approaches. Similarly, dynamic models with time-slices may also be considered as "event-driven", where the only events are associated with the end of each time-slice.

Because much of advanced technologies in the transportation arena are geared at the dynamic behaviour of transportation users and at the control or analysis of transient phenomenons, dynamic transportation models seem unavoidable within this context. Network users indeed have clearly different strategies in peak hour as opposed to off-peak, in week-ends as opposed to week days, in day as opposed to night, and the list of examples can easily be made longer. Furthermore, techniques like intelligent incident detection and interactive or automated route guidance are only meaningful in a dynamic context. It is remarkable that, although appreciated by many experts, this conclusion has not always been applied in its full requirements.

We conclude this paragraph by noting that the difficulty of evolving from a static to a dynamic model varies very much with the specific area of transportation considered. If dynamic route choice models are becoming "of age", the transition is much more difficult in demand analysis or land-use interactions, for instance.

3.3 A paradigm for city-wide modelling: The four stages model

Once the decisions about the territory and time representation taken, the model is already well shaped: it contains zones and a network for travelling from one zone to another. We should then use this structure for describing the traffic itself. We now focus our attention to a very classical methodology for city- or region-wide modelling originally proposed by Wilson in the beginning of the seventies [72]. This method proceeds in four successive stages:

- *Traffic generation.* In the first stage, each centroid of the model is considered and the number of trips leaving the associated zone, a well as the

number of trips ending in the zone are determined. This is the *generation stage.*
This stage classically uses statistical techniques for predicting the "emissivity" of a zone, that is the number of trips that depart from that zone. These techniques are based on the repartition of inhabitants (households) in coherent socio-economical groups, whose displacement behaviour can be analyzed. Similarly, the "attractivity" of a zone, i.e. the number of trips ending in the zone, can be established for various causes of displacement (the *functions*: work, shopping, school, etc.). Data is typically obtained from special surveys.
- *Traffic distribution.* The second stage then consists in calculating the repartition of these trips in (origin, destination) pairs. One speaks of the *distribution stage*, whose result can be placed in a big table: each entry of the table contains a number of trips leaving the zone corresponding to the row of the table and ending in the zone corresponding to its column. This is the well-known *origin/destination matrix*, or O/D matrix.
- *Modal choice.* The third stage models the competition between the various transportation modes (private cars, train, busses, metro, cycles,...) and establishes, for each (origin, destination) pair, the proportion of the trips using the different modes.
- *Traffic assignment.* Finally, the fourth stage, called the *assignment stage*, determines the actual routes in the network that are being used for each of the trips and each of the modes during the considered modelling period.

This four stage methodology has often been criticized because of its sequential character. For example, the choice between car and bus for a trip may depend on the congestion level on the roads or in the public transportation system, level which is not known until after the fourth stage. Congestion can also encourage drivers to change their destination, or even, in extreme cases, to forego their trip altogether. More generally, the four *successive* stages are often integrated and interdependent in practice.

We note that we haven't specified typical methodologies for the last three of the four stages. A survey of these methodologies is the object of the next section.

We finally note that the four stage methodology can also be applied, even partly, in more local studies, like the analysis of a set of intersections. In such a case, one would typically estimate the O/D matrix from measurements, say, which correspond to Stage 2, and then assign the corresponding demand on the considered network, which correspond to Stage 4.

Unfortunately, using the four stage modelling approach is not always a reasonable methodology. For instance, if the problem under consideration contains user information systems, the inherent sequentiality of the technique is then completely unrealistic, because information flow from one of the four stages to previous ones is very common in this context. For example, home-based information systems on expected travel time using private or public transportation are designed to modify the modal choice (Stage 3) as a function of network performance (Stage 4). Congestion dependent road pricing schemes have a similar effect, but are also anticipated to alter the transportation demand (Stages 1 and 2) itself.

Although a number of dynamic models have been proposed for traffic assignment (Stage 4), very much less is known about dynamic city-wide demand models, and even less on the necessary integration of both. This is an area where "classical" transportation modelling truly needs improvement. The situation is less critical for more local studies, as there exist methodologies for estimating a dynamic demand in this case (see Section 4.6 below).

4. Methods in transportation modelling and their connections with Operations Research

We now examine some classical methods and concepts whose application in transportation modelling is very widespread. Again, we have no ambition to cover the field: the reader is referred to the excellent encyclopaedia [22] or to [8] for further analysis.

4.1 User's rationality and utility functions

One of the most important question in modelling human behaviour is to decide on what is considered to be the "motor" or driving force behind the choice of a particular action. This very central question will probably always be open for debate. However, practical models have to make some kind of choice at this level. The main dimension along which existing methodologies can be ranked is the assumption that transportation users make, at least to some degree, *rational choices*. It is commonly assumed that using a transportation network always implies a certain cost for the user (economists talk about "disutility"): not only travelling takes time, but it could also cost in several other ways. The ticket price for the bus or the metro, the petrol consumed by our vehicle, the number of stops along the route, the accident risks, the inconvenience and stress caused by congestion...can also add to the mere time spent to our perception of transportation costs. Society as a whole also supports other costs associated with transport, as pollution, security, insurances, medical care, noise and other nuisances caused by traffic. As a result, many transportation modelling techniques take the point of view that these costs should be minimized, which is a classical linear or, more commonly, nonlinear optimization problem. Of course, the problem may also involve constraints (such as satisfying demand or not loosing travellers on the network). In practical instances, the problem also depends on how exactly the costs are defined.

The notion of global cost associated with displacement is typically taken into account by the choice of a *cost function* or *performance function* that determines, for each link of the network, a "cost" as a function of the traffic on that link. Units for this cost vary from model to model, and are chosen in relation with the final purpose of the modelling. For example, the quantity of exhaust gas of a certain type can be used if one wishes to measure the pollution impacts, while transportation time can be used if one is interested at asserting the level of service of public transport... There is however a remarkable consistency between all these choices, irrespective of the units chosen: the cost function is typically increasing as traffic along the link increases, slowly first but faster and faster for higher traffic levels. For example, the travel time along a link is relatively independent of traffic for low traffic levels, but can increase substantially for more congested levels, up to the point where not a single vehicle can move because everybody is stuck in a traffic jam!

It should be noted that the choice of a single performance function associated with the parts of the transportation network implies a complete and uniform rationality of the users in the measure where each users minimizes his or her disutility. This assumption has often been criticized as irrealistic. One way to overcome the difficulty is to introduce some degree of stochasticity in the users' choice criteria. We will briefly discuss these aspects below.

Finally, optimizing a single "utility function" (even containing some stochastic component) often raises the question of comparing various goods or services of qualitatively different nature, a very difficult task. For instance, it is often necessary to estimate the monetary value of time (VOT) for users or users' classes. Some concepts in multi-objective optimization constitute an interesting alternative at this level (see [10], [43] or several papers in [8] for an example) and will be developed further in this volume by Tanczos.

4.2 Physical analogies

The first generation of concepts in transportation modelling is widely based on analogies with the physical world. Traffic is considered as a flow (see above) or an electrical current. Various methods have been proposed to exploit these analogies. We will briefly describe here two "physical" concepts that are still of major importance in methods for traffic distribution, the second stage of the classical four stage model.

The first idea is that transportation demand between two centroids might be modelled using the law of *gravitation*, in which the intensity of the attraction force between two bodies is proportional to the masses of these bodies and inversely proportional to the square of the distance separating them. Transposed in transportation demand models, the trip demand from a given centroid i to another j is then proportional to e_i, the "emissivity" of i, and to a_j, the "attractivity" of j, and also proportional to some decreasing "deterrence" function f of their (shortest) distance d_{ij}. More formally, the number of trips q_{ij} between centroids i and j is then approximated by

$$t_{ij} = k\ e_i a_j f(d_{ij}). \tag{4.1}$$

In the strict gravitational framework, one chooses $f(d_{ij}) = 1/d_{ij}^2$.

Another somewhat similar idea is to use the concept of *entropy* in a physical system. In this approach, the distribution model attempts to recover a particular trip distribution such that its likelihood is maximized amongst all equiprobable demand configurations. An easy mathematical development then shows that

$$t_{ij} = k\ e_i a_j e^{-\sigma d_{ij}}, \tag{4.2}$$

where σ is a free model parameter. We immediately note the strong link between the entropy and the gravitational approach (see [63] or [9], for instance, for a discussion of this relationship). One then typically distinguishes (see [16]) between "singly constrained" distribution models, where the emissivity of all centroids is assumed to be known, and "doubly constrained" models, where constraints are put on both emissivities and attractivities. The choice of a suitable model for traffic distribution might be a delicate question (see [16]). For instance, it is often reckoned that singly constrained models are more appropriate for short-term studies or for studies involving non-work functions.

Needless to say, these physical analogies typically lead to the solution of optimization problems. Entropy is maximized and one performs a fit for a gravitational model. The use of operations research is thus again present.

We finally note that the use of these concepts in a dynamical context is somewhat difficult.

4.3 The assignment of traffic

We now turn our attention to some more mathematical techniques that are mostly used within the fourth stage of the general model described above, that is the assignement of traffic within each modal network, given the demand.

4.3.1 Shortest paths. The first popular idea is that network users choose the route that is *shortest* (in time or in any more general cost measure), while assuming that the cost associated with each link is constant, i.e. independent of the traffic on that link. Finding such a shortest path is maybe one of the oldest operations research problems. It has been widely studied. The reader is referred to [65] or [3] for further details and references.

This technique has the advantage of being readily interpretable and also extremely efficient from the computational point of view. One indeed knows very efficient algorithms for computing shortest paths in a network, even when the number of nodes and links is very large (over 100000 nodes is quite feasible). Traffic assignment based on this idea is known as the *all-or-nothing* assignment, because all trips between a given (origin, destination) pair are assigned to the same single route, a rather crude approximation.

Although mostly used in the static context, dynamic versions of shortest paths algorithm do exist (see [10] for instance). Their efficiency is considerably less than their static counterparts, but again the growth in computer power makes them more and more attractive for practical use.

Along the same line of thought, one may also consider the question of reconstructing the costs associated with routes in an unsaturated transportation network, as they are perceived by the network users. This is important if one wishes to take perception (in some ways a form of behaviour) into account. A first approach has been proposed and tested in [14]: the idea is to reconstruct the perceived delay associated with each link of the network from the observation of the paths actually taken (assuming that users choose the shortest route between their origin and their destination). This method, which in fact solves the *inverse* shortest path problem, is akin to the idea of using "mental maps" [13] in the process of route planning (see also [14], [46], [49] and [50]). Technically speaking, the network under study is represented by an oriented graph in which a set of shortest paths is known; the question is then to infer the value of the time delays associated with each arc of the graph and differing as little as possible from a set of a priori known costs (derived, for instance, from the knowledge of the geometrical characteristics of the road). Applying this methodology to urban situations, it is important to explicitly consider the delays at signalized junctions, and not to restrict the analysis to the estimation of the delays on the links only. This can be achieved by using a graph that contains detailed arcs to represent the various "turns" in a junction. The resulting problems, where the unknown costs on these turns are correlated (for instance by traffic light phasing), can also be solved reasonably efficiently, as shown in [15].

4.3.2 Wardropian traffic equilibrium. Probably the most famous and most commonly used assignment method is aimed at improving this approximation by assigning trips to more than one route between a given (origin, destination) pair. It is based on the notion of *traffic equilibrium*. Following the famous statement made by Wardrop in 1952 [68], "travel time on all used routes is smaller or equal to that that would be experienced by a single vehicle on any unused route". In other words, no network user can, in an equilibrium situation, improve its trip cost by a unilateral decision. One easily sees that a consequence of this assumption is that all used routes between an origin and a destination have the same (minimum) cost and that all routes of higher costs are unused. This concept is obviously substantially more complex because it implies that the choice of a route between two zones depends not only on the possible routes between these two zones, but also on the load of the network, resulting from the traffic between all other O/D pairs! This concept is called "User Equilibrium" because the criterion used is the travel cost for the network user. The work of Beckman, McGuire and Winston [5] was amongst the first to use this principle in a very long line of research (see [19], [26], [27], [30], [62], [16] or [58], for further discussion and references). It is important to note here that a number of the most efficient methods to compute the User Equilibrium are based on the observation that *the equilibrium conditions are in fact the optimality conditions of a well defined continuous linearly constrained optimization problem*, providing yet another link with operations research. This crucial link is very well developed in [16] and [58].

One can also consider the "System Equilibrium" which would occur if one chooses all routes to minimize the total transportation cost for all users (or for society). These two equilibrium concepts are unfortunately different, because the advantage of all may require that some users deliberately choose a route that is less favourable for them... The (alas) unrealistic nature of this assumption implies that User Equilibrium is most often used for describing the behaviour of users in the network, while the System Equilibrium is sometimes computed to obtain an "ideal" traffic situation.

The difference between User and System Equilibrium is best illustrated by the famous Braess paradox (see [13], [29], [21], [35], ...). This example shows that, in the presence of congestion, the building of a new road may result in a deterioration of the network performance, if one assumes that each user optimizes its own travel time. Generally speaking, it indicates the importance of an in depth analysis of planned changes in the transportation network. It also shows, a contrario, that traffic restrictions (like pedestrian zones, one-way streets or road pricing) can have a beneficial impact on the global situation.

4.3.3 Equilibrium and variational inequalities. It is important to note that the equilibrium method has been extended to handle situations more general than pure traffic assignment: one can talk of "equilibrium" between various transportation modes or of "equilibrium" between transportation demand and transportation supply (as specified by the network). The first of these extensions typically requires a modification of the equilibrium concept, because the cost of public transportation cannot in general be compared to that of using a private car. Similarly, entropy ideas are usually introduced in the second extension for improved realism. We observe that both these approaches answer, albeit in part, to the criticism on the excessive sequentiality of the four stages model, because two or more phases are now interdependent and solved for simultaneously. The reader is referred to [16], for instance, for a further discussion of the many extensions of the equilibrium methodology.

From the point of view of the connections with operations research, one of the most interesting extensions of the equilibrium concept is to problems where the interactions between parts of the model ruin the immediate translation of the equilibrium calculation into an optimization problem. This happens for instance if several classes of users are considered on the network, or if there are interactions between the flows on different arcs. One then focusses more directly on the equilibrium conditions, which can be considered as a particular case of *variational inequalities*. The numerical treatment of variational inequalities in this context has also been an area of research in between operations research and transportation modelling, and is the object of the paper by Ferris in this volume. One can also refer the reader to the work by Dafermos and followers, [19], [20], [54], [55], or to [58] or [67].

4.3.4 Stochastic models. Both complete equilibrium assignment and the simple shortest paths technique assume that all network user see the cost of their route in a uniform way. The variability inherent to human decision processes has very little room here, which can be judged unrealistic (see [36], for instance). As an answer to this criticism, several stochastic variants of shortest path techniques have been proposed with the aim of re-introducing some variability while maintaining the assumption that the cost associated with a link is independent of traffic: the idea is to assign a random utility to "reasonable routes" or even to individuals links in the network.

The first of these techniques, introduced by researchers like Dial around 1970 [23], consists in finding the set of reasonable routes between an origin and a destination. A route is said to be reasonable in this context if the users steadily increases its distance from the origin and decreases its distance to the destination when following the route. The shortest route is of course reasonable, but the set of reasonable routes is typically larger. The traffic is then spread over these routes, using a logit distribution function. In another approach, one typically averages the flows obtained on a given link during successive applications of the shortest path assignment method applied in a network whose link costs are drawn from a normal distribution with fixed mean. This "successive average" method correspond to applying the probit technique to the route choice problem.

Similarly, one can introduce the notion of stochastic user equilibrium (SUE), in which each users minimizes its own *perceived* travel time in the network, where the perceived travel time is assumed to be a stochastic perturbation (of known distribution) of the real (measured) travel time. A good introduction to this technique is provided in [16], for instance. We briefly summarize the relations between assignment methods in Table 4.1.

We note that the stochastic nature of travel time can be interpreted in two ways. The first and more traditional interpretation is that the random perturbation of travel time represents the variation of perceived travel times with respect to real ones, induced by the variation of perception from a user to the next. Since these perceptual variations are essentially unavailable for measurement[1], they are assumed to be random. The second interpretation is that the user itself assigns a random component to travel time, for instance to represent aleas as waiting times at signalized intersections and variations in travel speed due to weather conditions or other random events on the network. In this second interpretation, the user is not trying to optimize its travel time as he perceives it, but is instead trying to take into account

[1] See the comment on inverse shortest path in Section 4.3.1 for a possible approach.

	constant costs	traffic dependent costs
deterministic approach	all-or-nothing	equilibrium (UE)
stochastic approach	reasonable paths or averaging	stochastic equilibrium (SUE)

Table 4.1. Relations between some traffic assignment techniques

random variations of network performance, or, more generally, of the environment, that are beyond his or her control. This attitude is typical of the vast class of operations problems known under the name of stochastic programming. For instance, one may also consider the problem of traffic assignment with stochastic demand... Stochastic models in transportation will be the subject of the paper by Louveaux in this volume. Reference books include [25] and [40].

4.3.5 Dynamic traffic modelling. For the reasons discussed in Section 3.2, one may wish to include the time dimension in a traffic assignment model. This effectively comes in three main flavors.

The first approach is to generalize the concept of equilibrium to time-dependent equilibrium. The first steps in this direction were done by Merchant and Nemhauser in [52] and [53] and were followed by a long line of contributions. See [73], [74], [70], [71], or the recent book by Ran and Boyce [61] and the references therein, for instance. A debate remains amongst transportation modellers on the validity of the dynamic equilibrium concept, mostly because the assumptions of full information and full rationality, which are necessary in this context, may be judged even more unrealistic than in the static case. Models of this type are often of the continuous type and are based on the classical concepts of optimality conditions in nonlinear programming or variational inequalities, two branches of operations research that we have already mentioned.

The second approach attempts to replace the definition of an equilibrium situation by a simulation of the evolution of traffic over time, without any reference to a "steady-state". Traffic is then considered as a pure dynamical system[2]. The difficulty is then to define what replaces the concept of equilibrium in the mechanism for specifying route choice for the network users. One

[2] In the sense of the mathematical theory of dynamical systems: the way the model behaves at a given step only depends on the previous steps of its history.

may again try to minimize some variant of travel time (subject to the available knowledge of the network conditions) or one may define a more general behavioural theory for users, as is done for instance in the PACSIM model discussed below by Cornélis and Toint. A good reference for the questions raised in this context is [3]. Simulation models are often of the discretized type. The subject will be discussed in this volume by Cornélis and Toint and by Mahmassani. Because simulation is another classical subject in operations research, models of this type provide yet another link to this field.

Finally, one may consider dynamic traffic models of networks less complex than a complete city. This is especially useful in traffic control, where one tries to optimize traffic in a complex intersection, or to control the access to a motorway in order to avoid saturation and the resulting delays. Questions of this type will be discussed by Papageorgiou in this volume.

4.4 Routing and related problems

Another important question in transportation modelling is often to compute routes for vehicles or users subject to constraints more complicated than those present in the simple assignment problem. These may include constraints on goods delivery, passenger transfer, sequence of visites sites, time windows or vehicle capacity, just to cite a few of the most commonly occuring cases. The determination of a route satisfying all constraints then become another classical problem of operations research and combinatorial optimization, the routing problem. It has applications as diverse and as pervasive as school bus routing, post delivery, garbage collection, road maintenance planning, demand sensitive public transportation systems, or "just in time" freight delivery. Problem of a similar combinatorial nature also arise in other contexts related to transport, as train time-tabling or traffic light setting. As can be seen from these examples, the applications of combinatorial optimization in the transport arena are plenty. The interested reader is referred to [8] for a nice collection of recent contributions in this area, and also to the papers of Laporte and Toth in this volume.

4.5 Discrete choice techniques

Another powerful tool for transportation modellers is the theory of *discrete choices*. The main idea here is to assume that transportation users, when faced with a decision, can only choose between a limited set of possibilities, hence the name of discrete choice. The typical situation is that of modal choice (stage 3 in the four stages model), where this set might include private car, public transportation, cycling or walking, for instance. With each of these possible choices is associated a utility function, and it is assumed (again) that the user maximizes his utility (another operations research problem).

The difficulty arises from the fact that these utility functions are usually not accurately known and cannot be measured easily. Moreover, the factors (attributes) that influence these utility functions (such as individual variability in sociological status, history, etc) are also beyond a complete analysis. To circumvent this problem, the utility functions are then constructed such that they contain a random part, which is supposed to represent the contribution of all these unknown factors (see our discussion of the stochastic user equilibrium assignment). One then talks about *random utilities.* A random utility then merely gives the probability that a given choice will be made by the considered user.

The first major discrete choice model is the *logit model,* in which the probability of chosing a given alternative k among K possibilities is given by

$$p_k = \frac{e^{d(k)}}{\sum_{i=1}^{K} e^{d(i)}}, \tag{4.3}$$

where $d(i)$ is the deterministic part of the utility associated with the i-th choice[3]. As can be seen from this formula, this (sometimes irrealistically) assumes independence of the possible choices. The logit model is very widely used in all stages of transportation models: the modal choice is often represented by a logit model and the choice of reasonable routes (see below) can be made using this approach as well. The popularity of the logit model is strengthened by its ease of use in multinomial cases, that is when the number of possible choices for the user exceeds two. In this case, one often "cascades" the choices as a sequence of successive decisions, each of which being modelled by (4.3). One then talks about *nested logit* modelling. The definition of these models may involve some problems of their own, that are connected to operations research in a loose way (see [6]).

An alternative model, the *probit* model, is based on the more general assumption that the utility functions are normally distributed but not necessarily independent. This model is however more difficult to implement because it leads to difficult numerical questions and also requires additional information on the choices (the covariance matrix), compared to the logit approach.

The reader is referred to [1] for a comprehensive coverage of discrete choice theory in the framework of transportation modelling. This subject is covered by the paper of Bierlaire in this volume.

4.6 Activity chains

We now briefly introduce another methodology, *activity chain* modelling, which is related with discrete choice models in that it attempt to capture the behaviour of transporation users. One of its main objectives is to help in the development of inherently dynamic and behaviourally sensitive demand analysis tools.

[3] This formula is based on the assumption that the random terms in each utility function are independently and identically distributed Gumbel variables. The Gumbel distribution is an analytic approximation of the normal distribution.

The notion of an activity chain is nearly self explaining: this is the sequence of activities planned for one or more individuals of a household at a given time in the day and week. Given such chains (the more one knows the better), it is then possible to deduce the transportation demand by considering which successive activities do not take place at the same localisation. The examination of many such activity chains then leads to an intrinsically dynamic and behaviour orientated demand modelling, resulting in more classical dynamic O/D matrices. The advantages of this approach are mainly

- its intrinsic dynamic nature,
- its very adaptable framework (it covers, for instance, multipurpose trips, a common reality but gaping hole in many other modelling approaches),
- its sensitivity to behavioural variations across households, because different schedulings of the same activities are possible.

Of course, this technique is not without its drawbacks.

- It requires a large amount of data (travel diaries, mostly) which is often difficult to obtain.
- The structure of the activity chains is not always known a priori, which introduces yet another level of modelling, that of activity scheduling.
- Its practical treatment by reliable software tools is still in infancy.

Despite these serious inconvenients, activity chain modelling remains one of the most promising techniques, especially when one wishes to take the behaviour of the users into account, such as in models of information systems. References in this area include [38], [4], [37], [66] and [51].

In the context of activity based modelling techniques, another component of the traditional operations research scene plays an important role. As activities of a household or individual happen in time, there is a clear need for scheduling techniques in the establishment of the activity chains themselves. However, the author is not aware of many research contributions aimed specifically at bridging the gap between activity chain models and scheduling techniques.

4.7 Other continuous optimization and estimation methods

We close this overview of the methodologies used in transportation research by mentioning that, as is the case in many scientific applications, one often tries to fit a given model to a set of observations. For instance, one might try to fit a OD-matrix describing the transportation demand to observations of traffic flows or centroids' emissivities. Or one might wish to calibrate an assignment model to adequately represent the observed network flows. This calibration is usually carried out by making the model's output as close as possible to otherwise known measures of the real situation[4], a classical parameter fitting problem in operations research.

Once calibrated on the real situation (often called the *reference case* or the *do-nothing case*), the model can then be used to analyze diverse alternative scenarii, differing from the reference case in a well defined manner. The confidence one has in the obtained results for these alternative scenarii is clearly conditioned by the quality of the model calibration on the reference case, and thus, albeit indirectly, by the quality of the method used in the calibration phase.

Among the many methods for observation fitting, the least squares approach is often preferred (see [16] or [12] for an example of application to the distribution stage). One should also mention here that some inherently dynamic methods for local demand analysis can also be viewed in the framework of the least squares approach, in that they try to exploit correlations between flows penetrating into and leaving the study zone (see [18], [41] for instance).

Other optimization problems arise in data calibration. For instance, the parameters of a (nested) logit model are typically computed using a maximum likelihood estimator, itself resulting in a specific unconstrained continuous optimization problem (see [5]). And the list is far from complete...

5. Some limits of transportation modelling as an assessment tool

In this section, we briefly discuss some limitations that should be kept in mind when using transportation models as an assessment tool.

The first consideration is that a model can only really be used if it has been validated in real applications, and when its behaviour has been shown to be robust enough to allow reasonably safe extrapolation. While this is often the case for classical models, this is much more rarely the case for newer and more advanced models (as can of course be expected). This remark is especially true for dynamic models, where validation is a much harder task. Hence the necessity of pilot experiments and associated model appraisal. The "Traffic Appraisal Manual" of the english Department of Transport (1994) or the work by Leurent in his thesis [44] attempt to set up "rules of good practice" in this domain, but much remains to be done. One could mention at this point the conclusion drawn by Julien and Morellet in [39]: "Finally, the real debate on the choice of the type of modelling should not deal with what these or those models can do theoretically, but rather with the more or less important hazards to which the calibration of these models can be

[4] It is therefore important for a modeller not to use all the information he has at his disposal to build the model, but to keep some of it for this calibration phase.

subjected". Although the author feels that this conclusion is probably too severe, it also has its part of truth...

Correlative to the validation question is that of data collection. The more complex the model, the more data it typically requires, hence the more difficult to apply it with its full predictive power. This limitation has to be recognized, but does not necessarily prevent a cognitive use of the models as a tool to gain intuition and understanding of a practical situation. In particular the collection of data related to transportation user behaviour is a challenging task.

A final and insidious limitation of the modelling methodology one finds in practical applications is the fact that, often, limited funding for the transportation study results in a partial modelling, where important aspects of the methodology (as public transport modal choices) are grossly simplified or neglected altogether. The model users obtain of course what they are paying for, but a too restrictive financial appraisal of the modelling phase might, and sometimes does, lead to substantial qualitative errors in the conclusions! This type of error is also considered by [44].

6. Conclusion

We have developed a general outline of the context, basic tools and methods in transportation modelling, from which links with operations research have been pointed out, as well as with the subject of other papers of this volume.

It is hoped that the presentation has made clear the great diversity of research areas in transportation which can and should benefit from the developments and algorithms of continuous and discrete optimization, as well as other branches of operations research. The joint adventure of the two fields has already lasted for the best of forty years, and the multiplicity and importance of the remaining questions is such that one can safely anticipate a much longer fruitful cooperation.

References

1. R+D in advanced road transport telematics in Europe (DRIVE90). Technical Report DRI-201, DG XIII, EEC, Brussels, March 1990.
2. M. Acutt and J. Dodgson. The impact of economic policy instruments on greenhouse gas emission in australian transport. In D. Henscher, J. King, and T. Oum, editors, *Transport Policy, volume 3 of the Proceedings of the 7th World Conference on Transport Research*, pages 321–334, Kidlington, UK, 1996. Pergamon.
3. R. K. Ahuja, T. L. Magnanti, and J. B. Orlin. *Network flows, Theory, Algorithms, and Applications*. Prentice-Hall, Englewood Cliffs, New Jersey, USA, 1993.

4. P. Stopher amd M. Lee-Gosselin. *Understanding travel behaviour in an era of change.* Pergamon, Kidlington, UK, 1996.
5. M. J. Beckman, C. B. McGuire, and C. B. Winston. *Studies in the economics of transportation.* Yale University Press, New Haven (USA), 1956.
6. M. E. Ben-Akiva, A. de Palma, and I. Kaysi. Dynamic network models and driver information systems. *Transportation Research A*, 25(5):251–266, 1992.
7. M. E. Ben-Akiva and S. R. Lerman. *Discrete Choice Analysis: Theory and Application to Travel Demand.* MIT Press, Cambridge, USA, 1985.
8. L. Bianco and P. Toth, editors. *Advanced Methods in Transportation Analysis.* Springer-Verlag, Heidelberg, Berlin, New York, 1996.
9. M. Bierlaire. Evaluation de la demande en trafic : quelques méthodes de distribution. *Annales de la Société Scientifique de Bruxelles*, 105(1-2):17–66, 1991.
10. M. Bierlaire. A robust algorithm for the simultaneous estimation of hierarchical logit models. GRT Report 95/3, Department of Mathematics, FUNDP, Namur, Belgium, 1995.
11. M. Bierlaire, T. Lotan, and Ph. L. Toint. On the overspecification of multinomial and nested logit models due to alternative specific constants. *Transportation Science*, (to appear), 1997.
12. M. Bierlaire and Ph. L. Toint. MEUSE: an origin-destination estimator that exploits structure. *Transportation Research B*, 29(1):47–60, 1995.
13. D. Braess. Uber ein Paradoxon der Verkehrsplanung. *Unternehmensforschung*, 12:256–268, 1968.
14. D. Burton and Ph. L. Toint. On an instance of the inverse shortest path problem. *Mathematical Programming, Series A*, 53(1):45–62, 1992.
15. D. Burton and Ph. L. Toint. On the use of an inverse shortest paths algorithm for recovering linearly correlated costs. *Mathematical Programming, Series A*, 63(1):1–22, 1994.
16. E. Cascetta. Estimation of trip matrices from traffic counts and survey data: a generalised least squares approach estimator. *Transportation Research B*, 18(4/5):289–299, 1984.
17. L. Clement, D. Peyrton, and M. Frenois. Review of existing land-use transport models. Technical Report 58, CERTU, Lyon, France, 1996.
18. M. Cremer and H. Keller. A new class of dynamic methods for the identification of origin-destination flows. *Transportation Research B*, 21(2):117–132, 1987.
19. S. Dafermos. Traffic equilibrium and variational inequalities. *Transportation Science*, 14:42–54, 1980.
20. S. Dafermos. The general multimodal network equilibrium problem with elastic demand. *Networks*, 12:57–72, 1982.
21. S. Dafermos and A. Nagurney. On some traffic equilibrium theory paradoxes. *Transportation Research B*, 18:101–110, 1984.
22. A. de Palma, P. Hansen, and M. Labbé. Commuters' paths with penalties for early or late arrival. *Transportation Science*, 24(4), 1990.
23. R. B. Dial. A probabilistic multipath traffic assignment algorithm which obviates path enumeration. *Transportation Research*, 5(2):83–111, 1971.
24. R. M. Downs and D. Stea. *Maps in minds.* Harper and Row, New York, 1977.
25. Y. Ermoliev and R. J.-B. Wets, editors. *Numerical Techniques for Stochastic Programming.* Springer-Verlag, Heidelberg, Berlin, New York, 1988.
26. M. Florian, editor. *Traffic equilibrium methods.* Springer-Verlag, New York, 1976. Lecture Notes in Economics and Mathematical Systems 118.
27. M. Florian. Mathematical programming applications in national, regional and urban planning. In M. Iri and K. Tanabe, editors, *Mathematical Programming: recent developments and applications*, pages 57–82, Dordrecht, NL, 1989. Kluwer Academic Publishers.

28. R. L. Francis, L. F. McGinnis, and J. A. White. *Facility Layout and Location*. Prentice-Hall, Englewood Cliffs, New Jersey, USA, 1992.
29. M. Frank. The Braess paradox. *Mathematical Programming*, 20:283–302, 1981.
30. N. H. Gartner. Optimal traffic assignment with elastic demands: a review. *Transportation Science*, 14:192–208, 1980.
31. M. Germain, Ph. Toint, and H. Tulkens. International negotiations on acid rains in Northern Europe: a discrete time iterative process. In A. Xepapadeas, editor, *Economic policy for the environment and natural resources: techniques for the management and control of pollution*, pages 217–236, London, 1996. Edward Elgar Publishing.
32. M. Germain, Ph. Toint, and H. Tulkens. Financial transfer to ensure cooperative international optimality in stock pollutant abatement. Technical Report 97/4, Department of Mathematics, FUNDP, Namur, Belgium, 1997.
33. R. G. Golledge and R. J. Stimson. *Analytical Behavioural Geography*. Croom Helm, New York, 1987.
34. J. D. Griffiths, editor. *Mathematics in Transport Planning and Control*. Clarendon Press, Oxford, UK, 1992.
35. Å. Hallefjord, K. Jørnsten, and S. Storøy. Traffic equilibrium paradoxes when travel demand is elastic. Working paper, University of Bergen, Bergen, Norway, 1990.
36. R. Hammerslag. Dynamic assignment in the three dimensional timespace: mathematical specification and algorithm. US-Italy Joint Seminar on Urban Traffic Networks, June 1989.
37. D. Hensher, J. King, and T. Oum, editors. *Travel Behaviour*. Pergamon, Kidlington, UK, 1996. volume 1 of the Proceedings of the 7th World Congress on Transport Research.
38. P. Jones, editor. *Developments in Dynamic and Activity-Based Approaches to Travel Analysis*, Aldershot, UK, 1990. Avebury.
39. H. Julien and O. Morellet. What kind of relationship between model choice and forecast traffic for long distance trips? 5th International Conference on Travel Behaviour, Aix-en-Provence, 1987.
40. P. Kall and S. W. Wallace. *Stochastic Programming*. J. Wiley and Sons, Chichester, England, 1994.
41. H. Keller and G. Ploss. Real-time identification of O-D network flows from counts for urban traffic control. In N. H. Gartner and N. H. M. Wilson, editors, *Transportation and traffic theory*, pages 267–284, New-York, 1987. Elsevier. Proceedings of the Tenth International Symposium on Transportation and Traffic Theory, July 8–10 1987, MIT, Cambridge, Massachusetts.
42. A. Khasnabis and B. Chaudry. Transportation-land us interaction: the US experience and future outlook. In D. Henscher, J. King, and T. Oum, editors, *Transport Policy, volume 3 of the Proceedings of the 7th World Conference on Transport Research*, pages 21–30, Kidlington, UK, 1996. Pergamon.
43. F. Leurent. Cost versus time equilibrium over a network. *Transportation Research Record*, 1443:84–91, 1994.
44. F. Leurent. *Pour certifier un modèle*. PhD thesis, Ecole National des Ponts et Chausées, Paris, 1996.
45. M. J. Lighthill and G. B. Whitnam. On kinematic waves ii: a theory of traffic flow on long crowded roads. *Proceedings of the Royal Society, London*, 229A, 1955.
46. R. H. Logie and M. Denis. *Mental Images in Human Cognition*. North-Holland, Amsterdam, 1991.
47. R. F. Love, J. G. Morris, and G. O. Wesolowsky. *Facilities location. Models & methods*. Elsevier (North Holland), New York, 1988.

48. J. L. Madre and J. Maffre. Toujours plus loin ...mais en voiture. *INSEE Première*, 417, 1995.
49. A. M. MacEachren. Travel Time as the Basis of Cognitive Distance. *The Professional Geographer*, 32(1): 30–36, 1980.
50. A. M. MacEachren. *How Maps Work: Representation, Vizualization and Design.* Guilford Press, New York, 1995.
51. McNally and Recker. *Activity-Based Models of Travel Behaviour: Evolution and Implementation.* Pergamon, Kidlington, UK, 1996.
52. D. K. Merchant and G. L. Nemhauser. A model and an algorithm for the dynamic traffic assignment problems. *Transportation Science*, 12:183–199, 1978.
53. D. K. Merchant and G. L. Nemhauser. Optimality conditions for a dynamic traffic assignment model. *Transportation Science*, 12:200–209, 1978.
54. A. Nagurney. Computational comparisons of algorithms for general asymmetric traffic equilibrium problems with fixed and elastic demand. *Transportation Research B*, 20(1):78–83, 1986.
55. A. Nagurney. An equilibration scheme for the traffic assignment problem with elastic demand. *Transportation Research B*, 22(1):73–79, 1989.
56. G. L. Nemhauser, A. H. G. Rinnoy Kan, and M. J. Todd. *Optimization*, volume 1 of *Handbooks in Operations Research and Management Science.* North-Holland, Amsterdam, 1981.
57. M. Papageorgiou, editor. *Concise Encyclopedia of Traffic and Transporation Systems*, Oxford, UK, 1991. Pergamon Press.
58. M. Patriksson. *The Traffic Assignment Problem, Models and Methods.* VSP, Utrecht, NL, 1994.
59. S. H. Putman. *Integrated urban models*, volume 1. Pion limited, 207 Brondesbury Park, London NW2 5JN, 1983.
60. S. H. Putman. *Integrated urban models*, volume 2. Pion limited, 207 Brondesbury Park, London NW2 5JN, 1991.
61. B. Ran and D. Boyce. *Modelling Dynamic Transportation Networks.* Springer-Verlag, Heidelberg, Berlin, New York, 1996. Second edition.
62. K. Safwat and T. Magnanti. A combined trip generation, trip distribution, modal split and trip assignment model. *Transportation Science*, 22:14–30, 1988.
63. A. Sen and T. E. Smith. *Gravity Models of Spatial Interaction Behaviour.* Springer-Verlag, Heidelberg, Berlin, New York, 1995.
64. Y. Sheffi. *Urban Transportation Networks.* Prentice-Hall, Englewood Cliffs, USA, 1985.
65. R. E. Tarjan. *Data Structures and Network Algorithms.* SIAM, Philadelphia, USA, 1983. CBMS-NSF, Regional Conference Series in Applied Mathematics.
66. H. Timmermans. *Activity Based Approaches to Transportation Modelling.* Pergamon, Kidlington, UK, 1996.
67. Ph. L. Toint and L. Wynter. Asymmetric multiclass assignment: a coherent formulation. In J. B. Lesort, editor, *Transportation and Traffic Theory*, pages 237–260, Oxford, U.K., 1996. Pergamon.
68. J. Wardrop. Some theoretical aspects of road traffic research. *Proceedings of the Institute of Civil Engineers, part II*, 1:325–378, 1952.
69. F. V. Webster, P. H. Bly, and N. J. Paulley, editors. *Urban Land-use and Transport Interaction, Policies and Models.* Avebury, Aldershot, UK, 1988.
70. B. W. Wie, T. L. Friesz, and R. L. Tobin. Dynamic user optimal traffic assignment: a control theoretic formulation. presented at the US-Italy Joint Seminar on Urban Traffic Networks, Capri, 1989.
71. B. W. Wie, T. L. Friesz, and R. L. Tobin. Dynamic user optimal traffic assignment on congested multidestination networks. *Transportation Research B*, 24(6):431–442, 1990.

72. A. G. Wilson. *Entropy in urban and regional modelling.* Pion, London, 1970.
73. S. Yagar. Dynamic traffic assignment by individual path minimization and queing. *Transportation Research*, 5:179–196, 1970.
74. S. Yagar. Emulation of dynamic equilibrium in traffic networks. In M. Florian, editor, *Traffic equilibrium methods*, New York, 1976. Springer-Verlag. Lecture Notes in Economics and Mathematical Systems 118.

PACSIM: A New Dynamic Behavioural Model for Multimodal Traffic Assignment

Eric Cornelis and Philippe L. Toint

Transportation Research Group,
Department of Mathematics,
Facultés Universitaires ND de la Paix,
B-5000 Namur, Belgium

Summary. The present paper provides a brief introduction to PACSIM, a new model for multimodal traffic assignment. This model is distinguished by the introduction of several important features: full dynamism based on an event-driven paradigm, complete modelling of the network of information provided to network users, multilevel network representation, multimodal capabilities and the use of an explicit parametrisable behavioural theory. The paper discusses the main motivations and design concepts and presents an illustrative application of PACSIM to advanced incident warning. Some perspectives are finally outlined.

1. Introduction

Asserting the usefulness of traffic assignement models has become a very common-place argument over the past twenty years, but it is only comparatively recently that the need for complex dynamic models has really emerged (see [4]). This need is correlated with the desire to study the impact of various advanced information systems for network users. These systems propose to improve the congestion levels in large urban areas by providing timely and accurate information to the drivers ([3], [1], [17]). Since classical equilibrium type models typically assume that the user's information is complete, accurate and timely, the new proposals seemed to be irrelevant for their application, which, in turn, highlighted the inherent strong assumptions of such models (for instance, see [11] for a deeper discussion of this subject). A major research effort resulted world-wide, whose purpose was and still is to build traffic assignment models in which these asssumptions of perfect information are absent.

Developed partly with the support of the IMAURO project during the DRIVE I European program and partly with the support of the impulse program "Transport and Mobility" (Services of the Belgian Prime Minister), the PACSIM model can be considered as one of the attempts to build traffic assignment models in which the assumptions on the information available to users and on the actual behaviour of these users are relaxed and brought closer to reality. The main purpose is to provide a sensitive traffic modelling technique which could then be used to assert the qualitative effects of advanced driver information systems (ATIS[1] in the USA, ATT[2] in Europe). It is not our purpose here to describe the PACSIM model in detail (a preliminary detailed description may be found in [12]). We instead concentrate on some of the modelling concepts, with the aim to show how these are correlated with the major design challenges in this area.

2. A few desirable model features in the ATIS/ATT framework

When thinking of models adapted to assess the impact of advanced users information systems, it seems natural to search for directions and features which would make such tools viable. We now examine some of these.

2.1 A dynamic nature

Because much of ATIS/ATT is geared at the dynamic behaviour of transportation users and at the control or analysis of transient phenomenons, dynamic transportation models seem unavoidable within this context. Network users indeed have clearly different strategies in peak hour as opposed to off-peak, in week-ends as opposed to week days, in day as opposed to night, and the list of examples can easily be made longer. Furthermore, techniques like intelligent incident detection and interactive or automated route guidance are only meaningful in a dynamic context, as they involve "en-route" modifications of the user's route, which is totally impossible to represent with static approaches.

It is remarkable that, although appreciated by many experts, this conclusion has not always been applied in its full requirements. Because of the difficulty in building intrinsically dynamic models, static models whose parameters were made dependent on time have often been used. Whether this methodology is in fact coherent is clearly questionable, and heavily depends on the other factors in the chosen methodology. The crucial question is in the manner in which information (traffic or other) is propagated along the time dimension and the assumptions made on this propagation. But modellers and model users interested in ATIS/ATT should bear in mind that it isn't usually enough for a model to depend on time to become a truly dynamic model.

2.2 A flexible and standardized behavioural mechanism

In transportation models, the notion of global cost associated with displacement is typically taken into account by the choice of a *cost function* or *performance function* that determines, for each link of the network, a "cost" as

[1] Advanced Traffic Informations Systems.

[2] Advanced Transport Telematics.

a function of the traffic on that link. Units for this cost vary from model to model, and are chosen in relation with the final purpose of the modelling. For example, the quantity of exhaust gas of a certain type can be used if one wishes to measure the pollution impacts, while transportation time can be used if one is interested at asserting the level of service of public transport... There is however a remarkable consistency between all these choices, irrespective of the units chosen: the cost function is typically increasing as traffic along the link increases, slowly first but faster and faster for higher traffic levels. For example, the travel time along a link is relatively independent of traffic for low traffic levels, but can increase substantially for more congested levels, up to the point where not a single vehicle can move because everybody is stuck in a traffic jam!

It should be noted that the choice of a single performance function associated with the parts of the transportation network implies a complete and uniform rationality of the users in the measure where each user minimizes his or her disutility. This assumption has often been criticized as irrealistic. One way to overcome the difficulty is to introduce some degree of stochasticity in the users' choice criteria. Finally, optimizing a single "utility function" (even containing some stochastic component) often raises the question of comparing various goods or services of qualitatively different nature, a very difficult task. For instance, it is often necessary to estimate the monetary value of time (VOT) for users or users' classes. Some concepts in multi-objective optimization (see [18] for a good introduction) could present an interesting alternative at this level, but this approach is not widely used (see [10] for an example).

The use of utility function within the ATT framework has its supporters and detractors. The main arguments in favor of this approach are

- the possibly high disaggregation level allowed by the methodology,
- the easier calibration of the resulting models (at least for moderate disaggregation levels) compared to other alternatives,
- the existence of good related mathematical and software tools (see [12] and [5]).

Arguments that have been used against the technique are as follows.

- The behaviour of humans cannot be safely represented by the optimization of a utility function, mostly because many users do not know their utility function and also because they are, in complex cases, mostly unable to perform a proper optimization, especially when the decision process is quick as it is in traffic.
- The variation of perception among users' classes is badly represented by assuming random variations.

These considerations have led to modelling systems where behaviour is not represented anymore by the optimization of utility function, but by application of adapted *behavioural rules*. These rules are not assumed to reflect the (undetermined) rationality of the user, but are just considered as *descriptive* of behaviour, as opposed to resulting from deeper processing. This kind of approach is clearly very inspired by artificial intelligence techniques, where the inadequacy of utility functions has often been pointed out. This rule-based paradigm is very attractive in many ways (at least to the authors) but present the difficulty of the necessary statement of sufficient and meaningful rules. This is a very difficult task, mostly because relevant behavioural theories are still very much in the development stage, and certainly no general agreement has been reached as to which theories do best represent the behaviour of network users in what are potentially extremely diverse situations. Therefore, there seems to be little point in incorporating one particular behavioural theory in a traffic model, and exclude other promising approaches at the same time. New models should therefore provide the capacity to use a particular *behavioural theory as a model parameter*. Furthermore, as several theories are likely to be of interest, the specific way in which models handle behavioural theories must be extremely flexible and adaptable.

Several possibilities exist to fulfill these goals, ranging from parametrized cost functions (probably too restrictive) to complete representation of the theory using knowledge representations systems, as developed in the area of artificial intelligence.

The authors believe it is important that efforts are made towards a coherent way to represent behavioural theories, hopefully leading to some *standardization* in this research area. Furthermore, the representation format should allow simple and error free transmission to other researchers and, in the best case, be directly exploitable as such by modelling programs. These requirements are indeed crucial for these theories to play a really practical role in modelling methodology, even if they now appear to be somewhat idealistic. We refer the reader to [15] for a discussion of a potentially useful approach.

2.3 A behaviourally coherent traffic metaphor

The specific constraint on aggregation level appearing in the ATIS/ATT context is that of behavioural coherence. As is clear from above discussions, the modelling of the transportation system in this context imposes that special care is given to user's behaviours, both in the short term (in immediate decision like making en-route choice) but also in the longer term (attitudes). It is therefore important to base the models on a metaphor that provides this behavioural coherence.

It is quite difficult to think of a behaviourally coherent transportation flow, for instance, because the idea of flow implies some kind of uniformity in the pattern, leaving no room for differences within the flow. Although one could think of models describing the flow of several fluids together, this remains, to the authors' knowledge, rather speculative. One is therefore naturally led to more disaggregate metaphors, involving groups of users with similar behaviour. Many such metaphors have been proposed at a meso level, as *platoons* or *packets*. They are adequate for ATT sensitive models in direct measure of their behavioural coherence. The situation is finally most complicated when considering ATT sensitive public transportation modelling, since, in this context, the only reasonable "unit" seems to be the individual user, which directly leads to micro-models.

As a consequence of this discussion, it seems that, in first approach, micro-models are more adapted to behaviourally coherent simulation. Unfortunately, micro-models tend to be more limited in spatial coverage, because of their intrinsic higher complexity. There is therefore a need to develop meso-models whose spatial coverage can be much wider but whose behavioural coherence has to be specially emphasized.

2.4 A flexible network representation

A very flexible transportation network representation is also extremely desirable in our "behaviourally driven" methodology. Indeed, users can only decide on route and mode choice within the framework of their knowledge of the network. But one knows from the research in mental maps (see [13]) that this knowledge might present relatively severe distortions from reality. It also typically involves some level of network abstraction (only major links are perceived in distant or unfamiliar districts).

The behavioural route or mode decision can then be broadly split into two phases (see [3]), both driven by behavioural theory: the user knowledge is asserted first, followed by some choice between the decisions that are possible in the restricted context of this knowledge.

It is therefore crucial that the network representation can easily be adapted and modified (according to a given behavioural theory) to better reflect the user's perception of the network. This modification will thus typically involve network abstraction and distortions of various types (link suppression, cost modification, preferred routes, ...). This requirement in turn imposes the use of specialized and flexible data structures.

2.5 The information network

The impact of new advanced technologies in transportation has an important consequence on the spatial description of models. The main observation is that the object of study is no longer restricted to the movements of users in the transportation network, but also covers the exchange of information across its various parts: sophisticate route guidance or incident detection as well as relatively classical traffic light synchronization involve that exchange very clearly. Furthermore, the efficiency of advanced ATT is often said to be dependent on the quality, reliability, accuracy of the exchanged information, and on its distribution method.

Typical examples of relevant factors are

- the location of traffic monitoring equipment (the actual locations of detection and measurement equipment on the network of course condition the final information in that it specifies the parts of the network about which information could be available),
- the location of information distribution mechanisms (conditioning in turn the locations at which re-routing can occur),
- the accessibility of information (defining the classes of network users accessing possibly different types of information),
- the strategies used to process information from the network before subsequent redistribution to users,
- the actual interconnection pattern (topology) of monitoring devices, information processing centres and distribution mechanisms.

Besides the classical transportation network, we therefore discover another network, which describes the information flow (the feedback loop) between the transportation network and its users. We call this network the *information network*. It is the authors' view that explicit modelling of this information network is an important part of the present research challenges in this area.

Of course, these information network models should be flexible, in order to adapt to the many different cases of potential interest. In particular, they should include the relevant factors mentioned above, but also parameters that influence their efficiency (communication links' or detectors' reliabilities, for instance) and thus also permit the modelling of information system breakdowns.

3. The PACSIM approach to behaviourally driven route choice and assignment

After reviewing some desirable features for ATIS/ATT sensitive traffic models, we survey some of the modelling choices that were made for the traffic model PACSIM in compliance with our above comments.

3.1 The handling of time

As discussed above, an explicit time dimension is also recommended for the type of models we have in mind. PACSIM is thus a dynamic model: it is event driven in the sense that the passing of time is viewed as the succession of a number of *events*. These events are, within PACSIM, of two distinct natures. The first type of event is associated with true physical event happening in the transportation network itself, as occurrence of congestion or bad weather conditions, traffic light changes, arrival of cars at intersections, road works or traffic accidents. The second type of event is associated to the transfer of information throughout the information network: detectors actuation and information transmission across the communication network (including the specification of delays and information life-time). These two types of events are then scheduled within PACSIM, according to parametrized priority rules[3].

3.2 Modelling the information network

We also stated above that an explicit model of the information flow from the network to its users is desirable. In accordance with this suggestion, PACSIM contains such an explicit model. This model takes itself the form of a network, that features four main types of nodes: detectors, traffic information centres, traffic control centres and information sockets. We now briefly review their function and relations.

The detectors' purpose is to monitor the status of the network and to detect any traffic event therein. For increased realism, detectors have individually adjustable sensibilities[4], reliabilities, reaction delay and can be positioned anywhere on the network.

Traffic information centres are nodes at which information on detected traffic events is centralized, screened according to parametrizable selection criteria, and then distributed to other traffic information centres, traffic control centres and information sockets.

Information sockets model the various devices that are used to transmit information to network users (radio beacons and dedicated broadcastings, variable message signs, ...). They come in different types, varying according to technology and capacity, but also in the amount of information they can store (memory), time during which information is considered up-to-date (persistence) and ways in which priority conflicts between different messages are handled. They can also be situated anywhere on the network.

Traffic control centres represent instances that decide on traffic management, control and optimization strategies, as opposed to mere information distribution. For instance, they can maintain sets of recommended routes for route guidance applications and fleet management. Recommendations and informations implementing these strategies are also distributed to network users via information sockets.

These four classes of "information network nodes" are themselves linked by communication lines whose reliability and delay can be parametrized. Furthermore, the complete information network configuration (topology and node/link parameters) can be dynamically updated during the design period.

Again, the authors realize that the description of the information network in PACSIM might not capture every possible variant of information processing that can be thought of in the context of advanced driver information

[3] Most of these rules give priority to "physical events" (such as accidents of traffic light changes) on decisions taken by users (route changes) at the same instant.

[4] That is the type of effect and/or event that they are able to detect (as congestion, accident, unfavourable weather conditions, ...).

systems. However, the present definition already covers a fairly wide ground. It is also easily expandable if new technologies have to be modelled, that cause or result in new information flow patterns.

3.3 Network abstraction

Our previous discussion also recommends the availability of network abstraction mechanisms, in order to reflect the different perceptions of a specific network location by users that are close to it and by users that are further away: close locations are perceived with more disaggregate detail while distant ones are perceived in a more aggregated fashion (where only main districts and major links matter).

According to this view, PACSIM uses a three level hierarchical network representation. The three levels correspond to the *main* network, the *strategic* network and the *detailed* network, from the more aggregate to the more disaggregate. Explicit mappings are defined that assign a node of a disaggregate level to a node of the aggregate level just above, assuring operational coherence to the overall structure. This "multi-level" representation of the network of course plays an important role in the definition of the users' knowledge of the network but also in other parts of the model, as described in the next section.

3.4 Packets

The traffic metaphor used in PACSIM is *based* on behavioural coherence, therefore complying with the above arguments closely. PACSIM indeed considers traffic as composed of the movements of *packets*, which are groups of vehicles moving on the network with a coherent behaviour. They are created at network locations corresponding to origins of transportation demand at a given time, travel through the network and disappear at their destination. The behavioural coherence of the packets is ensured by the requirement that vehicles belonging to the same packet must have in common

1. their capacity to access driver information systems and the traffic knowledge provided through this channel,
2. their mode of transportation,
3. their trip purpose,
4. their destination in the network.

Although the authors realize that these characteristics might not be sufficient to describe behaviourally coherent classes of users, their specification however constitutes a first attempt in this direction.

3.5 The behavioural mechanism

The behavioural mechanism that drives PACSIM is organized along two complementary stages, broadly corresponding to the two phases for behavioural decisions outlined above. The user perception of the network is first established, and a routing decision is then taken within that restricted framework. We now briefly describe these two stages.

The network as perceived by the user (the packet), the *perceived network*, is first defined using the multi-level network representation, according to the following steps.

1. A perceived network is built, that contains a detailed view of the links and nodes in the surroundings of *anchor points*, with the level of detail fading away with distance. These anchor points represent parts of the network that are well known to the user (see [14]). At present, the origin and destination of the trip are typically chosen as anchor points. This procedure makes full use of the multi-level network structure.
2. Behavioural rules (see [6]) applicable to the packet then modify this restricted network, according to the packet characteristics and the information it has accumulated in past events. This modification can be of two possible natures:
   - the topology of the perceived network may be modified,
   - the relative costs of the perceived network links may be modified.

   In the first case, links can, for example, be removed from the network because they do not belong to the packet's knowledge of the area; links can also be added or reversed as a consequence of dynamic traffic management decisions (as tidal flows) which are communicated to the packets via driver information systems. The modification of costs allows to reflect behavioural preferences (heavy goods vehicle drivers tend to prefer major links, for instance) or again to take available broadcasted information on network congestion into account.

Once the packet's knowledge of the network is established, the behavioural rules are further applied to determine the *route structure* as a sequence of subroutes. These subroutes are finally computed and assigned to the packet. PACSIM actually uses shortest path computation in the modified network for this very last step. It also features a further option to randomly perturb the network cost to represent individual stochastic variations within a single behaviourally coherent class of network users (in a spirit close to that of stochastic assignment, as described in [8] and [16] for example).

We mentioned above that certain behavioural rules are applicable to packets. The set of all these rules constitutes a behavioural theory of route planning, which is used as input parameter by PACSIM. These rules are specified in a formal language that has been specifically designed for the purpose, and which is described elsewhere (see [15]). We only stress here that this representation technique has several decisive advantages: formulated in such a formal language, the behavioural theory can be perspicuous, unambiguous, concise and machine readable. We also note that rules expressed in this language cover variations in behaviour potentially influenced by a wide range of parameters as availability, quality, accuracy and timing of user's information, location in the network, trip purpose, transportation mode, time of the day,...

The inherent behavioural route choice of PACSIM imply that a mechanism is provided for the drivers to memorize some past informations for later use. However, routing information cannot be based on this information only, as has been already observed: in particular, a packet would not be able to start its route in the network because it is just, at this stage, beginning to accumulate information. Some degree of anticipation based on experience is thus necessary. PACSIM provides a mechanism (called the *background network*) to build up such long term dynamic experience of the network, which can be judged admissible in a limited sense (see [11]). This provides the requested anticipation of traffic effects in the route planning decision process, at the condition that the knowledge of the anticipated effects can be acquired with experience. This long term experience is common to all drivers in the present version of PACSIM, but it might be envisaged to disaggregate it further to better reflect the differences in network experience between different users. At present, PACSIM uses a classical Wardrop equilibrium traffic assignment for computing the background network costs.

As we noted above, it is not clear what behavioural theory will actually prove to be most useful. Hence, the theory that is presently used by PACSIM (see [12]) is not aiming at completeness or even correctness: the best that can be hoped for is that something of the *character* of driver response to advanced information systems has been elucidated. One of its purposes is to allow the testing of the behavioural theory representation, particularly with regard to flexibility and efficiency. Better theories emerging from ongoing research projects within the DRIVE program and elsewhere can then be used and tested later without further modification to PACSIM.

3.6 The multimodal aspects

In urban areas, a non-negligible part of the displacements uses public transport. As a consequence, it is reasonable to expect an urban traffic model like PACSIM to take this aspect into account. The method chosen is to include in the model a relatively detailed simulation of the public transport system along with the simulation of private traffic. The same methodology is used, in that the public transport model is dynamic (event-driven) and allows for both multimodal trips (that is trips using both private and public transportation modes) and the provision of advance information systems for users of public transport. The model is of course based on a description of the public transport network, which includes stops, lines, timetables,... as expected.

PACSIM determines the (multi-) modal choice for trips using a behavioural theory for users. In the scenarii tested so far, this theory has always been simplified to the maximization of a user dependent utility function, as is classical in several other techniques (see [2] for a general view of discrete choice models, or [5] for the description of HieLoW, an efficient package for calibrating such models). The utility functions definition of course allows for the representation of "captive" users, that is users for whom only one choice is possible (either they don't have access to a car and therefore have to use public transport, or their origin is not served by any public transportation line). The authors are once more aware that this mechanism may not fully represent the true behaviour of users, and that a more extensive theory is probably desirable. However, the data collection associated with the establishment of such a theory is far from obvious and has not yet been effectively performed in the application cases. It would be extremely easy to include a more elaborate theory within PACSIM if one is available, or, maybe more importantly, if one is interested in measuring the impact of specific changes in user behaviour.

Another important aspect of the multimodal character of PACSIM is that the two transportation modes (cars and busses) are interdependent. In almost all cases (except for dedicated bus lanes), busses and cars share the same road network, and the congestion caused by cars also implies delays for busses. Conversely busses also contribute to link saturation and congestion for private car traffic. This feature has for instance proved useful in the assessment of public transport time tables in the presence of peak-hour congestion.

It must be noted that the packet metaphor had to be adapted for trips using public transport. Indeed, it doesn't make sense anymore to consider packets of vehicles, but one must instead consider packets of users (passengers). In the same manner as vehicle packets are moved within the network, passenger packets are moved from their origin to a public transport stop, where they wait before embarking on a bus (say). Importantly, multimodal trips are explicitly modelled: for instance users can drive from their origin to a parking place and then board the public transportation system, possibly after some waiting time. This feature has allowed the test of park-and-ride strategies with PACSIM. In these tests, a discrete choice utility model was again used to determine the fraction of car users who effectively use the park-and-ride facilities.

3.7 Testing the impact of drivers' information scenarii on congestion

We now describe one particular application of PACSIM, which emphasizes its ability to model the effect of advance incident detection and warning systems.

This application was set in the city of Namur, for which PACSIM had been satisfactorily calibrated[5] for a period covering the morning peak hour.

In the application that we describe here, PACSIM was used to test the impact of two classes of drivers' information scenarii on urban congestion in the framework of incident detection and advance warning. The main idea was to simulate, during the morning peak period (7h30 to 9h am), a traffic incident on one of the two important bridges of the city of Namur, crossing the Meuse river and leading to the city centre. The incident occured at 7h55, blocking the bridge until 8h40. The main question was then to compare advance warning strategies using variable message signs (VMS) or specialized radio channels (RDS). The criteria used for the comparison were the congestion levels both on a road close to the incident and on the complete city network.

In a first group of scenarii, we considered the use of one or two VMS situated at strategic sites of the right side of the river. One location (called Chaussée de Marche) carries the traffic coming from the south-east (with important traffic streams coming from Luxembourg) and the other (called Rue de Dave) carries the traffic coming from the south along the right bank of the river. The overall network congestion (total number of vehicles in the modellized network) is shown as a function of time in Figure 3.1 for the following five scenarii:

1. no incident (and no information),
2. incident without users' information,
3. incident and users' information (1 VMS on Chaussée de Marche),
4. incident and users' information (1 VMS on Rue de Dave),
5. incident and users' information (2 VMS).

One clearly sees that the impact of the tested information scenarii on overall congestion is negligible. We also note that, when the VMS are operated, congestion is higher just after the incident is cleared out. This is explained by the fact that the informed drivers now use longer routes and thus contribute to network congestion for a longer period.

We also present, in Figure 3.2, the impact of the same five strategies on the number of vehicles on a street (Avenue Bovesse) very near the incident and leading, on the right bank of the river, from the unperturbed bridge to the other. The "Rue de Dave" location gives extremely good results as the informed users can modify their route and take the unperturbed bridge before contributing to the congestion on the studied link. The other location (Chaussée de Marche) has little or no influence on the saturation of the studied link since informed drivers who planned to use the blocked bridge could choose an alternative route which does not include the studied road.

[5] In particular, the results produced by PACSIM were compared in detail with the application of a traditional stochastic assignment model (see [7]) applied during an unperturbed modelling period.

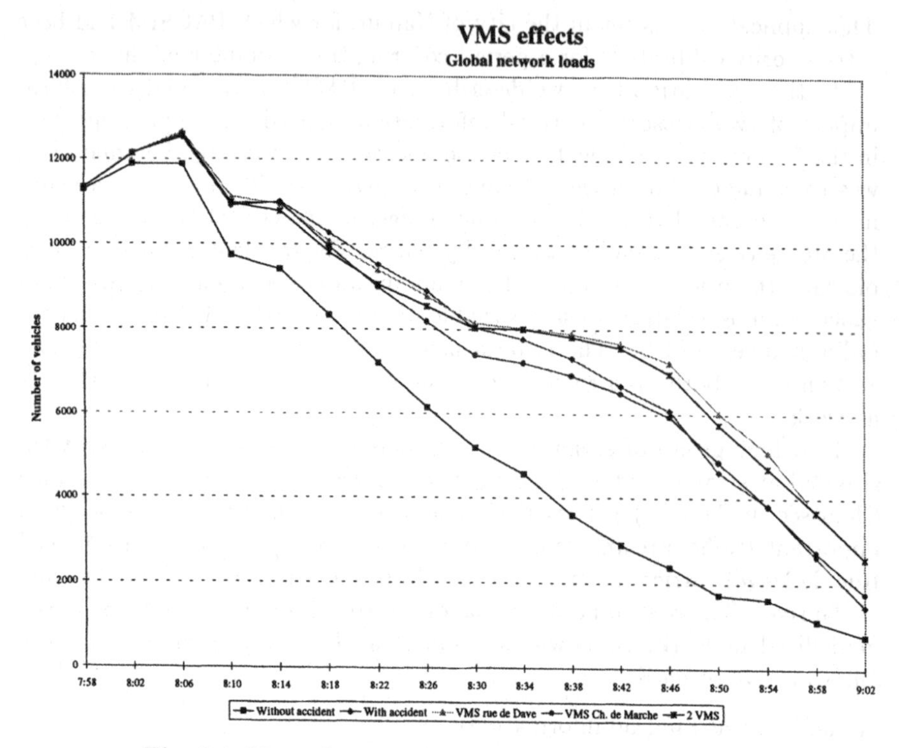

Fig. 3.1. Network congestion for different VMS scenarii

This leads us to not unexpected conclusion that the effects of VMS information strategy are quite local and crucially depend on the localisation of the sign(s).

A second class of scenarii features radio (RDS) information systems, for which we assumed three different levels of market penetration (10%, 20% and 50% of the drivers being equipped of RDS receivers). We also assumed that equipped drivers always took the information supplied to them by this channel into account, which amounts to a slight redefinition of the market penetration levels. The congestion levels on the city network and street near the incident are again shown in Figures 3.3 and 3.4 respectively, as a function of time and RDS market penetration.

It is interesting to observe that the benefits on global congestion only appear for the highest RDS market penetration level (50%). In the other cases, the number of rerouted users is not large enough to compensate for the local congestion caused by the accident. We see that the RDS information strategy has more global effects, compared to warning using VMS, as users are reached all over the network and not only if they happen to pass a given VMS location.

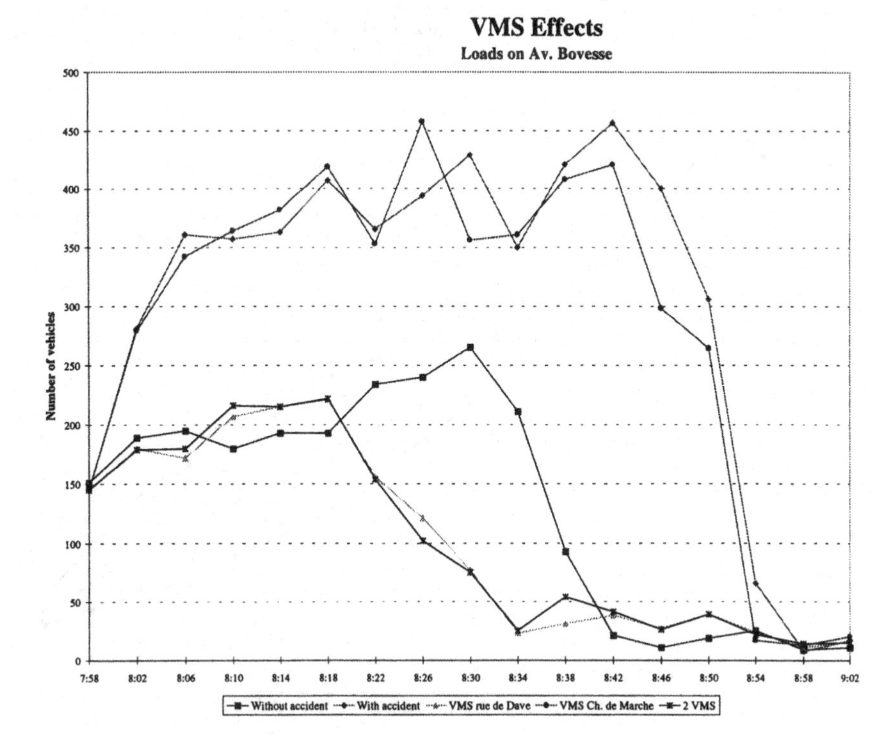

Fig. 3.2. Street congestion for different VMS scenarii

Of course, the experiments with PACSIM may be (and have been) extended in several other directions. For instance, one could measure the impact of the delay between the incident occurence and the moment where the information is distributed to the users (in the tests described above, a very short delay of 75 seconds was assumed). PACSIM was also used to calculate an optimal location for the VMS on the network (given some specific incident scenarii). Unfortunately, no a posteriori checks could be made because the city effectively didn't implement the system under test in the field. Further work on a posteriori validation thus remains to be done.

3.8 Bus schedule verification

In the second application that we describe here, the multimodal character of PACSIM was used to verify the realism (or lack thereof) of the bus timetables supplied by the bus company. Table 3.1 shows, for a specific line in the city of Namur, that the supplied timetable only corresponds to a situation with very low congestion level, in which bus traffic is not hindered by that of private cars.

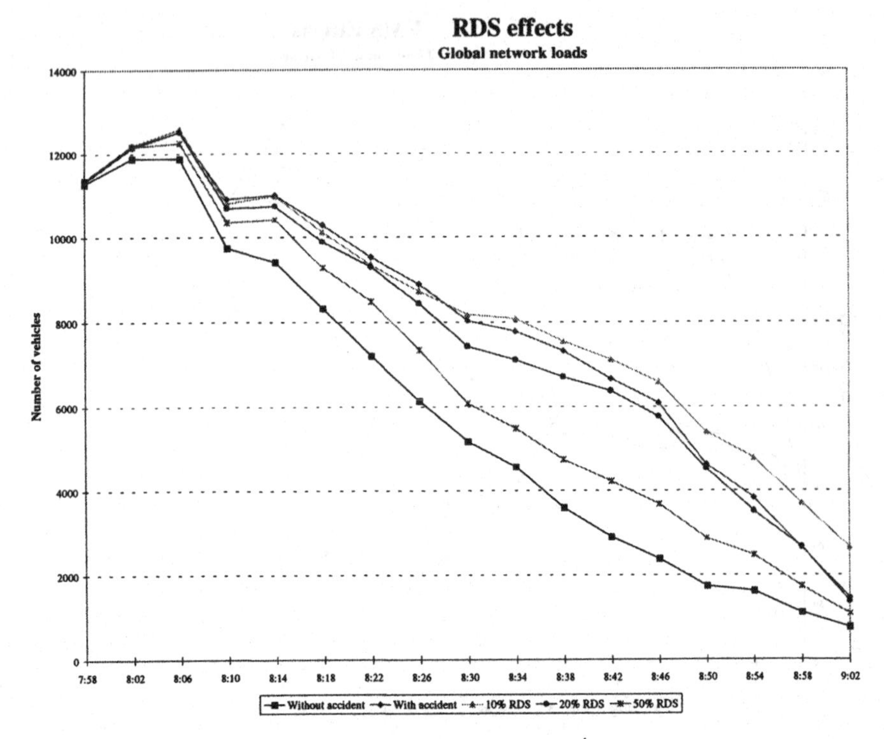

Fig. 3.3. Network congestion for different RDS market penetration levels

4. Perspectives

Further validation of PACSIM is of course highly desirable, including its use in the context of a real advanced information system. Projects are currently under discussion that would allow this development to take place, notably in Switzerland.

On the modelling front, current efforts in the development of PACSIM aim at coupling the traffic model with a gas pollutant emission model. The detailed level of the simulation provided by PACSIM indeed allows relatively complex emission models to be used. In particular, the number of stops and the engine heating time for a given vehicle can be taken into account. This opens interesting perspectives on the possible use of such a model (and of its behavioural component) for the appraisal of strategies for sustainable mobility.

Acknowledgments

The authors are indebted to P. Manneback, Ph. Dehoux and D. McArthur, who all contributed to what PACSIM is today. Thanks are also due to the Eu-

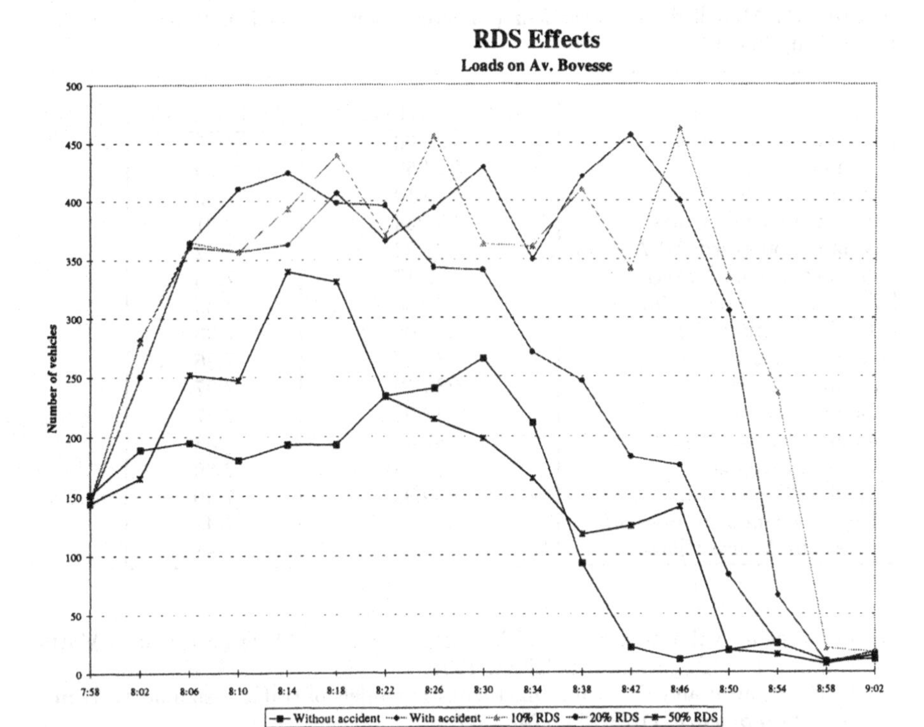

Fig. 3.4. Street congestion for different RDS market penetration levels

ropean DRIVE program and the Belgian "Transport and Mobility" impulse program which both provided partial support for this research.

References

1. R. Arnott, A. de Palma, and R. Lindsey. Does providing information to drivers reduce traffic congestion? *Transportation Research A*, 25(5):309–318, 1991.
2. M. Ben-Akiva and S. R. Lerman. Disaggregate travel demand and mobility choice models and measures of accessibility. In D. Hensher and P. Stopher, editors, *Behavioral Travel Modeling*. Croom Helm, London, 1979.
3. M. E. Ben-Akiva, A. de Palma, and I. Kaysi. Dynamic network models and driver information systems. *Transportation Research A*, 25(5):251–266, 1992.
4. M. E. Ben-Akiva. Dynamic network equilibrium research. *Transportation Research A*, 19:429–431, 1985.
5. M. Bierlaire. HieLoW : un logiciel d'estimation de modèles logit emboîtés. *Les cahiers du MET*, 2:29–43, novembre 1994.
6. P. H. L. Bovy and E. Stern. *Route choice: wayfinding in transport networks*, volume 9 of *Studies in Operational Regional Science*. Kluwer Academic Publishers, Dordrecht, The Netherlands, 1990.

Table 3.1. Modelled vs theoretical (company supplied) arrival times at various stops along line 12

STOPS	EMPTY NETWORK	TIMETABLE
Namur Gare routière	7:27:00	7:27
Namur rue de Fer	7:27:52	7:29
Namur rue de l'Ange	7:30:12	7:30
Namur place d'Armes	7:30:54	7:31
Namur boulevard Ad Aquam	7:31:32	7:32
Jambes avenue Materne	7:32:17	7:33
Jambes place de Wallonie	7:34:20	7:35
Jambes station Nord	7:35:05	7:35
Jambes viaduc	7:35:30	7:36
Jambes chaussée de Liège	7:36:10	7:37
Jambes Sarma	7:37:02	7:37
Jambes haute Enhaive	7:37:04	7:38
Jambes l'Atelier	7:39:33	7:39
Jambes Erpent Val	7:39:35	7:40
Erpent chemin de Bossime	7:40:10	7:41
Lives-sur-Meuse village	7:41:27	7:41

7. E. Cornélis. Validation de PACSIM. GRT Report 94/11, Department of Mathematics, FUNDP, Namur, Belgium, 1994.
8. C. F. Daganzo and Y. Sheffi. On stochastic models of traffic assignment. *Transportation Science*, 11(3):253–274, 1977.
9. A. Daly. Estimating "tree" logit models. *Transportation Research B*, 21(4):251–268, 1987.
10. A. de Palma, P. Hansen, and M. Labbé. Commuters' paths with penalties for early or late arrival. *Transportation Science*, 24(4):276–286, 1990.
11. Ph. Dehoux and Ph. Toint. Some comments on dynamic modelling, in the presence of advanced driver information systems. In G. Argyrakos, M. Carrara, O. Cartsen, P. Davies, W. Mohlenbrink, M. Papageorgiou, T. Rothengatter, and Ph. Toint, editors, *Advanced Telematics in Road Transport*, pages 964–981, Amsterdam, 1991. Commission of the European Communities - DG XIII, Elsevier.
12. Ph. Dehoux, P. Manneback, and Ph. L. Toint. PACSIM: a dynamical traffic assignment model. ii: functional analysis. GRT Report 90/na, Department of Mathematics, FUNDP, Namur, Belgium, 1990.
13. R. M. Downs and D. Stea. *Maps in minds.* Harper and Row, New York, 1977.
14. R. G. Golledge and R. J. Stimson. *Analytical behavioural geography.* Croom Helm, New York, 1987.
15. D. McArthur. A rule language for describing driver behavior in a IRTE. In G. Argyrakos, M. Carrara, O. Cartsen, P. Davies, W. Mohlenbrink, M. Papageorgiou, T. Rothengatter, and Ph. L. Toint, editors, *Advanced Telematics in Road Transport*, pages 1488–1498, Amsterdam, 1991. Commission of the European Communities - DG XIII, Elsevier.
16. Y. Sheffi. *Urban Transportation Networks.* Prentice-Hall, Englewood Cliffs, USA, 1985.
17. D. P. Watling and T. van Vuren. The modelling of dynamic route guidance. *Transportation Research, Part C*, 1C(2):159–182, June 1993.

18. P. L. Yu. Multiple criteria decision making: Five basis concepts. In G. L. Nemhauser, A. H. G. Rinnoy Kan, and M. J. Todd, editors, *Optimization*, volume 1 of *Handbooks in Operations Research and Management Science*, pages 663–700, Amsterdam, 1989. North-Holland.

Automatic Control Methods in Traffic and Transportation

Markos Papageorgiou

Dynamic Systems and Simulation Laboratory,
Department of Production Engineering & Management
Technical University of Crete, Chania, Greece

Summary. The paper presents some basic notions and application procedures of Automatic Control and Optimisation methodologies, and outlines current and potential applications of these methodologies to traffic and transportation systems. After definition of some basic notions and control structures, a regulation problem is presented along with possible solution methods. Practical procedures when applying automatic control and optimisation methods in real-time are discussed in some detail. A comparison of common features and differences of various network process control problems is presented before proceeding to an outline of applications of Automatic Control and Optimisation methods to various transportation domains including motorways, urban road networks, automated highway systems, rail, air, and maritime traffic. Some general conclusions and promising future research topics are also provided.

1. Introduction

Automatic Control comprises those theoretical methods and practical procedures that enable the development of technical systems capable of accomplishing autonomously certain prespecified tasks. Figure 1.1 illustrates the basic elements of an automatic control system. The process (e.g. traffic flow in an urban network) includes all technical or physical phenomena that should be influenced according to specific goals. The dynamic process changes depend upon:

- Some external quantities that are assumed independent of the dynamic evolution of the process (e.g., in the case of urban traffic, the external quantities may be the traffic lights, the traffic demand, the origin-destination pattern, the incidents, the environmental conditions, ...)
- The proper behaviour of the process according to its technical and/or physical nature (e.g. the travel times or the queue storage of vehicles on an urban link, the flow or storage capacity, ...).

The external quantities may be classified in:

- Inputs, whose values may be selected from an admissible control region (e.g. the traffic lights, the variable message signs, ...).
- Disturbances, whose values cannot be manipulated but may possibly be directly measurable via appropriate devices (e.g. traffic demand), or may be estimated or predicted via appropriate algorithms (e.g. traffic demand, origin-destination pattern, ...).

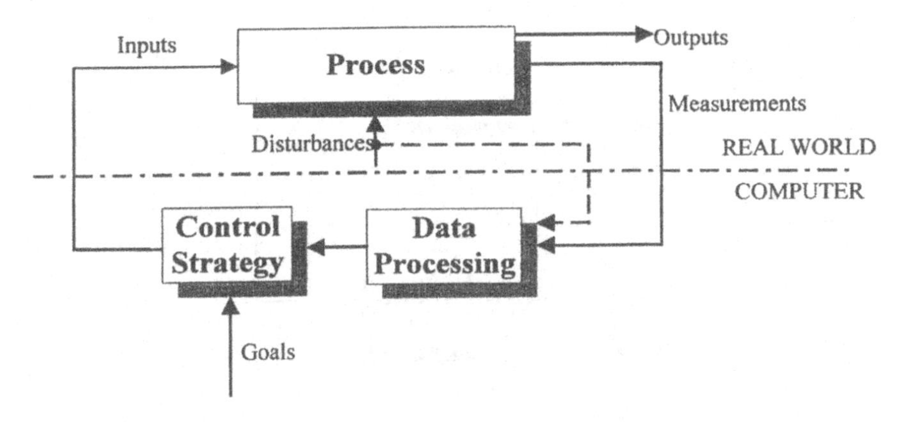

Fig. 1.1. Basic elements of an automatic control system

The process outputs are the quantities chosen to represent the behavioral aspects of interest (e.g. the outputs of urban traffic may be the total travel time, the queue lengths, ...). The data processing block in Figure 1.1 comprises the estimation and/or prediction tasks, based on real-time measurements of internal process quantities or disturbances. It is the task of the control strategy:

To specify in real time the process inputs, based on available measurements/estimations/predictions, so as to achieve the prespecified goals regarding the process outputs despite the influence of various disturbances.

If this task is undertaken by a human operator, we have a manual control system. On the other hand, in an automatic control system, this task is undertaken by an algorithm (the control strategy). In modern and/or complex systems, the control strategy as well as the data processing algorithms are typically implemented in a digital computer.

In the example of urban traffic control (Figure 1.2), the control strategy should calculate in real time the traffic light switchings based on available measurements/estimations/predictions, so as to minimize the total travel time for any demand values, any origin-destination pattern, and even in presence of incidents.

It is not too difficult to analyse any automatic control system within the frame of Figure 1.1, even if the corresponding processes may be of very different nature, such as robots, airplanes, chemical processes, water or gas networks, ... The specification of inputs and measurements for a particular automatic control system is closely related to technological issues. On the other hand, the data processing and particularly the control strategy blocks contain the system's "intelligence", i.e. its capability to face automatically and efficiently any situation arising due to the impact of the disturbances.

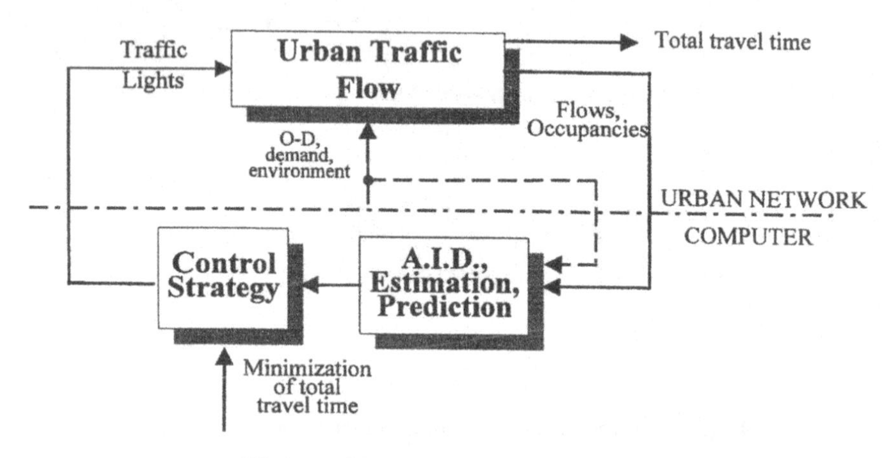

Fig. 1.2. Urban traffic control system

At this point it is necessary to make a distinction between the notions of control strategy and mathematical model. A mathematical model of a process comprises a number of equations that describe, with a more or less limited accuracy, the proper (internal) process behaviour in the considered context. Hence, a mathematical model, fed with input and disturbance values, may be employed to calculate with a certain accuracy the corresponding output values or other internal quantities. For example, an urban traffic flow model, fed with the values of the traffic lights, of the traffic demand, of the origin-destination pattern, ..., may be used to calculate the corresponding queue lengths and travel times in all network links. Dynamic models describe the time development of the process phenomena. For one and the same process, there may be several useful models with different levels of resolution, accuracy and complexity. For example, for the urban traffic flow, existing models are microscopic or macroscopic, static or dynamic, stochastic or deterministic,...

On the other hand, a control strategy is an algorithm that makes the decisions regarding the control actions that should be applied at each instant in time, i.e. it calculates the input values. As we will see later, a control strategy may be designed on the basis of a mathematical model or may even explicitly include a mathematical model as a mean of assessing in real time the efficiency of this or that control action. Despite these connections, it appears important to distinguish clearly between the tool that imitates the process behaviour (mathematical model) and the tool that makes control decisions (control strategy).

How should one design a control strategy? One possibility could be to try to imitate (to model) the behaviour of a (real or hypothetical) human operator (expert system approach). Another possibility is to attempt to understand (to model) the process behaviour and then to apply systematic methods (automatic control approach) that lead to an adequate control strategy. The next sections will provide some basic information regarding the automatic control approach.

Automatic Control exists as an independent discipline since some 50 years. During this period in time, automatic control engineers have developed and refined a number of methods for the systematic design of efficient, reliable and robust control strategies, and they have applied these methods to a high number of processes (space, defense, robotics, chemical processes, traffic, environment, ...). The basic philosophy and the importance of these methods are related to their general applicability: they are not particular heuristics valid just for a specific process, but general methods applicable to any process that can be described by certain types of mathematical models, regardless the physical process nature (robot, airplane, traffic, environment, ...). This general approach reaches its limits if, for a specific process:

- There is a lack of understanding of the process behaviour, i.e. no adequate model available
- Certain complexity limits are exceeded
- The process behaviour is of a discrete or event-oriented or combinatorial nature.

Under these conditions, that have become more and more frequent in recent years, continuing efforts for developing general, efficient, and systematic methods did not always reach the maturity required for successful practical applications.

Before proceeding into more details regarding the automatic control approach, it is useful to present some features of the basic structure of an automatic control system as represented in Figure 1.1. The control system is characterized by a closed-loop structure, whereby the calculation of inputs is effectuated on the basis of measurements of process internal quantities which, by their turn, are influenced by the inputs. What could be an alternative structure? Assume availability of a process model of the type

$$y = f(u, d) \tag{1.1}$$

where the vectors y, u, d include, respectively, the outputs, the inputs, and the disturbances, while f is a generalized operator, e.g. a number of differential equations. Assume also (see Section 3 for a more general case) availability of desired output values y_d. If (1.1) is invertible, one obtains

$$u_d = f^{-1}(d, y_d) \tag{1.2}$$

which corresponds to an open-loop control strategy (Figure 1.3) that makes no use of process measurements. The advantage of this structure, compared to the closed loop, is the rapidity of control action: the strategy reacts immediately to disturbance variations without waiting for the disturbance impact to become visible in the internal process variables. Moreover, if the process

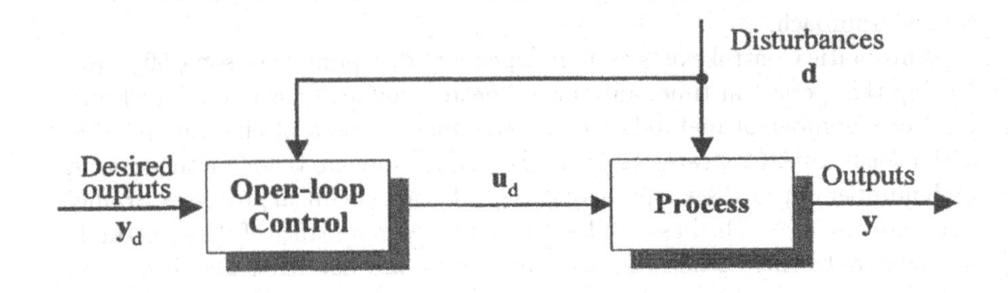

Fig. 1.3. Open-loop control system

itself is stable, there is no risk of obtaining an unstable system as in the closed-loop case.

Unfortunately, the disadvantages of open-loop control are much more important than the advantages. In fact, if no real process measurements are utilized, like in the open-loop structure, the real process state is never known, i.e. there is no way for the control strategy to know whether the real outputs y are actually close to y_d, hence the control strategy will not react if y is far from y_d. This uncertainty may originate from:

- The limited accuracy of any mathematical model
- The presence of non-measurable disturbances.

Because of this inherent uncertainty, any automatic control system is forced to include a closed-loop structure. Nevertheless, if measurements/estimates/predictions of some major disturbances are available, they may be used to ameliorate the control system efficiency as indicated in the dashed signal line in Figure 1.1.

2. The regulation problem

The regulation problem is a special case of the control system of Figure 1.1, whereby the control goal is to lead and maintain the process output y near prespecified corresponding desired values y_d that are called set values. Moreover, it is usually assumed that the real outputs y are measurable in real time.

Example 2.1: Temperature regulation in a room. The process includes the heating actuators and the heat transfer phenomena within the room. The output is the internal temperature while the input is the heating valve position. The main disturbances are the external temperature, opened doors or windows, the number of persons in the room, ... The goal of the regulation is to maintain the internal temperature close to the chosen set value despite the disturbance variations.

Example 2.2: Isolated ramp metering, see Figure 2.1. The process is the motorway traffic flow downstream of the on-ramp. The output is the downstream occupancy rate, while the input is the ramp flow. The major disturbances are the upstream flow, the environmental conditions, the drivers behaviour, the percentage of trucks, a possible vehicle stop, ...

Example 2.3: Coordinated ramp metering of several motorway on-ramps, see Figure 2.2. Compared to the last example, there are several outputs y_1, y_2,... and several inputs u_1, u_2, ...

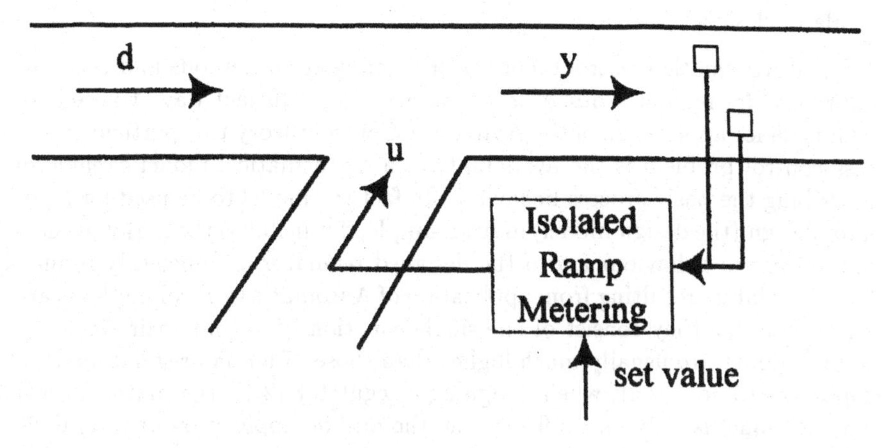

Fig. 2.1. Isolated ramp metering

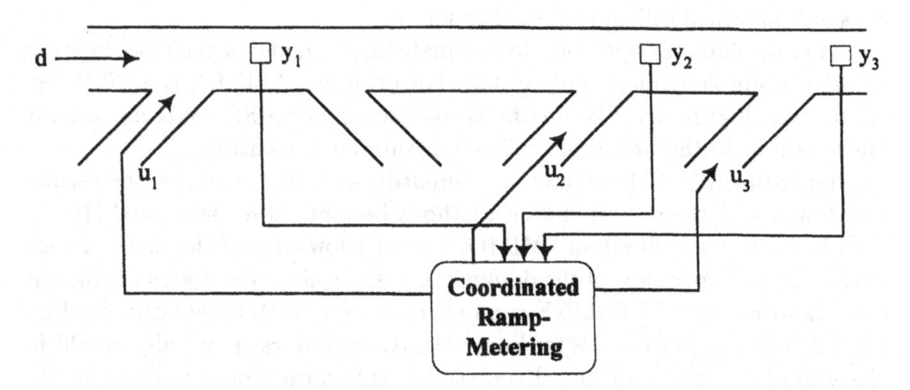

Fig. 2.2. Coordinated ramp metering

Regulation problems call for a regulator, i.e. a formula

$$u = R(y, d) \tag{2.1}$$

that guarantees $y \approx y_d$ despite the presence of disturbances. The major performance criterions for a regulator are:

- **Stability** (above any other consideration)
- **Rapidity** of response in case of a change of the set value or of the disturbances
- **Stationary accuracy**, i.e. $y \approx y_d$ under stationary conditions
- **Robustness**, i.e. preservation of the control performance even if the real process behaviour is not identical to the mathematical model used for regulator design.

The Automatic Control theory offers a number of methods and theoretical results for regulator design in a systematic and efficient way. A necessary condition for application of the Automatic Control theory to a particular process control problem is the availability of a mathematical model capable of describing the basic process behaviour. In fact the model to be used for regulator design (the design model) may be simple, if it includes the major aspects of the process behaviour and if the designed regulator is sufficiently robust. Most regulators resulting from application of Automatic Control methods are very simple, as they consist of one single equation (2.1), but their efficiency and reliability are usually much higher than those of human regulators. It is important to note that, when designing a regulator (2.1), the mathematical process model is only used off-line, i.e. the on-line application of (2.1) does not include any model equations.

Most Automatic Control methods are applicable to linear models, while the nonlinear control theory is less developed. In many cases, it is possible to linearize a nonlinear model (e.g. around the set values) before the regulator design, which may call for special measures during control operations in order to avoid practical difficulties, see Section 5.

The regulator design for SISO (single-input-single-output) processes is relatively simple (though not trivial). For example, ALINEA is a SISO regulator, see Figure 2.1. The methods used for linear SISO cases are usually those taught in the first basic course on Automatic Control.

Regarding MIMO (multiple-input-multiple-output) processes, the regulator design and the corresponding methods become more elaborated [1], [7], [13]. In both cases, SISO and MIMO, a good knowledge of the methodology and a certain experience of the designer are essential for the design of efficient control strategies. METALINE is an example of a MIMO regulator, see Figure 2.2. The design methods for linear MIMO regulators are usually taught in the second and more advanced courses on Automatic Control. These methods are the linear-quadratic (LQ) optimization, pole assignment methods, decentralized control, ... Particular attention should be paid to the robustness properties of the designed regulators via recently developed, powerful methods and tools.

Further methods for particular classes of regulators are available within Automatic Control theory, like, for example, non-linear regulators (for non-linear processes) and adaptive regulators (whereby the regulator parameters are adjusted automatically in real-time by suitable mechanisms in order to account for process uncertainties or for time-varying process behaviour)[2].

3. Optimal control strategies

We will now address the general case of the control system of Figure 1.1 under the assumption that the control goal can be expressed as the minimisation (or maximisation) of a quantity J (the objective function or optimisation criterion). This quantity may, for example, correspond to a system output y that depends on the inputs u and on internal process variables x

$$J(u,x) = y(u,x) \rightarrow Min. \tag{3.1}$$

In the case of urban traffic, this criterion typically corresponds to the minimization of total travel time in the considered network.

If there are two or more competing subcriteria (or outputs) $y_1, y_2, \ldots$ to be minimized, one may construct the overall criterion J as a weighted sum

$$J(u,x) = \alpha_1 y_1(u,x) + \alpha_2 y_2(u,x) + \ldots \tag{3.2}$$

where the weighting parameters α_j are chosen so as to satisfy

$$\alpha_1 + \alpha_2 + \ldots = 1, \quad \alpha_j > 0. \tag{3.3}$$

These parameters express the relative importance of each subcriterion.

The internal variables x depend upon the inputs u and disturbances d according to a mathematical model. A great part of dynamic processes may be described by a state space model that has the general form

$$\dot{x} = f(x,u,d) \tag{3.4}$$

where the vector x comprises the state variables. Note that, if the initial condition, i.e. the value of x for time $t = 0$, is known

$$x(0) = x_0 \tag{3.5}$$

and the time trajectories $u(t)$, $d(t)$, $t \in [0,T]$, are also given, the differential equations (3.4) may be resolved to deliver the corresponding state trajectory $x(t)$ over the same time period $[0,T]$.

The choice of inputs u is usually limited due to physical or technical constraints that define an admissible control region via a set of inequalities

$$h(x, u, d, t) \leq 0 \quad \forall t \in [0, T]. \tag{3.6}$$

The optimal control problem may then be expressed in the form of a mathematical optimisation problem as follows:

Problem P1: Given the initial condition (3.5) and the disturbance trajectories $d(t)$, $t \in [0, T]$; find the input trajectories $u^*(t)$, $t \in [0, T]$, that minimize the criterion J subject to the model equations (3.4) and the constraints (3.6).

What is the relationship between this optimization problem and the control strategy of Figure 1.1 or Figure 1.3? We first note that the future values of the disturbances are assumed known in P1 which implies availability of corresponding predictions. Second, the required solution of P1 is a trajectory $u^*(t)$ to be calculated before $t = 0$ and to be applied during the period $[0, T]$ in an open-loop manner. Hence, if the model (3.4) or the disturbance predictions are not accurate, the real state variables and the real process outputs will be accordingly different from the optimization results. Moreover, it should be noted that the control $u^*(t)$, $t \in [0, T]$, resulting from P1, is only optimal for the particular initial condition considered in P1.

In order to obtain a closed-loop solution, that also considers available disturbance predictions, one should consider the following optimisation problem:

Problem P2: Given the disturbance trajectories $d(t)$, $t \in [0, T]$; find a function R

$$u(t) = R[x(t), t], t \in [0, T] \tag{3.7}$$

that minimizes the criterion J subject to the model equations (3.4) and the constraints (3.6).

The main differences between P1 and P2 (and between their respective solutions) are the following:

- The solution of problem P2 is not a trajectory $u^*(t)$, $t \in [0, T]$, but a function $R(x, t)$, called the control law, that may be executed in real time by use of state measurements x (closed-loop solution).
- The solution of problem P2 is independent of the initial condition and hence applicable anywhere in the space (x, t).

Figure 3.1 illustrates the difference between both procedures in the space (x, t). In the case of P1, the problem solution delivers a control trajectory $u^*(t)$, $t \in [0, T]$, and the corresponding state trajectory $x^*(t)$, $t \in [0, T]$ (Figure 3.1a). If an unexpected instantaneous disturbance appears during this time period, the real state trajectory $x(t)$ will be different from $x^*(t)$. In the case of problem P2, on the other hand, the problem solution is represented by the function (3.7) that reacts in an optimal manner at any point in the space (x, t), which is indicated in Figure 3.1b by the small arrows. Hence, even if an unexpected instantaneous disturbance appears, the control reaction will be optimal by following the arrows. Clearly, in absence of any unexpected disturbance both solution will be identical for a given initial state. It should be noted, however, that in case of an unexpected constant (not instantaneous) disturbance, even the closed-loop control may become non-optimal. One way to partially overcome this inconvenience is to introduce a suitable dynamic relationship between $u(t)$ and $x(t)$ instead of the static function (3.7).

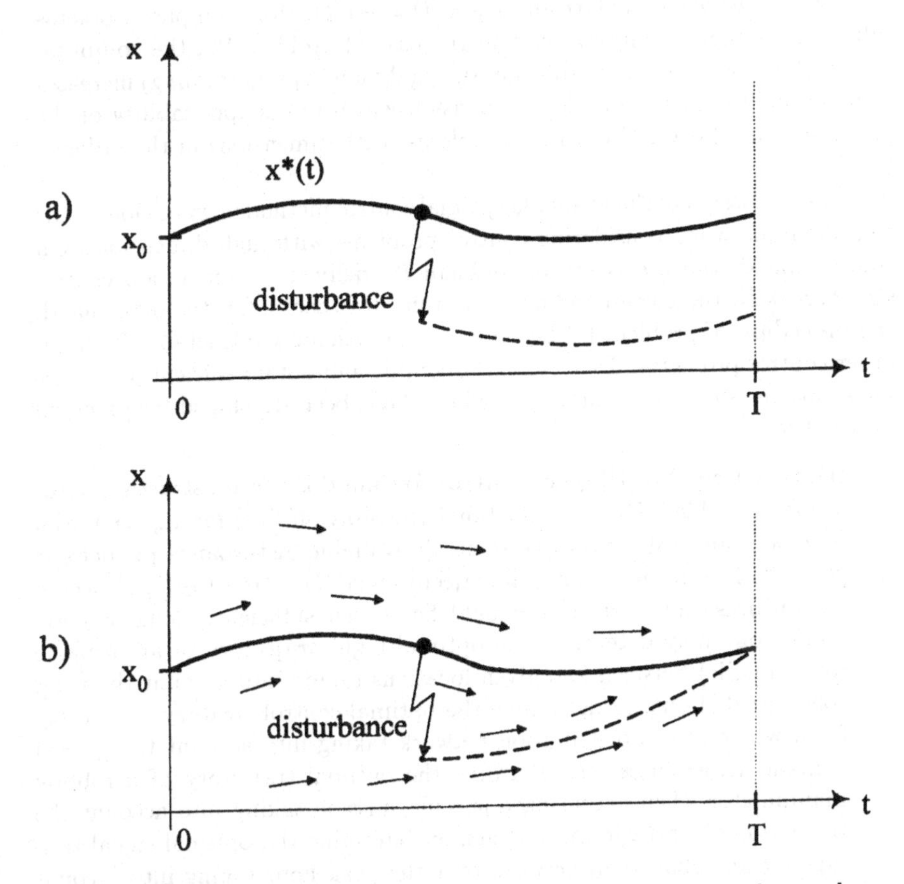

Fig. 3.1. Illustration of open-loop (a) and closed-loop (b) optimal control

What are the available methods within the Automatic Control and Optimisation theories for solution of problems P1 and P2?

Problem P1 may be resolved analytically (using Pontryagin's Maximum Principle [3], [5], [12], [23], [24], [25]) for problems of simple structure (e.g. linear model) or low dimensions. In this case the control trajectory may be given analytically, e.g. $u^*(t) = e^{-t}$. In most practical situations, however, the solution can only be found by implementing a corresponding numerical optimisation algorithm (e.g. a feasible direction method) in a digital computer whereby the corresponding computational effort increases polynomially with the problem dimension. Hence, the utilisation of powerful algorithms allows for the solution of large-scale problems (e.g. with hundreds of state variables) within less than 1 CPU-min.

With regard to problem P2, an analytical solution is also feasible only for problems of simple structure (e.g. linear model, quadratic criterion, no constraints) or low dimensions [4], [14], [15], [23]. In this case, the function (3.7) will be delivered analytically, e.g. $u(t) = -x(t)$. More complex problems call for a numerical solution. But in the case of problem P2, the computational effort for a numerical solution (using dynamic programming) increases exponentially with the problem dimension which limits applicability of the procedure to relatively low order problems, with dimensions in the order of 8 or 10.

This discussion of the available optimal control methodology obvious leads to a dilemma when considering control problems with high dimensions. On the one hand, the procedure of problem P1 delivers an open-loop control structure with the corresponding important drawbacks. On the other hand, the procedure of problem P2 becomes computationally intractable for high-order control problems. To avoid this dilemma and obtain efficient and feasible solutions, some suboptimal procedures have been developed for practical applications:

1. **Hierarchical Multilayer Control**: Within this control structure, a solution of problem P1 is specified first (possibly off-line) for a given initial condition and taking into account all available disturbance predictions [8], [18]. This leads to optimal trajectories $u^*(t)$, $x^*(t)$, $t \in [0, T]$, where the optimisation horizon T should be chosen sufficiently long. For example, one may determine the optimal flight trajectory of an airplane (or a missile) between two given locations taking into account the wind speed predictions; or determine the optimal control strategy over a day for a water or gas distribution network taking into account the typical demand trajectories; or determine the optimal trajectory of a robotic manipulator when executing a specific mission taking into account the prediction of obstacle movements; or determine the optimal signal timings for an urban road network over the peak hour taking into account the demand predictions. However, if the thus determined optimal control $u^*(t)$ is applied in real time, the actual trajectory $x(t)$ may be far from the theoretically calculated $x^*(t)$ because of:
   - limited model accuracy within P1
   - limited prediction accuracy regarding the disturbances
   - occurrence of unexpected disturbances.

   In order to avoid the deviations $x(t) - x^*(t)$, one may define a regulation problem (see Section 2) using $x^*(t)$ as a desired trajectory (set value). If the process under control is nonlinear, a linearisation around $x^*(t)$ may facilitate the application of linear Automatic Control methods for the design of a subordinated regulator (Figure 3.2) that has the task to react

(i.e. to modify $u^*(t)$) in a closed loop, using real-time measurements $x(t)$, so as to guarantee $x(t) \approx x^*(t)$ in a quick, stable, and robust way.
In summary, a hierarchical multilayer structure first calculates the trajectories $u^*(t)$ and $x^*(t)$ for a sufficiently long time horizon taking into account all available predictions, whereby $x^*(t)$ will be mathematically optimal but physically suboptimal due to different sources of uncertainty. Then, a classical regulator has the task to achieve $x(t) \approx x^*(t)$ despite the various uncertainties and unexpected disturbances.

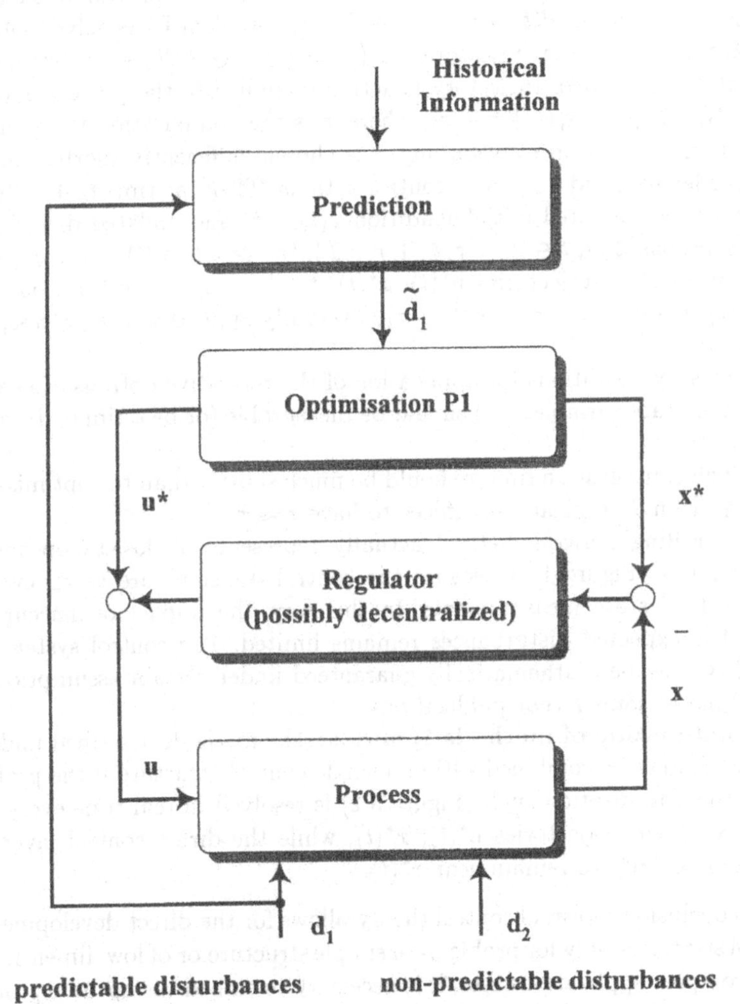

Fig. 3.2. Multilayer control structure

2. **Repetitive Optimisation (with rolling horizon)**: For some applications, the prediction of the main disturbances may only be sufficiently accurate over a relatively short time horizon. In these cases, it makes little sense to solve problem P1 over a long time horizon, because the solution trajectories $u^*(t)$, $x^*(t)$ might be inadequate as the result of inaccurate predictions. Hence, the optimal control problem P1 should better be embedded in a closed-loop control structure, whereby its numerical solution is effectuated in real time in the following way. At time t_o, based on a measured initial condition $x(t_o)$ and on available disturbance predictions $\tilde{d}(t)$, $t \in [t_o, t_o + T]$, the problem P1 is solved numerically to obtain the trajectories $u^*(t)$, $x^*(t)$, $t \in [t_o, t_o + T]$, but only a part of the control trajectory is actually applied to the process, namely $u^*(t)$, $t \in [t_o + \tau, t_o + \tau + \bar{\tau}]$, where τ is the computation time for the numerical problem solution and $\bar{\tau}$ is chosen sufficiently shorter than T in order to avoid "myopic" control actions. Then, at time $t_o + \bar{\tau}$, based on a new measured initial condition $x(t_o + \bar{\tau})$ and updated disturbance predictions $\tilde{d}(t)$, $t \in [t_o + \bar{\tau}, t_o + \bar{\tau} + T]$, the problem P1 is solved again to obtain the trajectories $u^*(t)$, $x^*(t)$, $t \in [t_o + \bar{\tau}, t_o + \bar{\tau} + T]$, but only $u^*(t)$, $t \in [t_o + \bar{\tau} + \tau, t_o + 2\bar{\tau} + \tau]$ is actually applied to the process, and so forth.
Necessary conditions for application of the repetitive optimisation are:
   - The state variables $x(t)$ should be measurable (or be estimated) in real time
   - The computation time τ should be much shorter than the optimisation horizon T; typically it suffices to have $\tau \approx \bar{\tau}$.

   The rolling horizon method actually represents a closed-loop control structure (Figure 1.1) because the control decisions are taken every $\bar{\tau}$ based on state measurements. In this way, the impact of inaccuracies and unexpected disturbances remains limited. The control system stability may be mathematically guaranteed under certain assumptions as proved in some recent publications.
3. **Combination of methods 1. and 2.**: The methods described under 1. and 2. may be combined within a single control structure if the problem of the optimisation layer (Figure 3.2) is resolved in real time every $\bar{\tau}$ to provide the trajectories $u^*(t)$, $x^*(t)$, while the direct control layer still regulates $x(t)$ to remain near $x^*(t)$.

In conclusion, optimal control theory allows for the direct development of control strategies only for problems of simple structure or of low dimension. In more complex application cases, it is necessary to embed the optimal problem solution in a (possibly hierarchical) control structure so as to circumvent the accumulation of unavoidable uncertainties.

It should also be noted that all above procedures are also applicable to discrete-time optimal control problems based on a discrete-time dynamic model

$$x(k+1) = f[x(k), u(k), d(k)] \tag{3.8}$$

instead of (3.4), where k is the discrete-time index.

The quality of decisions delivered via solution of an optimal control problem is clearly superior to a human operator's performance. This superiority becomes even more pronounced for complex problems. In some cases, the decision delivered by the optimal control problem solution might even appear strange or inexplicable at first view, calling for a more careful analysis of the results so as to understand and appreciate their sensible background and express them in human reasoning terms. Thus, in these successful applications, the optimisation algorithm may surprise its human designer by proposing non-typical but still optimal solutions. What is the origin of this "intelligence" of a piece of software? It is a general theory, that does not rely on empirical and questionable heuristic rules, but allows for an exhaustive but efficient search in the space of all feasible decisions and the selection of the best (optimal) ones.

4. Optimisation theory

Optimal control problems may be understood as a particular area of the broader Optimisation Theory. The basic problem in optimisation theory is to specify, within a given space, an optimal element that minimizes a criterion value $J \in R$ subject to prespecified constraints [23]. According to the nature of the searched space and of the constraints, one may distinguish different classes of optimisation problems with corresponding solution methodologies, as for example:

- **Static Optimisation**: One looks for a vector x in the Euclidean space R^n. Typical constraints are equality constraints $c(x) = 0$ and/or inequality constraints $h(x) \leq 0$. Solution methods are largely based on the notion of gradient leading to local or global minima (or maxima) depending upon the particular application [9], [10], [16], [19], [27].
- **Dynamic Optimisation**: This is the class, where optimal control problems belong to. One looks for real valued trajectories $x(t)$, $t \in [T_1, T_2]$, $x \in R^n$. The equality constraints may include differential equations which typically correspond to the dynamic process model. Problems of this class are handled using the variational calculus (Pontryagin's maximum principle) or dynamic programming.
- **Combinatorial Optimisation**: In this class, the search space typically includes a finite (countable) number of elements. In problems of this type, the extremely useful notions of gradient and infinitesimal calculus do not exist. As a consequence, the numerical solution of many of these problems (NP hard) faces the combinatorial explosion, i.e. the exponential increase of the computation effort with the problem dimension. Nevertheless, for

many combinational problems, there are optimal or suboptimal algorithms with polynomial complexity [20].
- **Stochastic Optimisation**: This class includes static, dynamic, or combinatorial problems whereby the constraints may include stochastic variables with known probability distribution. In view of this well-defined uncertainty, the optimisation goal is to minimize the expected value of a criterion $J \in R$.
- **Game Theory**: This is the class of optimisation problems including two or more (instead of one) decision makers with competing objective criteria.
- **Multicriteria Optimisation**: These problems consider simultaneous minimisation of more than one (typically conflicting) objective criteria. The solution of these problems is not represented by a single optimal element within the problem space, but by a set of elements (Pareto set) that contains all non-inferior elements. An element is non-inferior if there is no other feasible element that has all criteria values lower.

Optimisation methods are extremely efficient in terms of their proposed problem solution, as they deliver - by definition - the best. Applications include:

- Parameter estimation (e.g. for a model), on-line or off-line [17], [26]
- State estimation (in real time) [6], [11]
- Planning (off-line)
- Control (on-line)
- Prediction (on-line or off-line).

For some problem types, the solution may be calculated analytically, which is an advantage for practical implementation of the results. In case of an algorithmical solution by use of a digital computer, the issue of the required computation effort becomes important, particularly for real-time applications.

In several potential application cases, the problem under consideration is known in general verbal terms, i.e. not with mathematical precision. The formulation of a problem of this type in mathematical terms is neither a trivial act nor unique. The same application problem may be presented in one or another form, depending upon the introduced objectives, constraints, assumptions, and simplifications. The decision on how to present a practical problem in mathematical form should be taken under consideration of the corresponding computational efficiency of available solution algorithms, the implications of the introduced simplifications, the relevance of the chosen objective criterion, ... It is useless to oversimplify a problem to enable application of a particular efficient algorithm, because the delivered theoretically accurate solution may be practically irrelevant; on the other hand, an exaggerated formulation accuracy may exclude the application of optimisation methodologies (e.g. due to intractable computation effort), allowing only heuristic solutions of inferior quality. For these reasons, the formulation of a general application problem in mathematical terms will be more adequate if the designer has sufficient knowledge both of the various optimisation methodologies and of the particular application area. It should be noted, however, that in some areas it appears hardly possible to present a particular application problem in the form required by the optimisation methods, in which case heuristics may be the only alternative approach.

When a particular problem is presented in form of a mathematical optimisation problem, it is necessary to investigate the required computational effort for its solution. Limits may originate from:

- Very large dimensions
- Combinatorial explosion
- Real-time application.

In conclusion, the Optimisation Theory offers a number of efficient methods applicable to a great variety of problems prevailing from almost all scientific and technical disciplines. Because the solutions delivered by the optimisation methods are the best possible, this approach should be considered before any other. An optimisation approach is not possible when the nature of the considered problem does not allow for an adequate mathematical formulation, or when the computational effort for an acceptable solution accuracy is too high.

5. Heuristics

The theories of Automatic Control and Optimisation deliver systematic and efficient solutions for well-defined problems. However, for most practical applications it appears necessary to complement these solutions with various kinds of heuristics so as to take into account practical aspects that were neglected in the theoretical elaboration, or to consider specifications that may be hardly expressed in the context of an optimisation problem. These heuristics are present almost in every control system but they are not necessarily mentioned in the corresponding scientific publications when they are only applicable to a specific control problem. In the following, a certain classification is suggested for the most common heuristics that appear in control systems.

1. **General Structural Heuristics**: In Automatic Control practice there is a number of control structures that have been proposed to respond to certain practical considerations. Two structure types were already presented in Section 3, namely the hierarchical multilayer control and the repetitive optimisation. We may call these procedures structural heuristics because their structure is not fully supported by theoretical results but is mainly based on empirical considerations. In many practical applications, these structures are applied without a possibility to theoretically

guarantee their stability or robust performance under all possible conditions. Often, these structures are only tested by use of simulations or in real life, for a limited number of possible scenarios.
Automatic Control includes several typical structural heuristics that are general enough to be applied to a great number of different problems. Another well-known structure of this type is cascade control. As an example, consider control of an underground rail vehicle that has to travel automatically from A to B following an optimal prespecified trajectory (set value) $x^*(t)$. The process input is an electric voltage that influences sequentially the current of the electric engine, the speed, and the position of the vehicle, that latter being the process output. However, a single control loop according to Section 2 may have a number of disadvantages, where the increased complexity is only one of them. It appears more reasonable to develop a cascade if closed loops according to Figure 5.1, whereby each control loop provides a set value for the next interior control loop. Each interior control loop is much quicker than the next exterior one and may hence be neglected when designing the latter. Besides the advantage of a simpler control design, cascade control allows to take benefit from intermediary measurements so as to react rapidly to unexpected disturbances that may appear at any of the subprocesses of Figure 5.1.

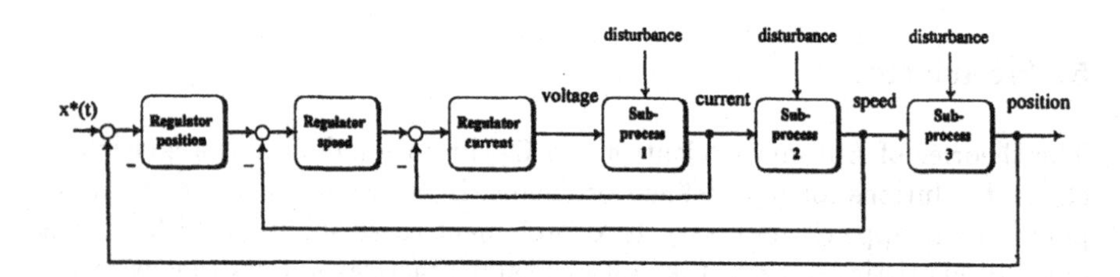

Fig. 5.1. Cascade control

2. **Surveillance and Emergency Heuristics**: Every control system includes a more or less high number of devices (measurements, communications, actuators, controllers, computers, ...) that may fail during the system operation. Failure of a control system device should first be automatically announced to allow for a quick repair and for further emergency actions by the system operators. On the other hand, it is necessary for the control strategy to continue - to the extend possible - its functioning, taking into account the equipment failures. This implies that the control strategy should include a surveillance and emergency decisions modul that enables a graceful degradation in case of failures. In view of the high number of devices included even in relatively simple control

systems, and of all possible failure combinations, the emergency modul typically includes a complex decision tree for automatic response to any possible failure situation. The development of this modul may be based on the proper answering of questions like: Is a measurement, provided by a failed detector, estimable from other measurements? Is a particular failed device absolutely necessary for the control system functioning? Is it necessary to modify the system's main goal in view of the current failures? If the original control goals cannot be retained due to failures, is it possible at least to guarantee a safe system functioning?

In many applications, a device failure or the occurrence of an unexpected disturbance may not be directly detectable. Hence an automatic diagnosis (or indirect detection) modul is necessary in order to take into account these exceptional events. A typical example of this kind of diagnosis in road traffic is the automatic incident detection. For the development of automatic detection tasks, the theories of optimisation and of signal processing offer a number of systematic and efficient methods. Nevertheless, for many applications of this type, a mathematical formalisation may not be evident, in which case appropriate heuristics may be the only remaining approach.

3. **Specifications Heuristics**: In some control problems, the operational specifications may require several distinct operational states in dependance of the current system situation or of the time-of-day. As an example, for an isolated ramp metering system (Figure 2.1) one may provide the following operational specifications:

   The system should be out of operation between 22:00 and 6:00h. Between 6:00 and 22:00h the traffic lights should be set to flashing orange if the mainstream occupancy downstream of the ramp is higher than 10% and lower than 15%. Beyond 15%, the regulator should calculate the phase durations for green and red. If the occupancy at the upstream end of the ramp exceeds 50%, a maximum green phase is imposed in order to avoid a ramp queue interference with the surface street traffic.

   Operational specifications of this kind lead to the definition of distinct control states, each with its corresponding particular control actions. The switching decisions between these states (depending on real-time measurements and/or time-of-day) are automatically taken on the basis of suitable heuristics. This is a non-trivial task for complex control systems with many distinct control loops and distinct control states, because the applied heuristics should guarantee proper, complete, unique and contradiction-free decision making for all possible combinations of measurement values.

4. **Other Heuristics**: Many practical control systems include various other kinds of heuristics, for example in order to tackle nonlinearities or input constraints ... It should be noted that some of these heuristics may be

avoided by theoretically sound and efficient Automatic Control procedures.

In conclusion one may state that, despite the variety of theoretical results and algorithms offered by Automatic Control and Optimisation, it seems almost inevitable to complement a rigorously designed control strategy by various heuristics, aiming to address particular practical aspects. For simple systems, the development of these heuristics may be relatively easy. But for complex systems with dozens of control inputs, sub-loops, and measurements, it is desirable to tackle this problem in a more systematic, theoretically founded way. Some concepts and tools in this direction is provided by a special branch within Automatic Control theory, namely the theory of Discrete Event Dynamic Systems.

6. Automatic control application procedures

As already mentioned, Automatic Control became an independent scientific theory only some 50 years ago. Nevertheless, control problems were encountered even before 1940 in several technical domains. The oldest regulators (integrated in water distribution works) appeared in ancient Egypt and Greece, but it was only after the Watt vapor engine that regulators could be physically distinguished from the process under control, and it was only in this century that the concept and the effect of feedback loops was explicitely studied (e.g. in the context of electrical circuits), largely understood, and intentionally employed for system regulation.

Until 1940, regulation problems and their respective solutions remained at a low scientific level and addressed only particular application needs. The engineers in charge of water works, electrical, chemical, or mechanical systems invented independently regulation mechanisms based on feedback loops, that stabilised, mostly without theoretical justification, the corresponding quantities, without really understanding how and why these control systems worked. It is only after 1940 that one begins to realize that the behaviour of processes of completely different nature may be very similar once expressed in mathematical equations. As a consequence, the development of regulation methodologies could be based on general equations like (3.4), without the need to consider the particular properties of individual processes.

This general view leads to the birth of Automatic Control as an independent discipline, whose results are generic enough to be applied to many, apparently different, practical problems. Nevertheless, the ties of Automatic Control and its application domains remain. On the one hand, more and more Automatic Control engineers are in charge of developing practical control systems in different application areas. On the other hand, the practical applications and the new problems they reveal, indicate the requirements for further theoretical developments. It should be noted, however, that a notorious gap has always been claimed between theory and practice of Automatic Control, i.e. a certain inertia of penetration of theoretical results in various application domains.

The necessary steps when developing a control system for a particular application are typically the following:

1. **Modelling/Identification** is the phase of development of a mathematical process model. The model may be deduced from according laws of physics, chemistry, ... (deductive way); or it may be induced from experimental results (inductive way) showing the processes response to selected input signals; or via a combination of both approaches. Frequently it is necessary to derive more than one models for the same process, e.g. a simple control design model and a more accurate simulation model.
2. **Control System Configuration**: If not provided by the control problem, one has to select the variables, their locations, and the correspond technologies for measurements and actuators. These decisions are neither easy to make nor negligible, as they may have a major impact on the control system performance, independently of the employed control strategy, particularly for large-scale processes.
3. **Control Strategy Design** is effectuated on the basis of the control design model, using the methodologies mentioned in earlier sections.
4. **Simulation Test**: Particularly for complex processes, it is advisable, convenient, and cost-effective to test in simulation the control strategy (or to compare the performance of various alternatives) for different scenarios before actual application.
5. **Implementation** of the control system including the measurement devices, the actuators, the communications and the control strategy. The latter may be implemented in an analog or (more and more frequently) in a digital way (computer) within a decentralised, centralised, or hierarchical structure.
6. **Experimentation/Validation/Evaluation**: This phase aims to test the proper functioning of the real control system (strategy, software, hardware, equipment), first in parts and then as a whole, and to evaluate its actual performance, before entering the operational (and completely autonomous) phase.

In most cases, these six steps must be partly iterated several times before the final system, with its desired performance characteristics, becomes operational. Even after this phase, elaboration of farther (theoretically or practically based) improvements is quite common, particularly for complex and large-scale systems.

For relatively simple systems, the mostly methodological steps 1, 3, 4, and partly 2 and 5 may be completed within a few days by an experienced and knowledgeable Automatic Control engineer. On the contrary, for complex and/or large-scale systems, several person-years may be required. In fact, the development of efficient control systems in areas like electric power distribution, communication networks, various military systems, various traffic control problems, requires the control system designer to acquire a profound understanding of the particular process behaviour, the relevant technologies, the operational objectives and constraints. In many cases, the forming of interdisciplinary development teams may be the most convenient approach.

7. Evolution of Automatic Control

The beginning of Automatic Control as an independent scientific discipline in the 1940's was essentially motivated by high-performance problems in the military and space areas. First methodological developments were based on the notion of feedback that was studied to a certain extend within electrical circuits. But the high-performance requirements of the above mentioned application areas necessitated the development of a general and solid Automatic Control theory to replace particular solutions and questionable heuristics used until that time.

The 1950's were mainly characterized by two parallel developments:

- On the one hand, development of a solid theory and a profound understanding of SISO control systems (single input, single output) including efficient and general methods of regulator design.
- On the other hand, establishment of methodological connections between Automatic Control and Optimisation, this latter having already a long history of some three centuries and comprising remarkable theoretical results. These developments have established the theoretical basis of optimal control, essentially via the work of Pontryagin (Maximum Principle) and of Bellman (Dynamic Programming).

At the same period, the developed Automatic Control methods are also applied in domains different than military and space, namely in the chemical process industry, industrial production, mechanical engineering, energy production and further areas where SISO-type control problems appear. The implementation of these control systems was based almost exclusively on analogue technologies (electrical, pneumatic, hydraulic, ...)

The 1960's are mainly characterized by the progress from SISO to MIMO systems (multiple input, multiple output) based on two partially competing and partially complementary methodological streams:

- The "american" stream (Kalman) that is based on the state-space process description (time domain) and introduces the fundamental notions of controllability and observability and celebrated design methodologies such as the Kalman Filter and LQ (Linear-Quadratic) Optimisation [6], [7].
- The "british" stream (Rosenbrock) developing a genuine generalisation of SISO methodologies for MIMO problems in the frequency domain (pole assignment).

These new methods are generally well-suited for the requirements of both high-performance applications (military, space, flight systems) and more traditional, less demanding application areas such as industrial control.

In the 1970's, two major developments have had an impact on the future of the Automatic Control discipline:

- The rapid developments in computer technology that enable the implementation of high-complexity control strategies; the broad use of microcomputers is of particular importance in this context.
- The consideration of large-scale applications (integrated industrial production, environment, traffic and transportation, ...) and the tendency to integrate decentralized sub-process control systems call for the development of new theoretical methods (Large-Scale System and Control Theory) which has not always been a successful endeavour.

The most important developments of the 1980's are, on the one hand, the appearance of a new, interesting application area, namely robotic systems; on the other hand, one begins to realize that the extremely useful tools of continuous dynamic system theory (state-space and frequency domain descriptions) are not applicable to a rapidly increasing class of applications of discrete nature, which calls for the development of new theoretical tools applicable to so-called discrete-event dynamic systems. In the same period, the issue of control system robustness for MIMO processes becomes a focus of important methodological developments and corresponding applications in many areas.

At the same time, the more and more complex tasks of surveillance, diagnosis, coordination, human-machine interface, emergency management, ... appear within large-scale environments. This calls for corresponding methodological tools and general computational environments to support the design and implementation of corresponding heuristics. To this end, some methods and tools from the Artificial Intelligence area find their way to corresponding applications where Automatic Control and Optimisation methods are hardly applicable.

Finally, some keywords characterizing Automatic Control developments in the 1990's are interesting control design methods for non-linear systems, application of artificial neural networks for control purposes, and adaptive control methods.

In conclusion and with regard to future developments, the following could be stated:

- Optimisation provides methods for making optimal decisions. Automatic Control offers design methods for autonomous, real-time, decisions-making systems.
- These methods cover a broad range of current and potential applications and should be the first approach to be attempted because they are efficient, easily transferable, flexible, and robust.

- There are still some domains where theoretically founded methods are hardly applicable; more or less systematic heuristic procedures should be applied in these cases.

In any case, if theoretically founded methods of Automatic Control and Optimisation are not applied, this should not be due to the ignorance of these methodologies, as it is frequently observed in many practical areas including traffic and transportation applications.

8. Overview of network process control systems

As already mentioned earlier, two fundamental characteristics of virtually any traffic or transportation system are the presence of a network structure and of some kind of flow therein. Based on both characteristics, it is not difficult to identify other domains of similar character: water, gas, or electric power distribution networks, sewer networks, communication networks (for telecommunication and/or data transfer). In all these processes with a meshed structure, a partially predictable and geographically distributed demand has to be satisfied by use of a corresponding large-scale network. In order to satisfy the demand, different kinds of flow control may be applied within the network so as to minimize certain objective criteria subject to capacity and storage constraints. More particularly:

- In water or gas distribution networks, the demand appears at flow destination locations. The flow origins are also distributed in the network and they provide, according to their capacities, the necessary amounts to cover the demand. Thus, the control problem is to transfer the water or gas from the origin sources to the demand locations (where it should be delivered with a pressure above a contractual threshold) so as to minimize
  - the cost of the required energy consumption
  - the network losses (that depend, in the case of water networks, from the pressure).

  subject to various storage (for reservoirs) and flow (for pipelines) capacity constraints.
- In sewer networks, the control problem is comparable to the above with two notable differences:
  - The demand is present at the network origins.
  - The main control goal is to minimize overflows so as to reduce their polluting impact on the receiving waters.
- In electric power distribution networks the problem is similar to the one of water or gas networks with two important particularities that may render the solution more difficult:
  - No storage possibility available.
  - Very rapid flow phenomena impose strict real-time requirements.

- In communication networks we roughly find again the water and flow network control problem but with more (and more important) particularities:
  - The demand is present at the network origins.
  - The flow particles have individual destination addresses, hence it is not sufficient to globally control the flow, but one must address sub-flows with particular destinations.
  - Rapidity of flow phenomena.
  - The control system's communications, i.e. the connection between measurement devices, controller and actuators, are part of the flowing medium (data packets!) in the same network. This virtually imposes a decentralized control strategy so as to avoid significant delays in the real-time control reactions.
  - The objective criterion to be minimised is related to the total time spent by data packets in the network.
- In urban and motorway traffic networks, the problem is again to satisfy a demand that appears at the network origins and contains sub-flows with different destinations, so as to minimize the total time spent in the network. A new particularity in this kind of networks is that the flow particles are individual decision-makers whose perceptions and goals are not identical with those of the control system. These individual decision makers may choose - within certain limits - their transport mode and time, their speed, their route, their traffic lane, ... This particularity has significant implications in the design of certain control measures such as route guidance, information providing systems, ...
- In air, maritime, and rail traffic networks we find again general characteristics of other network control problems. Major particularities include:
  - Very limited storage possibilities within the network.
  - Pursuing and monitoring each individual particle is at least as important as global flow control. Hence, some combinatorial and discrete problems accounting for departure/arrival times, routing, potential conflicts and delay of each particle are envisaged.

Although storage and queue-forming phenomena are common in most of these processes, the particular phenomenon of flow congestion that reduces the network's (or particular links') capacity and leads to a deteriorated infrastructure utilisation is mostly observed in communication and traffic flow networks of all kinds. In both cases a common resource (the network infrastructure) is used competitively by many users. In this context, the appearance of a congestion that reduces the network's capacity at the time (rush hour) it is most urgently needed is a paradox that has to be faced, reduced or avoided to the benefit of all users.

In communication networks the capacity reduction during periods of strong demand is due to buffer overflows leading to the real or assumed loss of corresponding data packets that are eventually re-sent from their respective origin nodes which further increases the demand and hence the buffer overflows, and so forth.

In urban, air, rail, and partially maritime traffic networks, capacity reduction is mostly due to node blocking (junctions, airports) or due to a blocking of downstream links. Finally, in motorway networks, a self-blocking phenomenon as an inherent flow characteristic may become apparent even within individual overloaded network links ("congestion from nothing"), independently of all other network links.

Naturally, measurement and actuator devices are quite different in different network processes. On the contrary, the control problem characteristics, and in some cases also the mathematical model equations (for example the partial differential equations governing water, gas, or traffic flow) may be very similar. As a consequence, envisaged or applied solutions may contain important common aspects as well.

In almost all network applications one may distinguish between the following typical operational states:

- Low demand (off-peak).
- Predictable high demand (rush hour).
- Unpredictable high demand (special events).
- Capacity-reducing or link-blocking or node-blocking incidents.
- Local device failures.

Historically these network processes were initially not controlled at all, or they included some kind of fixed control, often based on purely constructive means. Local control measures, that were later implemented in some of these network processes, addressed mostly safety rather than efficiency considerations, e.g. traffic lights at urban road junctions, that were initially introduced to enable the safe crossing of competing streams of vehicles and/or pedestrians.

The demand increase and the high economic and/or ecological cost of an infrastructure expansion eventually motivated the development of real-time control systems of increasing sophistication and, hopefully, increasing efficiency. The eminent need to increase system performance, particularly under non-predictable demand or incident conditions, on the one hand, and the methodological and computational developments on the other hand, have lead to according ameliorations of the control strategies employed. Although the sophistication level may differ according to the application and the particular network, some common tendencies may be observed in most network control systems:

- The development and implementation of a unique central control unit for a large-scale process has some disadvantages concerning reliability (in case of failures), complexity, and computational effort. For this reason, whenever possible, decentralized and/or hierarchical control structures may be preferable.

- In a lower, possibly decentralized direct control layer (Figure 8.1), one finds relatively simple control strategies for well-defined local subproblems, e.g. pump or gate control, road junction control, or motorway ramp control. The local control strategies may be of SISO or MIMO type, or they may involve optimisation problems of low dimension, often complemented with suitable heuristics. Moreover, first data compression and data processing, local diagnosis and emergency measures are often included in this control layer.
  The local control measures are coordinated via suitable instructions (e.g. set values) received from the superior control layer. In this sense, it is the basic task of each local controller to implement these coordinating instructions in its area of responsibility. Note that a decentralized structure of the control strategy does not necessarily imply a decentralized hardware implementation or a geographic decentralization as well. In case of communication link failures in a geographically decentralized control system, the local controller may continue to work autonomously, with pre-programmed set values.
- The superior global control layer is responsible for efficient operation of the overall system. In some applications, we may have a whole series of intermediary layers, with increasing responsibility domains. The global control strategy is typically represented by one (or more) MIMO-regulators of high dimension or by an optimisation problem of the overall process, that deliver set values for the inferior control layers. Optimisation methods are particularly suitable to the global control of network processes, as the corresponding objectives may be readily expressed in mathematical terms. Nevertheless, for very complex problems of combinatorial nature it may be useful to complement the algorithmical control strategy parts with adequate heuristic rules so as to increase the robustness, reliability, and operational transparence of the system. Other characteristics of the global control layer are:
  - The control strategy is based on aggregated process models and assumes the proper operation of the inferior control loops (except in cases of manifested failures).
  - The updating frequency of decisions is relatively low.
- The control system of a complex network process includes the tasks of surveillance of proper operation, of diagnosis of failures or incidents (that may not be directly visible in the measurement data), of archivation and statistical evaluation of the system performance, and of activation of emergency measures in case of failures or incidents. These tasks were fulfilled in the past exclusively by human operators, but they are gradually transferred to computers by use, e.g. of adequate Artificial Intelligence tools (like expert systems). Whatever the share of these tasks between human operators and computers, it is necessary to include in the system a human-machine

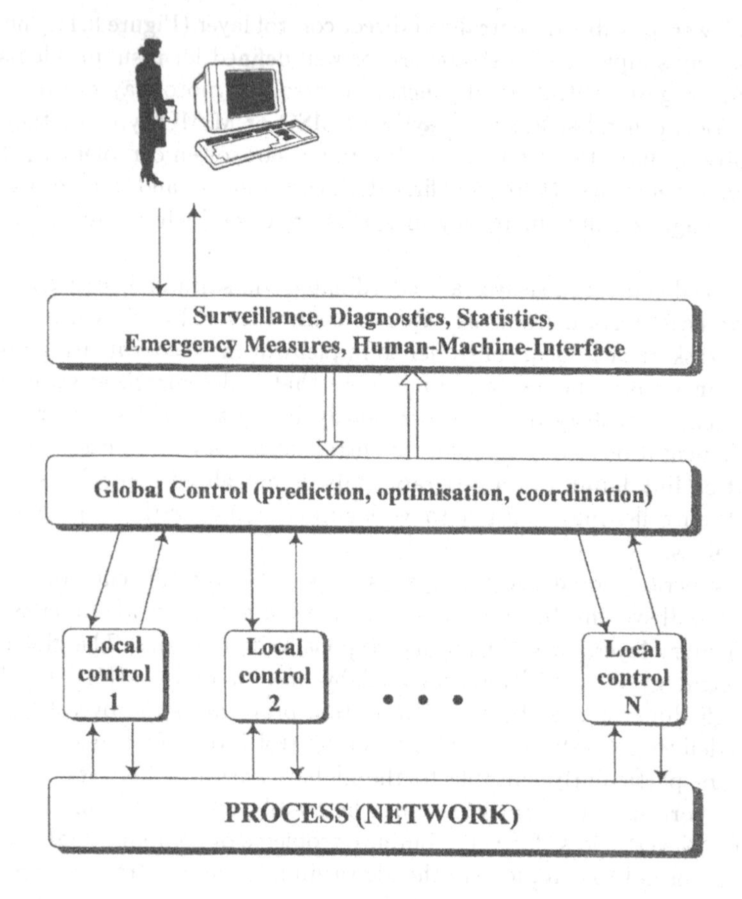

Fig. 8.1. Decentralized and hierarchical control structure

interface with appropriate graphical displays to enable the ultimate human supervision of the complete system from a central control room.

A particular characteristic of road traffic is the parallel presence of various distinct control measures for the same traffic network, e.g. signal control, individual or collective (via variable message signs) route guidance, drivers information systems, ... It is an important goal of current research and development efforts to integrate these parallel systems, whereby integration may address all involved elements such as the control strategy, the corresponding software, the measurement and communications devices, and, last not least, the administrative frame.

9. Control applications in motorway traffic

9.1 Modelling, Simulation, Prediction

Macroscopic modelling of motorway traffic flow considers the corresponding particular flow properties on the basis of aggregated traffic variables such as the traffic density (veh/km), traffic volume or flow (veh/h), and mean speed (km/h) [21]. For the macroscopic description of traffic flow on a motorway link or a motorway network, the classical mathematical tools of state differential equations (3.4) or difference equations (3.8) are perfectly suitable (see e.g. the equations of the model METANET).

Microscopic modelling considers the movement of individual vehicles in dependence of the movement of adjacent vehicles. Although the basic kernel of this description is expressed in form of differential equations (car-following model), there are a series of additional actions, like, for example, lane choice, lane change, merging at on-ramp and off-ramp areas, that can only be described via heuristic rules [22].

As it was mentioned earlier, the development of a mathematical model may be based either on the physical laws governing the process behaviour or on experimental approaches. In both cases, it is necessary to estimate unknown parameters included in the corresponding mathematical equations, and to validate the model accuracy versus real measurements. For this parameter estimation and model validation tasks, the application of suitable optimisation methods is an efficient, solid, and convenient approach.

Regarding the development of generally applicable simulation packages, the utilisation of modern software means, notably for the corresponding human-machine interface or for the integration of sub-models, is particularly attractive. Of course, the descriptive kernel of any simulation package is essentially identical to a (microscopic or macroscopic) mathematical model.

Short-term and medium-term prediction models for motorway traffic flow are also virtually identical to the utilized (typically macroscopic) traffic flow models; it is the particular way of using them in real time, feeding them with available real-time measurements, that distinguishes them from simulation tools.

9.2 Ramp Metering

For the development of isolated ramp metering strategies, application of SISO regulator theory (input: ramp flow, output: mainstream occupancy rate downstream of the ramp, see Figure 2.1) is perfectly suitable. In fact, the field evaluation of the SISO-regulator ALINEA with other heuristic isolated ramp metering strategies has demonstrated ALINEA's clear superiority in terms of efficiency, simplicity, flexibility, and robustness.

For coordinated ramp metering of several ramps along a motorway axis (Figure 2.2), the application of MIMO regulator theory, either directly (e.g.

the control strategy METALINE is based on straightforward application of LQ-optimisation) or in cascade, is an adequate approach. If a high number of ramp meters is included in an extended motorway network (e.g. an orbital motorway around a city center) then it appears necessary:

- to explicitely consider total travel time minimisation over the whole system
- to explicitely consider control constraints (including maximum allowable ramp queue lengths)
- to explicitely consider the processes nonlinear behaviour (e.g. in case of non-recurrent congestion).

In this case the optimal control approaches outlined in chapter 3 are perfectly applicable. Detailed investigations have shown that the corresponding problem P1 may be solved in real time even for quite large-scale networks. Further elaboration and field evaluation of sophisticated control strategies of this kind remain an interesting research and development subject for the near future.

9.3 Route Recommendation via VMS

Real-time route recommendation may be provided to motorway network users via road-side VMS (variable message signs), particularly under non-predictable (non-recurrent) congestion conditions, so as to minimise the individual and/or the total travel time. For large-scale motorway networks, this is a fairly complex control problem due to:

- The presence of multiple origins, multiple destinations, and multiple connecting routes.
- The interactions between sub-flows with different destinations that share the same network.

Despite its complexity, this problem can be expressed in rigorous control engineering terms and may be solved in a fairly general way either by employing a suitably designed MIMO regulator or via an optimal control approach.

9.4 Driver Information via VMS

Variable message signs are often used within motorway networks in order to inform the drivers about the current (or predicted) traffic situation (e.g. congestion length or travel time) in a limited network area. If the type of displayed information is always the same (e.g. in the Ile-de-France network, it is the current travel time in both links following a bifurcation) and if the information provided is always correct to the best of the system's knowledge, then there is no space for a control strategy. In fact driver information systems, though popular with the drivers (who may only trust themselves as decision makers) and with the road authorities (who may not wish to be blamed to provide incorrect route recommendation), are of limited value for efficient traffic control because:

- For VMS-space and readability reasons, only a tiny fraction of the traffic information available at the control center can actually be provided to the drivers, e.g. as in the Ile-de-France network mentioned above; in most cases this amount of information is not sufficient for the drivers to make optimal routing decisions, particularly if their trip includes several decision points (bifurcations).
- Even if drivers would have complete information of the current network conditions, it would be impossible for them to make an optimal decision in the few-second period between reading the message and reaching the bifurcation location.

9.5 Variable Speed Limitation

Variable speed limitation via VMS as well as other similar control measures (e.g. keep lane, congestion warning, ...), sometimes referred to under the general term lane control, aim at homogenizing traffic flow so as to reduce the risk of congestion and increase throughput. Currently the corresponding control strategies are based on heuristic rules [22]. More rigorous automatic control methods have been only occasionally suggested for this problem, mainly due to the lack of a proper quantitative modelisation of the impact of the control measures on traffic flow, but also due to

- the discrete character of the control input and
- the safety constraints imposing a gradual (not abrupt) variation of speed limitations in space and time.

9.6 Surveillance, Estimation, Detection

Data cleaning is a major task of control system surveillance. It comprises the sub-tasks of identification of erroneous real-time measurements and of their correction/replacement based on time/space extrapolation and/or other estimation and filtering techniques.

Real-time estimation of traffic variables that are not directly measured (for economic or technical reasons) is another important and interesting surveillance task. The general issue here is to use modelling and algorithmic "intelligence" in order to extract from real-time measurements more information than they explicitely include. Examples of this type of estimation problems are:

- Estimation of traffic density (that may only be measured by use of video sensors) from loop detector information.

- Estimation of the complete traffic state between two distant (e.g. 2 km or more) loop detectors.

These and further estimation tasks may be accomplished by use of appropriately designed Kalman Filters, typically based on macroscopic traffic flow models.

Another particular surveillance problem for motorway traffic is the automatic incident detection (AID). A high number of different methods were proposed for AID in motorway traffic based on loop detector or on video sensor information. An incident is an external event that reduces the motorway capacity, e.g. lane(s) blocking due to an accident or a failed vehicle. The proposed AID methods include algorithmic, heuristic, and mixed approaches that range from Kalman Filters and other statistical approaches to catastrophy theory and neural networks. Most of these methods include certain parameters (thresholds) that should be calibrated based on real data (prevailing from traffic situations with and without incidents). To this end, global optimisation methods are an efficient and convenient tool.

10. Control applications in urban road traffic

10.1 Modelling and Simulation

The general statements included in section 9.1 are largely applicable to modelling and simulation of urban road traffic as well. Two notable particularities are the following:

- Utilisation of microscopic models is more common in the urban context, particularly for high "friction" networks with frequent vehicle stops, parking, off-junction pedestrians, delivery vehicles ...
- The graphic interfaces of simulation packages may be more sophisticated in the urban context.

For both reasons, heuristic rules and advanced software tools are of higher importance compared to motorway traffic.

A particularly interesting problem in the context of urban traffic is traffic assignment. This problem addresses the description of the route choice behaviour of drivers with different origins and destinations in a network. The corresponding mathematical description is based on the celebrated Wardrop's Principle that states (in its simplest version) that all utilised routes connecting a particular origins-destination couple have the same cost (e.g. the same travel time). The treatment of various versions of the traffic assignment problem is essentially based on suitable optimisation methods. A subject of intense recent research is the extension of Wardrop's Principle and the development of corresponding tools and software packages to the dynamic case. To this end, well-known methods of optimal control and automatic control appear particularly useful.

10.2 Signal Control

The development of control strategies for traffic signals at urban junctions is a high-importance task within the traffic control area that has occupied many researchers during the last decades. The proposed control strategies may be classified according to the following characteristics:

- fixed-time versus real-time strategies
- isolated (one junction) versus coordinated (several junctions) strategies
- saturated versus non-saturated traffic.

The greatest part of proposed strategies of any of the above types are based on more or less sophisticated optimisation methods (e.g. the strategies TRANSYT, MAXBAND, SCOOT, PRODYN, OPAC, CRONOS). On the other hand, several less known (no publications available) operational strategies are largely based on heuristic rules.

Currently, the main research and operational interest in this area is focused on the elaboration of control strategies that belong to the most advanced but also most complex class: coordinated, real-time control of saturated traffic. The complexity of this problem is due to:

- The large-scale character (many junctions)
- The real-time treatment (limitations of computational effort)
- The partially combinatorial problem nature.

Because of these characteristics, the ultimate solution of the advanced signal control problem can hardly be obtained by the rigorous formulation of one single optimisation problem. New approaches that include (in a cooperative, possibly hierarchical structure) algorithmic parts, heuristic rules, and advanced software tools may lead to efficient solutions for this problem.

10.3 Individual Route Guidance

The dynamic, individual route guidance of equipped vehicles from a control center is an innovative and promising control measure that is currently in an experimental or early operational phase in several countries. When the percentage of equipped vehicles is low, the guidance problem is relatively simple and requires real-time calculation (based on real-time measurements and short-term predictions) of the time-shortest path for each node-destination couple in a network using appropriate shortest path algorithms. On the other hand, as the percentage of equipped vehicles increases, the problem becomes more complex because the guidance of a high number of equipped vehicles towards the corresponding currently-shortest paths may radically change the current or the predicted traffic situation on these and other paths, thus rendering the guidance decisions irrelevant. Automatic control methods (MIMO regulators) and optimal control methods are adequate tools to address the corresponding feedback control problem.

10.4 Estimation Problems

Several variables that are useful in the context of real-time signal control are not directly measurable via conventional measurement devices and should be estimated in real-time.

The knowledge of the current averages of turning movements at the network junctions is a prerequisite for most signal control strategies. Also the current queue lengths in the network links are required by many algorithms. As direct measurements of these variables are usually not available, their values must be estimated in real time based on a limited number of local loop-detector measurements. These estimation tasks are accomplished by use of well-known and efficient optimisation methods such as the least squares or the Kalman Filter. Note that utilisation of video sensors in the urban environment is likely to improve substantially the accuracy of these estimations.

10.5 Parking Control

Parking control systems inform drivers in real-time via VMS about availability of parking spaces in corresponding areas. The utilised strategies are large heuristic and do not seem to be of high interest for application of Automatic Control or Optimisation methods.

11. Other applications in traffic and transportation

11.1 Integrated Control

As already mentioned, the integration of parallel control systems of road traffic is a very recent and interesting endeavour in the general area of traffic control. There are different aspects of system integration:

- Integration of control strategies for different control measures
- Geographical control integration (e.g. urban and motorway traffic control)
- Integration of software tools (data bases, archives, algorithms, HMI, ...)
- Integration of devices (measurements, communications, hardware equipment, visualisation, ...)
- Integration (connection) of control centers
- Administrative issues of integration.

Automatic Control and Optimisation have a major role regarding the integration of control strategies. On the other hand, the integration of control systems increases the importance and complexity of the surveillance layer (Figure 8.1) which calls for corresponding advanced software tools.

11.2 Automated Highway

Automated Highway System (AHS) design and implementation requires the employment of most advanced automatic control and optimisation methods due to the very high safety and performance requirements and also due to the high degree of automation. Particular control problems in this context include:

- Distance regulation between vehicles (longitudinal control) aiming to replace the human driving regulators thus increasing safety, efficiency, and reliability; this is a typical regulation problem that is resolved by corresponding automatic control methods guaranteeing (among others) stability of the string of vehicles which is not the case with human drivers.
- Lateral regulation of vehicles to remain in their current lane is also a typical regulation problem, though with many alternative approaches depending upon corresponding technological questions, mainly with regard to measurement devices.
- Platoon forming, platoon splitting, lane change and similar activities are largely based on discrete decisions and corresponding communications between the actively or inactively involved vehicles. As these procedures may have severe implications in terms of safety, the application of methods (e.g. from discrete-event dynamic system theory) that guarantee complete, unique, and contradiction-free decision making appears indispensable.

11.3 Rail Traffic Control

Within rail traffic systems (underground, intercity, high-speed rail networks) the real-time control problems are essentially related to surveillance and safety issues. On the other hand, for off-line planning and scheduling problems (allocation of locomotives to trains, crew scheduling, time plan), combinatorial optimisation problems are mostly involved in the aim of optimal utilisation of the available infrastructure, particularly during peak hours [22].

An important safety-related task within rail traffic systems is collision avoidance (on a line or at a node). To this end, traditional measures that are based on robust electromechanical devices implementing simple but efficient logical (boolean) functions are quite broadly utilised. Within modern systems, however, the implementation of electronic devices (microcomputers) with high-redundancy architectures (to satisfy high reliability requirements) become increasingly common.

For surveillance of rail traffic from a central operation room, advanced telematic devices (radio transmission, satellite communications, ...), and HMI tools are increasingly employed. Major surveillance tasks include:

- Monitoring of the movement of each train in the network
- Verification of the proper functioning with respect to the time schedule
- Intervention in case of severe disturbances so as to re-normalize the traffic.

The task of re-normalisation of traffic is fairly complex and largely manually executed as yet. Involved real-time sub-tasks include:

- Prediction of the duration of an occurred incident that blocks a line
- New routing of affected trains in the network, if necessary
- Suitable time schedule modification to address the current abnormal situation.

The main goal of these actions is the traffic normalisation, i.e. the quick and smooth return to the initial time schedule. Recently, methods and tools from Automatic Control and Artificial Intelligence (Expert Systems) have been proposed for a partial automation of the corresponding procedures and/or as decision support for human operators.

Within high-frequency underground rail systems, the surveillance of the time schedule and the regularisation control measures in case of large deviations are quite important in order to avoid a destabilisation of traffic. The traffic system is inherently unstable under high demand conditions for the following reason. If, due to an initially insignificant disturbance, a particular train arrives with a little delay in a station, then the number of waiting passengers will be slightly higher than anticipated, which will increase the anticipated stopping period of the train in that station, thus increasing its delay when arriving to the next station, and so forth. At the same time, the time distance between the retarded train and the next train becomes increasingly shorter, as the next train is used by increasingly less passengers, hence it has increasingly shorter stop periods than anticipated, and so forth, until both trains actually follow each other, a particularly inefficient situation.

These unstable phenomena may appear both in manually driven and in automated underground systems and may be avoided quite efficiently by suitably designed control strategies.

11.4 Air Traffic Control

The rapid increase of air traffic demand during the last decades was more accentuated than in other transport modes. Air traffic control should guarantee a safe and efficient flow both near the airports and in the air, particularly under high density and/or under adverse environmental conditions. Like in rail traffic, surveillance and safety issues, both at a traffic level and for individual airplanes, are major control tasks [22].

A first problem to be solved off-line is specification of air links, e.g. over a country's or a continent's surface, that may contain many airports, so as to minimise the total flight length subject to various technical, operational and geopolitical constraints. Interesting combinatorial optimisation problems appear in this context. The problem of assigning a flight within a network of air links is another question within air traffic control, that becomes more difficult and interesting when it requires a real-time modification of the underlying plan due to external disturbances. This kind of problems are usually encountered at regional control centers.

At the airport level, the principal control objective is to specify in real time the optimal sequence and landing trajectories of all airplanes appearing in the corresponding responsibility zone so as:

- to minimize total delay
- to minimize the total fuel consumption
- to respect safety and other constraints.

This control problem becomes more complex under high density conditions (air queue formation). Moreover, the problem must be re-solved whenever a new airplane appears in the airport's zone. Despite availability of some optimisation algorithms, treatment of this task remains largely a responsibility of human controllers albeit with the support of automatic data processing algorithms and of visualisation interfaces.

The airport-level control problem corresponds to a local control structure. The coordination of actions between different airports usually improves air traffic performance regarding safety, delays, and fuel consumption. For example, allowing an airplane to start depending on the traffic situation expected in the destination airport at the time of arrival, may contribute to air queue reduction and hence to less fuel consumption. On the other hand, this coordination increases the problem complexity further, hence calling for utilisation of efficient algorithms.

11.5 Maritime Traffic

Main control concerns in maritime traffic are - like in rail and air traffic - safety and monitoring of ships from shore or a port. In cases of high density or of adverse environmental conditions, the control task is to specify in real time the optimal routes for each ship so as to guarantee efficient and safe passage. If there is a collision risk, rule-based detection algorithms activate an alarm. The basic tasks of maritime control are currently assumed by human operators, but optimisation-based or rule-based tools are gradually introduced in some cases. For example, on-board trans-ocean weather routing is increasingly based on a rolling horizon control approach using environmental forecasts over the next few days, and employing dynamic programming algorithms to minimize fuel consumption and/or arrival time.

12. Conclusions and future needs

Automatic Control and Optimisation include a number of general and systematic methods for the solution of practical problems. However the formal representation of a practical problem to enable efficient application of theoretical methods is neither trivial, nor unique, nor always possible. On the one hand, the complexity level should be sufficiently high to facilitate realistic solutions. On the other hand, this level should not exceed certain limits so as to facilitate the problem solution avoiding an exaggerate or untractable computational effort. Suboptimal, possibly decentralized or hierarchical solutions are in most cases a good compromise.

Even if comprehensive solutions based on rigorous methods appear impossible in some areas, it may be useful to identify subproblems leading to algorithmic modules within an environment that is dominated by heuristic rules and/or by human operators interventions.

Within the traffic and transportation area, the applications of Automatic Control and Optimisation are quite numerous. However, further development and application work is needed motivated by:

- the presence, currently, of naïve solutions that may be enhanced, generalized, or simply replaced
- the interest in an optimal use of the available infrastructure in view of a steadily increasing demand
- integration and coordination of subsystems
- appearance of new technologies for measurements (e.g. video sensors) or for control (e.g. route guidance)
- the increasing computational capabilities that allow for implementation of more complex strategies
- the availability of interesting research results regarding traffic control, that await their validation in practice.

Note that, in many cases, the efficient application of Automatic Control and Optimisation methods to complex problems may require the formation of interdisciplinary teams combining good knowledge of:

- the particular application area
- the available Automatic Control and Optimisation methods and tools
- the particular or general implementation devices
- the operators and users needs for the application area.

References

1. B.D.O. Anderson, J.B. Moore. Linear Optimal Control. Prentice-Hall, Englewood Cliffs, New Jersey, 1971.
2. K.J. Aström. Introduction to Stochastic Control Theory. Academic Press, New York, 1970.
3. M. Athans, P.L. Falb. Optimal Control. McGraw Hill, New York, 1966.
4. D.P. Bertsekas. Dynamic Programming and Stochastic Control. Academic Press, New York, 1976.

5. A.E. Bryson, Jr., Y.C. Ho. Applied Optimal Control. Ginn, Waltham, Massachusetts, 1969.
6. C.K. Chui, G. Chen. Kalman Filtering. Springer-Verlag, Berlin, 1987.
7. C.K. Chui, G. Chen. Linear Systems and Optimal Control. Springer-Verlag, Berlin, 1989.
8. W. Findeisen, F.N. Bailey, M. Brdys, K. Malinowski, P. Tatjewski, A. Wozniak. Control and Coordination in Hierarchical Systems. Wiley, New York, 1980.
9. R. Fletcher. Practical Methods of Optimization (2nd edition). John Wiley & Sons, Chichester, 1987.
10. P.E. Gill, W. Murray, M.H. Wright. Practical Optimization. Academic Press, New York, 1981.
11. A.H. Jazwinski. Stochastic Processes and Filtering Theory. Academic Press, New York, 1970.
12. D.E. Kirk. Optimal Control Theory. Prentice-Hall, Englewood Cliffs, New Jersey, 1970.
13. H. Kwakernaak, R. Sivan. Linear Optimal Control Systems. Wiley-Interscience, New York, 1972.
14. R.E. Larson, J.L. Casti. Principles of Dynamic Programming - Part I: Basic Analytic and Computational Methods. Dekker, 1978.
15. R.E. Larson, J.L. Casti. Principles of Dynamic Programming - Part II: Advanced Theory and Applications. Dekker, 1982.
16. L.S. Lasdon, A.D. Waren. Survey of Nonlinear Programming Applications. *Operations Research*, 28:1029–1073, 1980.
17. L. Ljung. System Identification - Theory for the User. Prentice-Hall, Englewood Cliffs, New Jersey, 1987.
18. M.D. Mesarovic, D. Macko, Y. Takahara. Theory of Hierarchical, Multilevel Systems. Academic Press, New York, 1970.
19. M. Minoux. Programmation Mathématique. Tome 1 et 2. Dunod, Paris, 1983.
20. C.H. Papadimitriou, K. Steiglitz. Combinatorial Optimization, Algorithms and Complexity. Prentice-Hall, Englewood Cliffs, New Jersey, 1982.
21. M. Papageorgiou. Application of Automatic Control Concepts to Traffic Flow Modelling and Control. Springer-Verlag, Berlin, 1983.
22. M. Papageorgiou, editor. Concise Encyclopedia of Traffic and Transportation Systems. Pergamon Press, Oxford, 1991.
23. M. Papageorgiou. Optimierung - Statische, dynamische, stochastische Verfahren für die Anwendnung. (2nd Edition) Oldenbourg, München, 1996.
24. A.P. Sage, C.C. White. Optimum Systems Control. Prentice-Hall, Englewood Cliffs, New Jersey, 1977.
25. M.G. Singh, A. Titli. Systems Decomposition, Optimisation and Control. Pergamon Press, Oxford, 1978.
26. T. Söderström, P. Stoica. System Identification. Prentice-Hall, New York, 1989.
27. H.M. Wagner. Principles of Operations Research. Prentice-Hall, Englewood Cliffs, New Jersey, 1975.

Mobility and Accessibility - The Case of Brussels

Hugues Duchateau

STRATEC
Boulevard A. Reyers, 156
B-1030 Brussels, Belgium

Summary. The consumers' search for greater mobility and the goods and service providers' search for the best accessibility have always shaped our territory. However, the growth of automobile mobility over the last thirty years is slowly suffocating the cities. Until now, the remedies have most of all tried to satisfy the need for mobility by acting on the supply of transport. This approach has proved to be all but ineffective.

A new, more global approach aiming to improve both mobility and accessibility which is being implemented in Brussels leads to the proposal of a new strategy divided into 5 lines of action :

- a regional development policy orienting the location of entities creating important movements of people into the city centre in the vicinity of the busiest nodes of the public transport networks;
- improvement of the links in public transports between the city centre and the suburbs (suburban metro/suburban express rail system);
- parking control;
- decrease of automobile traffic in residential areas and limitation of the congestion in the city centre by means of traffic-flow control measures;
- improvement of the travel conditions of pedestrians and cyclists.

1. The Concepts of Mobility and Accessibility

Mobility, meaning the possibility to move from one place to another, constitutes a freedom sought by all citizen, enabling them to maintain or expand choices in everyday life. It permits them to choose an employer or a workplace not located in the vicinity of their home, to go shopping where they want and where prices are best, as well as to spend their holidays at ever increasingly distant places thanks to the relative decrease in the cost of transport. This explains why mobility is quite a normal concern for any consumer. The private car meets this concern perfectly well, and this is the reason for its wide popularity.

However, thanks to an increase in revenue and to the mobility cars have provided them with, inhabitants of cities are now able to leave city centres to live in a more attractive atmosphere of less populated areas. By doing so, they are partly responsible for the depopulation of urban centres and for the congestion of their access roads.

Accessibility, meaning ease of access, is a concern more closely related to the production and distribution of products or services. In trade and non-trade sectors, players try to place their establishment in order to minimise transportation costs or to minimise the amount of time their clients spend travelling. The search for accessibility, acting as a mainspring, has shaped and still shapes our territory. At a time when public transport was the main answer to the mobility demand of the people, the competition for accessibility was responsible for the success of the town-centres for commercial and office establishments. More recently, the development of automobile mobility has re-oriented the search for accessibility by distributors of goods and services to more peripheral locations. This has led to the establishment of businesses, workshops, storage areas and offices areas along high speed roads and urban ring roads or in the vicinity of highway interchanges and, at the same time, to the decline of these activities in the city centres.

The combination of consumers' mobility in private cars and the search for accessibility on the producers' side therefore induced a process of delocalisation of housing as well as production and commercial functions, which contributes to strengthening the dependence on cars for satisfying the need for mobility.

The effects of this behaviour of a greater search for mobility by the consumers and for a better accessibility by the producers combine to multiply themselves and lead to the congestion of road infrastructure, to the deterioration of the service quality of public transport and to the worsening of the quality of life in the cities, thereby accelerating the process of destructuration and decline of the cities.

Up until a little while ago, the reactions of politicians responsible for urban transport policy most of all consisted in trying to satisfy or control mobility by measures affecting the offer of transport (successively, creation of high-speed urban roads, building of tubes and, after that, improvement of the performance of surface public transport modes).

These mobility strategies, while sophisticated, not only fail to take into account the pernicious evolution described above but even accelerate this evolution.

As a reaction to this, the idea is slowly emerging that strategies have to be based on more global perspectives including not only actions on the supply of transport but also actions aimed at influencing the demand and the trends in the urban players' choice of location.

This new approach leads to larger strategies aiming at reducing the need for moving about in private cars by improving the general accessibility to city centres and by orienting the location of functions and activities toward those centres.

The illustration below shows the steps that were taken in this direction in the elaboration of the new Land Use/Transport Development Plan of Brussels (IRIS Plan).

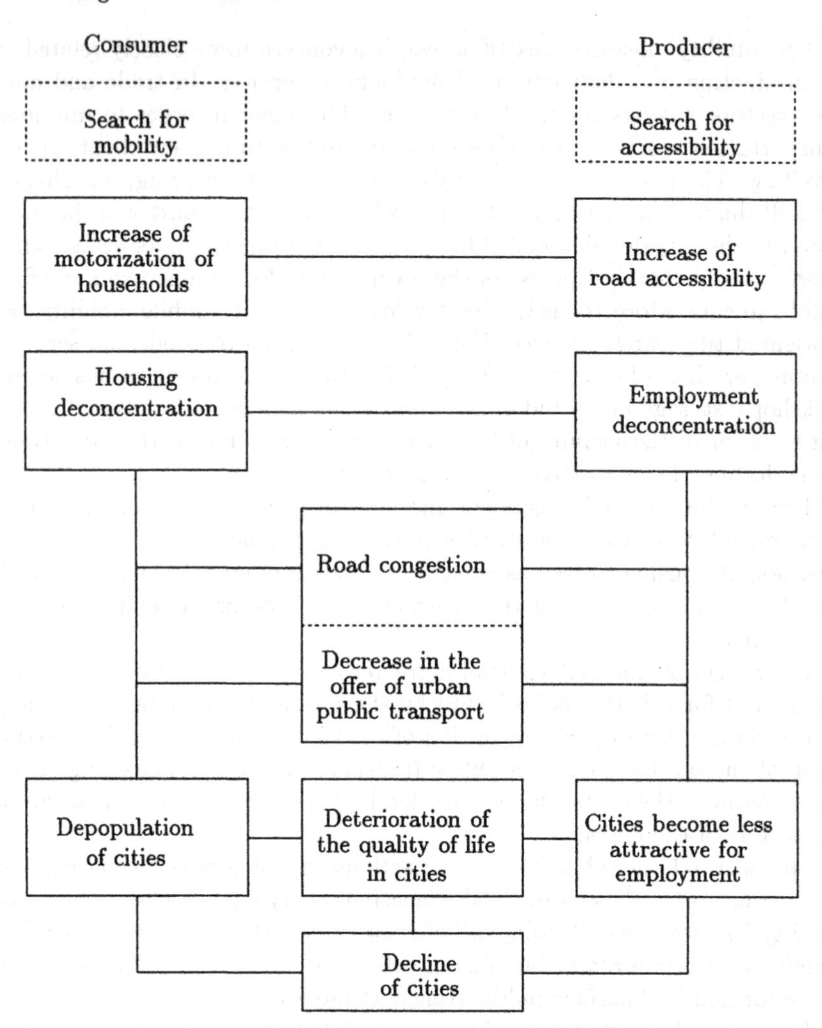

Fig. 1.1. The cycles of urban decline

2. Urban Mobility at a Deadlock

2.1 The Citizens' Frustrations

Inhabitants of cities, and particularly those of Brussels, its surroundings and, to a lesser extent, those of the rest of the country, feel that the transportation system which is at their disposal in order to travel around the city or to get to the city imposes higher and higher restrictions on their mobility. These restrictions vary according to the case concerned :

- Some have seen the time it took them to travel to work by car doubled in less than ten years due to the congestion of access roads to Brussels; moreover, once they arrive close to their destination, they cannot find a place to park their car.
- Others who used to take the train or tube to go to work now have to go in their own car because their employer moved from the city centre to a new building in the periphery that is not accessible enough by public transport.
- Children of some people do not have effective public transport at their disposal to reach their school, and their parents are thus obliged to drive them to school and back.
- Another category are households that do not live in the vicinity of shops or basic services and are obliged to use their car or public transport in order to carry out some trips they could make on foot or with their bicycle if facilities were better distributed.
- Finally, people who do not have a car because of their age, a physical disability or insufficient revenues, which means those who are called "captive" users of public transport are more and more limited in their possibilities to travel where they want when they want because of the suppression of services due to the increased scarcity of public transport users.

These drawbacks are not insignificant. They are responsible for a decrease in the quality of life in cities and the frustrations they cause cannot be ignored by political leaders.

2.2 Threats to Prosperity

The ease of contacts and exchange of goods and services is the basis of the urban society's prosperity. Any hindrance to these contacts and exchanges has a negative effect on this prosperity.

Many companies leave city centres for locations which are considered more accessible by their suppliers, clients and staff.

At the same time, the comparative advantages of housing in cities are decreasing, which reinforces the tendency of better-off inhabitants to look for a place to live in the surrounding area of the city, whereas the poorest inhabitants tend to accept housing left in the centres.

Because of the narrowness of the territory of the Brussels-Capital Region, this phenomenon extends beyond the limits of the region, entailing a loss of its substance and a relative impoverishment, as shown by the table below.

Table 2.1. Share of Brussels in the National Economy

	1980	1990
Share of Brussels in the Gross National Product	15.5%	13.4%
Part of Brussels in the global amount of taxable revenue of individuals of the country	11.6%	9.8%

These global trends, as well as the results of the more precise analyses, show that the Brussels-Capital Region is seriously endangered by an evolution similar to that of some North American cities whose centre has been completely deserted by a lot of companies as well as by the middle-class population or the most well-off.

2.3 Tendential Evolution

Mobility is a capacity all consumers try to develop in order to increase their freedom of choice. For it is thanks to this that an employee can prospect in a larger part of the employment market or that a household will be able to choose among more suppliers, goods and services.

The car is the perfect answer to this need, and the evolution of the motorization of households proves this success. Since 1975, the average annual increase in the number of vehicles on the road in Belgium has been 2.6%. On January 1st, 1991, the average number of private cars at the disposal of households per 100 inhabitants in the country amounted to 38. The trend of the evolution observed this last decade suggests that the growth of the motorization rate will continue in the coming years, although possibly at a lesser speed because of a progressive saturation effect in the demand for vehicles.

The general assumption in Belgium is that this saturation point will be reached at 50 vehicles per 100 inhabitants. It is relatively modest compared to the levels already reached in North America (nearly 60 vehicles per 100 inhabitants).

Of course, households also took advantage of the new form of mobility that a private car offers in order to change the way they choose where they live. For many of them, the comparison of the advantages and the price of a location in the old urban centre with those of a location in the suburbs has led to the choice of the latter.

The rise in population in the suburbs of Brussels shown in figure 2.1 [1] is a consequence of this choice.

The increase in mobility due to the car has also had indirect consequences on the suppliers of goods and services by extending their market areas and by modifying the conditions of competition that they impose on each other and by forcing them to reconsider their location criteria. For some of them, especially traders, the location choice is a question of survival because accessibility to their sale points is a critical element of competition. The increase of jobs in the periphery of the city shown in figure 2.2 indicates that decisions against the city were of large scale in the 80's. This contributed to the decline of the city.

In the last decade, the combination of these movements has led to an increase in commuting from home to work in Brussels from 276,000 units in 1980 to 322,000 units in 1990.

If the demographic and economic trends are confirmed, this commuting could rise to 357,000 in 2000 and to more than 400,000 in 2005 (see figure 2.3).

The combined effect of the deconcentration of housing and of employment and the increase in motorization of the population has led to a very important increase in automobile traffic both around and in the city. According to the surveys of the Ministry of Public Works, this increase has now reached an average rate of 5.8% a year for the years 1985 to 1990.

The situation as it was during the morning peak period at the beginning of 1991, has been reproduced with the help of traffic simulation models (see figure 2.4). If the deconcentration of employment and of housing and the increase in motorization continue as the trends show and all the other variables remain unchanged, especially as far as the offer of transport is concerned, the congestion of the road network will develop as shown in figure 2.5. The result of this is a doubling of automobile travel time during the morning peak period.

Such an evolution is, of course, not realistic because it would lead to an unbearable deterioration of the functioning conditions of the city:

- Urban economic players cannot accept such a situation because the worsening of their accessibility endangers their very survival; if nothing is done to change the situation, their reaction will be to leave the city for a more or less far peripheral location.
- The inhabitants will support neither the impediments to mobility due to congestion nor the increase in pollution that will result from it; their reaction will be similar to that of the economic players: those who can afford it will leave the city in huge numbers.

[1] Source: STRATEC, Regional Movements Plan, Trend Scenario, report (p), July 1992.

- As far as the regional public authorities are concerned, they too cannot stay put without any reaction to the threat of seeing a rise in the exodus of inhabitants and employment.

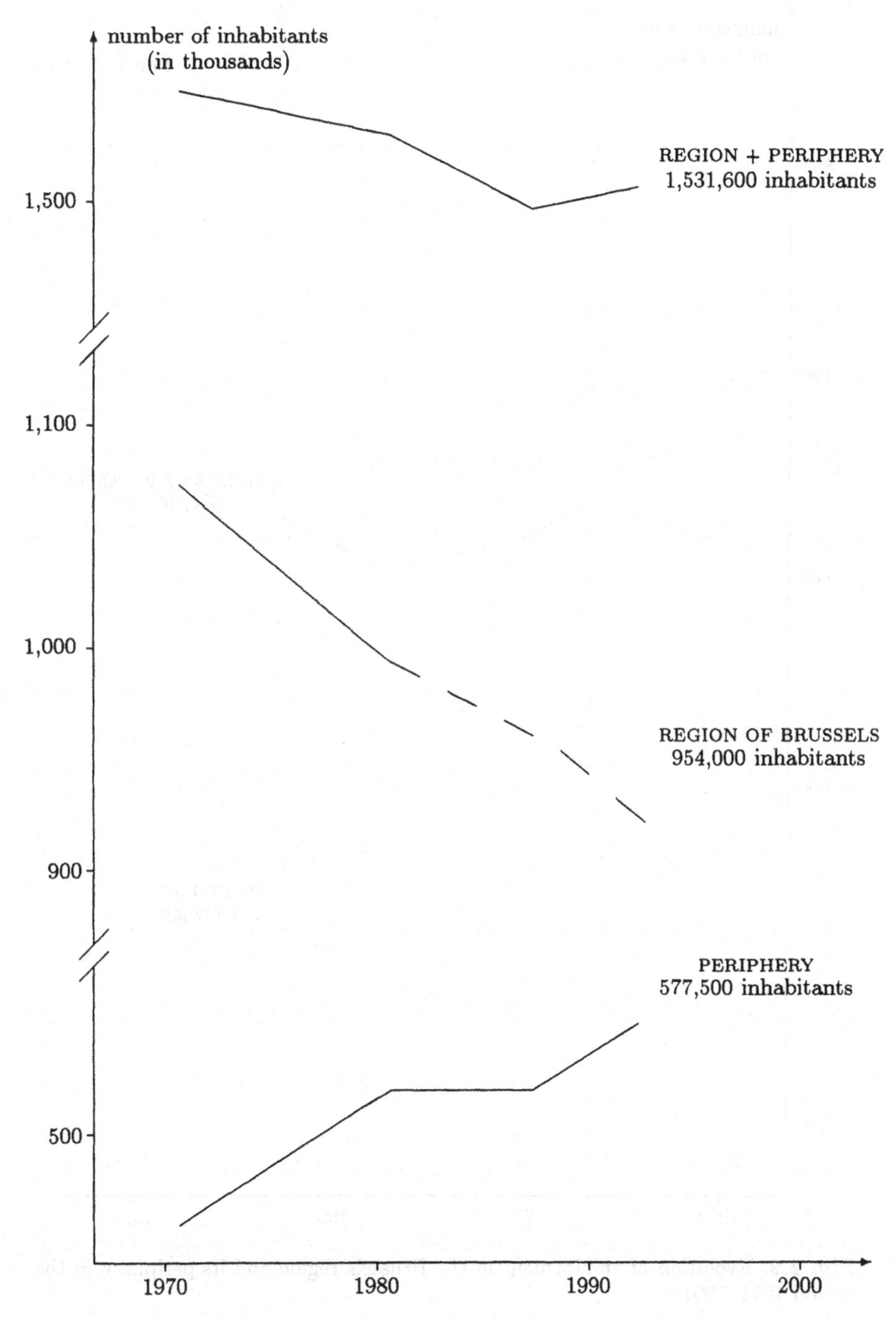

Fig. 2.1. Evolution of the population of the Brussels region and its periphery in the period 1971–1991.

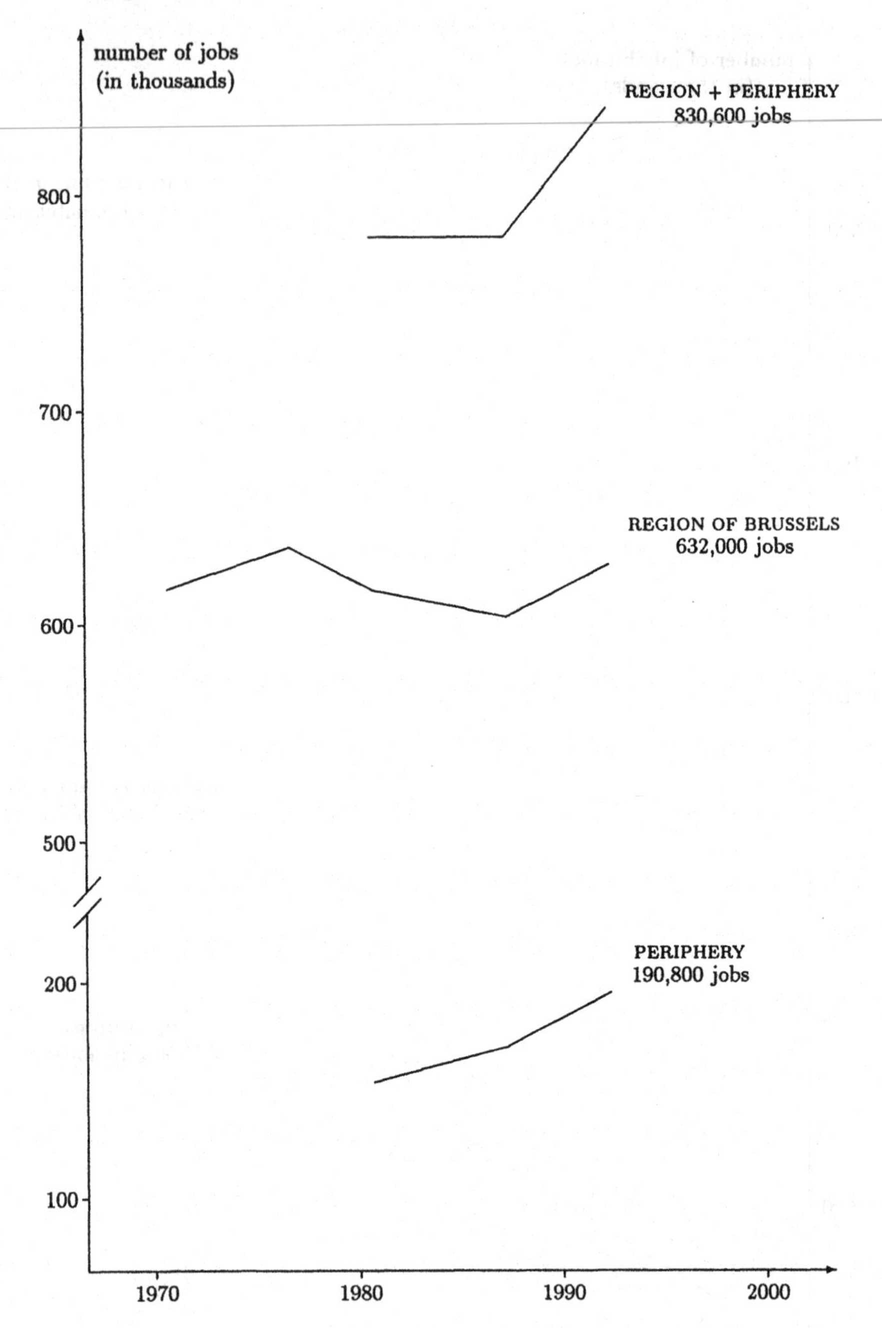

Fig. 2.2. Evolution of employment in the Brussels region and its periphery in the period 1971–1991.

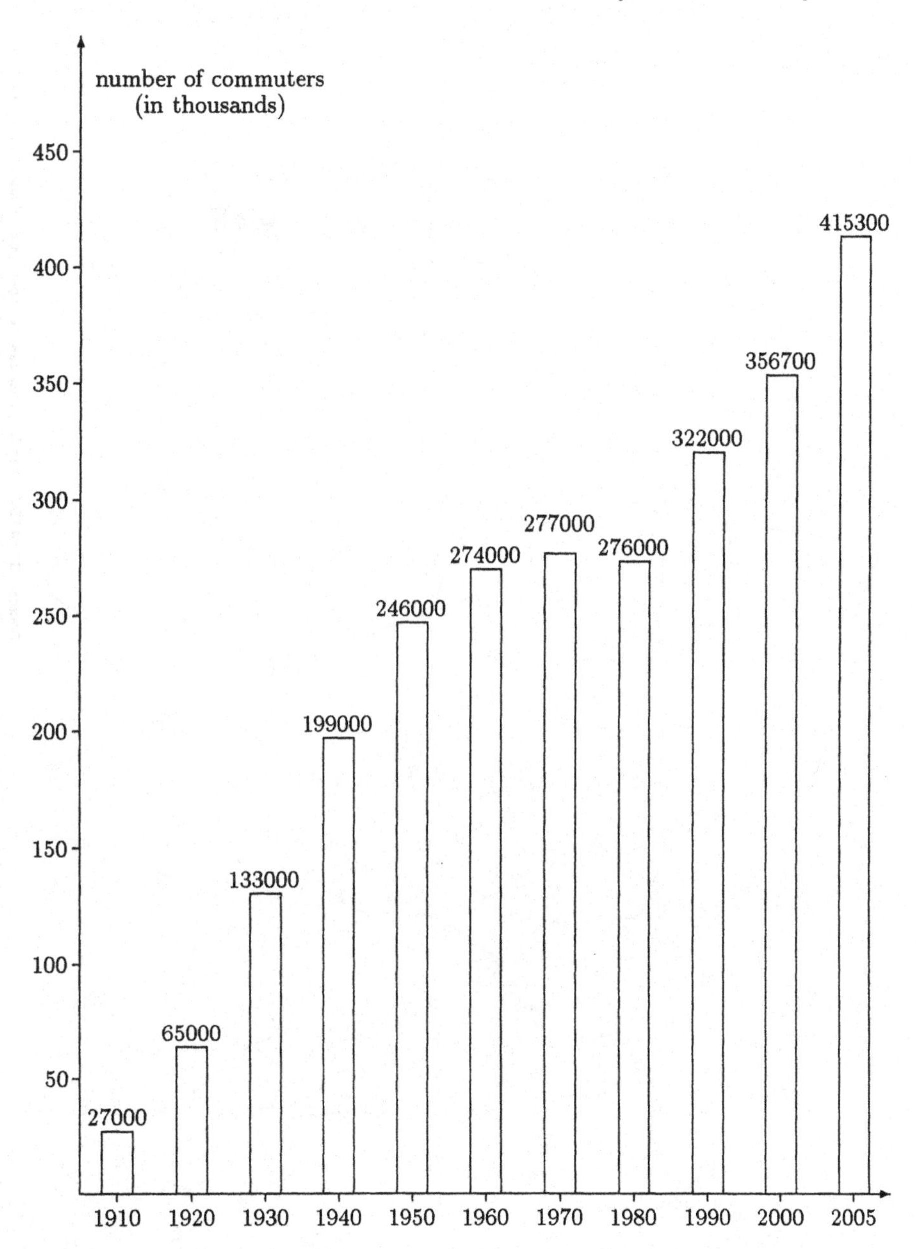

Fig. 2.3. Region of Brussels - commuting from home to work in the city.

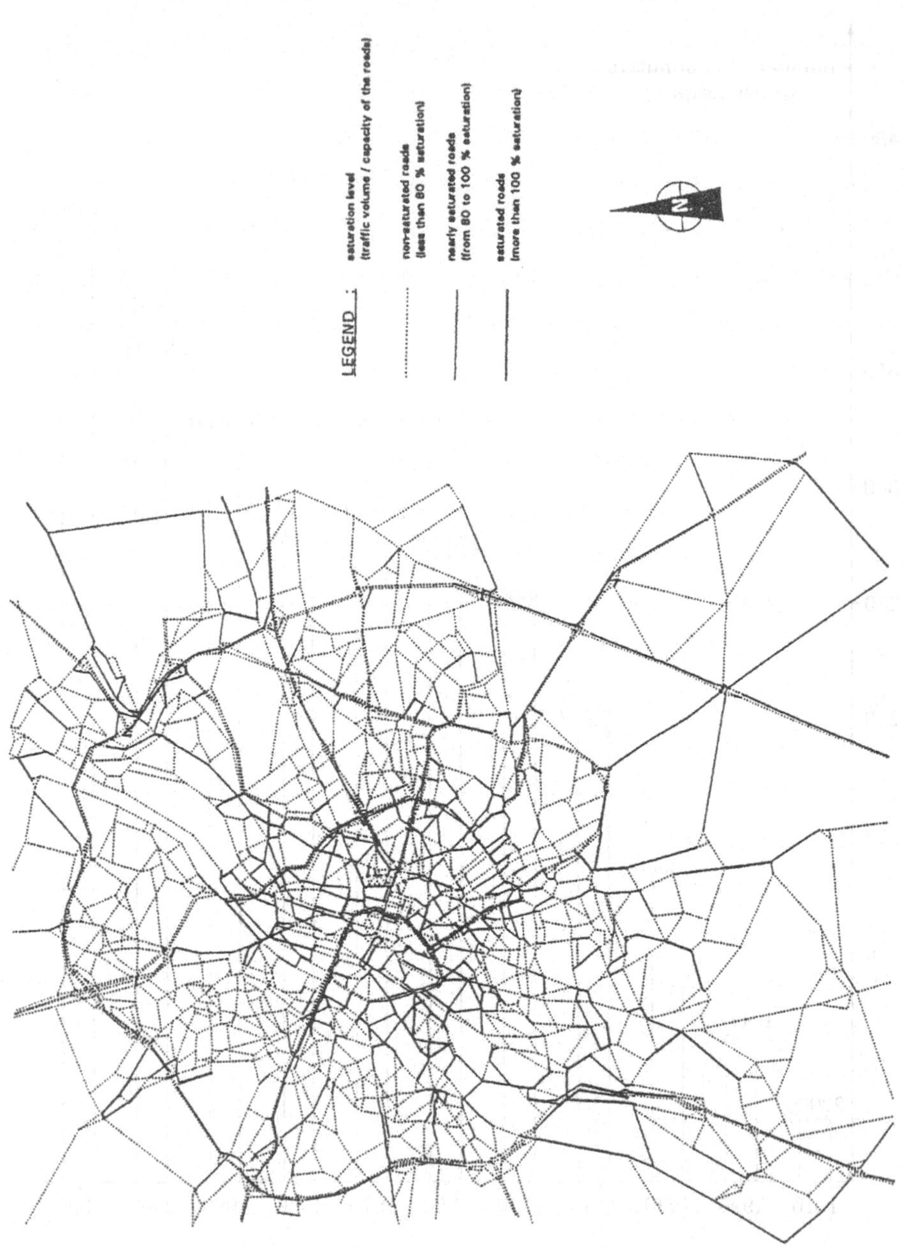

Fig. 2.4. Congestion of the road network in 1991.

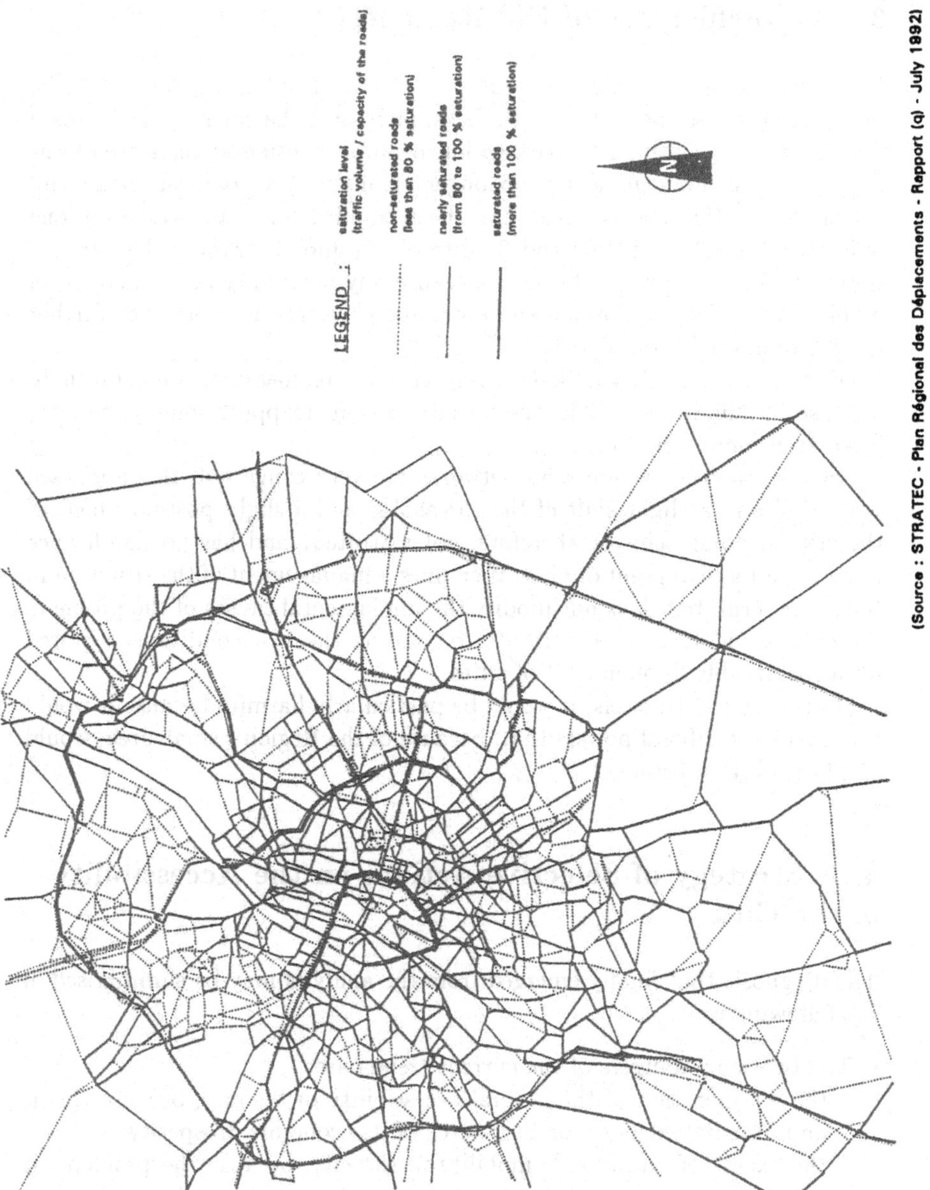

Fig. 2.5. Congestion of the road network in 2005.

3. The Inefficiency of Old Remedies

Restoring mobility by healing symptoms of road traffic congestion, that is, trying to suppress the traffic bottlenecks, increasing the capacity of the main road networks, creating new parking lots or implementing sophisticated techniques of traffic management, is no longer a realistic strategy in the long term.

These are the answers that have been applied for years with only one objective: increase the flow and fluidity of automobile traffic. They are responsible for their own inefficiency because any possibility of an increase in mobility is used by residential and economic players to increase even further their demand of automobile travel.

Experience has shown that in any city where measures were taken to increase the fluidity of traffic, the initial problems reappear some years later in an even more acute form.

Indeed, in cities where road networks are very congested, this approach only results in a slight shift of the thresholds and a slight postponement of the critical point. This is, therefore, not a strategy and has no significance but from a tactical point of view. It enables a management of the situation in the short term, but does not modify the fundamental causes of the problem. In any case, it does not contribute to creating the new conditions required for a sustainable development of cities.

In the case of Brussels, it would be particularly harmful because it would favour the centrifugal powers that, because of the Region's small area, would empty it of its substance.

4. A Strategy of Selective Actions on the Accessibility of the City

The diagnosis that finally emerged from the analysis may be summarised in the following way:

- The two main dangers of the current trends are:
  - the deterioration of the general accessibility of the city, because this is an immediate danger for the source of its economic prosperity;
  - the excess of automobile mobility in the city, because the problems it creates for the environment is a threat to the population but also to the economy.
- With regard to these two dangers, the reactions that can be considered by the Region's public authorities must be selective; these are:
  - selective improvement of accessibility by public transport;
  - selective restriction of accessibility by private cars.

The global strategy emerging from this diagnosis centres around five groups of action that respectively aim at:

- urban structure;
- public transport;
- car parking;
- automobile traffic;
- getting around on foot and by bicycle.

The actions proposed and their justification are listed below.

4.1 Actions on Urban Structure

An analysis of the mechanisms leading to an increase in automobile mobility shows that more and more city users are obliged to move around in their own cars because urban functions to which they must or want to have access (workplace, services, shops, hospitals, ...) are located in places where public transport does not go.

As far as Brussels is concerned, this conclusion was drawn from the observation of the modal distribution of trips during the morning peak period, as a function of whether the destination of these trips was in the Brussels-Capital region or more to the outside and whether the place of residence was located inside the Brussels-Capital Region, outside the Region at its periphery (outer suburbs comprising 33 townships) or in the rest of the country. Table 4.1 [2] compares data collected on:

- the Central Business District from the large central roads up to the Cinquantenaire park, accessible by train through the Central station, the Congrès station, the Quartier Léopold station and the Schuman station and by the tube through the central sections of lines 1A, 1B, 2 and 3;
- the inner suburbs between the large middle ring roads and the outer limits of the region, accessible by some suburban railway stations and by the ends of the tube lines.

The very large differences in the modal distribution shown in table 4.1 are principally caused by the difference of accessibility between the Central Business District and the inner suburbs.

The Central Business District is easily accessible by public transport, from everywhere in the city thanks to the tube, and from the whole country thanks to the railway; on the other hand, access by car is more difficult because of the congestion and the difficulty of finding a parking place.

By contrast, the inner suburbs are relatively easily accessible by road, among others because of the presence of the ring road, and is less easily accessible by public transport (the ends of the tube lines).

Table 4.1 shows that, with everything else remaining unchanged:

[2] Source: STRATEC, Regional Movements Plan, report (q), July 1992.

Table 4.1. The effects of the locations of origin and destination on the modal distribution of movements. Situation in 1991, average workday, 7 a.m. - 9 a.m.

Travel to the central business district			
Origin	Total amount	% in public transport	% in privately owned vehicle
Region	56900	51	49
Periphery	14200	42	58
Rest of the country	31900	71	29
TOTAL	103000	56	44
Travel to the inner suburbs			
Origin	Total amount	% in public transport	% in privately owned vehicle
Region	57900	29	71
Periphery	31300	13	87
Rest of the country	24100	35	65
TOTAL	113300	26	74
Travel across the whole region			
Origin	Total amount	% in public transport	% in privately owned vehicle
Region	271200	39	61
Periphery	94200	21	79
Rest of the country	113900	51	49
TOTAL	479400	38	62

- a certain grouping - together of - motorised trip destinations (workplace and higher education areas, among others) in the city centre would allow reducing the total number of trips by car to the Region; in this way, the transfer of 10,000 destinations of trips from the middle ring to the centre would allow reducing by 3,000 the number of trips by car to the city, as 74% of the trips to the inner suburbs are made by car against 44% for the trips to the Central Business District;
- in the same way, returning to live in the Region by a part of the population which today lives in the periphery, would have a positive influence as 79% of the trips from the periphery are made by car against 61% from the Region itself;
- on the other hand, it can be seen that commuting to Brussels from the rest of the country contributes to a lesser extent to the automobile traffic in the city than other trips.

These global observations, completed by ideas directed more specifically to the development of areas ideal for walking and cycling and with access to public transport, have led to the formulation of a series of recommendations which, via the Land Use Master Plan, can have determining effects on the needs of trips by car and on the problems these trips cause. These recommendations are listed in table 4.2.

Table 4.2. Actions on Urban Structures

1.		To improve the accessibility of workplaces:
	a.	Concentrate the employment which generates a high traffic of people per m2 in areas which are easily accessible by public transport.
	b.	Favour the development of dense residential areas along zones where there is a good public transport service.
	c.	Reserve dense housing areas within reach by foot of highly concentrated employment areas.
	d.	Facilitate Park & Ride and Kiss & Ride for the access to and from less densely populated areas.
	e.	Preserve the industrial railway lines in order to re-allocate them to passenger transport.
2.		To facilitate movement during the working day:
	a.	Accelerate processes of renewed use of the ground in the central areas in order to avoid an under-utilisation and to avoid "no man's lands" whose presence increases the need for motorised transport.
	b.	Organise a mix of mutually complementary activities that do not depend on car use, in the centre of the city.
3.		To facilitate accessibility of shops and services:
	a.	Promote small local shopping centres.
	b.	Maintain and re-vitalise the big commercial centres in the Central Business District.
4.		To facilitate accessibility to collective facilities:
	a.	Locate the regional facilities in the centre.
	b.	Promote the growth of local facilities.
5.		To preserve the accessibility of education establishments:
	a.	Maintain the network of neighbourhood education establishments.
	b.	Discourage the establishment of comprehensive schools at the periphery of urban areas.

4.2 Actions on Public Transport

The analysis of the behaviour of users who have the possibility to use either their car or public transport to get around show that the choices they make are indeed strongly linked to the respective characteristics of the travel they make in one of these two modes. More precisely, the choice of public transport to effect a given trip is most probable if the gap between the duration of the trip by public transport and the duration of the trip with a private car is small (see figure 4.1). Improving the quality of the service offered by public transport is therefore not a useless proposition.

In the case of Brussels, urban public transport offers a relatively good service to the centre from points of origin located in the Region, and the railway is also very effective from any city in the country. By contrast, the service of public transport from the periphery (see table 4.3 [3]) and generally from a distance of 35 km around Brussels is poor.

Table 4.3. Average duration of movement during the morning peak period (7 a.m.-9 a.m.) to the business centre of Brussels (in minutes)

	Average duration of trips		
Origin	By public transport (1)	By private car (2)	Difference (1) - (2)
Region	34.0	24.3	9.7
Periphery	62.5	44.1	18.4
Rest of the country	68.9	64.5	4.4

This is why measures leading to the creation of a suburban railway service are first on the list of objectives of the Regional Movements Plan of Brussels-Capital concerning public transport (see table 4.4 below). These measures should allow reducing the actual duration of trips by at least a quarter of an hour.

4.3 Actions on the Parking of Private Cars

When calibrating the model shown in table 4.1, it appeared that the gap between the trip durations was insufficient to explain the distribution of the observed modes of behaviour. Other elements contribute to determining the choice. The most important of these is parking: when time needed in order to find a parking place increases because the number of spaces is limited or

[3] Source: STRATEC, Regional Movements Plan, report (q), July 1992.

Table 4.4. Actions on Public Transport

a.	Improve the accessibility of the periphery by developing the suburban services of the National Railway Company (by creating an express suburban rail network).
b.	Improve the accessibility of the centre by railway through a more efficient use of the railway infrastructures originally aimed at transporting goods but which are at the moment unused (East and West orbital lines).
c.	Create Park & Ride infrastructures in areas where the deconcentration of housing makes it impossible to offer good public transport services.
d.	Take advantage of existing underground infrastructures (metro and premetro) which account for nearly 50% of the present volume of users of the urban transportation company.
e.	Improve the commercial speed of the surface urban network: * reserved rights of way; * construction measures near crossroads enabling public transport to have priority; * remote control of the traffic lights by public transport vehicles.

parking is not free, the probability that someone chooses public transport increases significantly.

This confirms the importance to be attached to the control of public parking in order to control automobile mobility. This all the more so as actions on parking can be made selectively: by way of a parking policy, one can aim to reduce the use of private cars for long trips (e.g. from home to work), without limiting this use for smaller trips (shopping, business, ...).

Proposed actions concerning parking are listed in table 4.5 below.

Table 4.5. Actions on Automobile Parking

a.	Reduce parking possibilities on streets in the city centre by severely limiting long-term parking and by re-allocating the public space thereby gained to short-term parking, to pedestrians and to lanes reserved for public transportation and to green areas or planting areas.
b.	Strengthen control on the duration of parking in alternating "even-day/odd-day" places.
c.	Implement an effective system of restricting parking along roads to residents living in housing areas without garages.
d.	Review the rules and regulations relating to the building of off-road parking spaces and vary their maximum number according to the number of jobs, the kind of activity and the level of service offered by public transportation networks in the local area.

4.4 Actions on Automobile Traffic

All actions proposed in tables 4.1. and 4.3. will lead to a global reduction of the pressure of the automobile in the city. This decrease must immediately be made to lead to a better control of the rest of the traffic and to a suppression of all inter-area transit traffic in residential areas and in shopping and leisure areas where the traffic is most often perceived as an aggression (noise, risks of accidents for the children, pollution, etc.) Actions of this type are listed under a, b and c of table 4.6 showing strategic actions on automobile traffic.

However, collective problems caused by automobile traffic congestion will not necessarily disappear, at least not until the implementation of a system of charging fees for the use of road infrastructures equal to the marginal cost of congestion.

Until the time such a user fee system is implemented, congestion must be brought under control, that is, limited to places where it causes the least problems. These are the aims of actions d and e in table 4.6.

Table 4.6. Actions on Automobile Traffic

a.	Extending to the whole region the concept of dividing the territory into zones inaccessible to motor transit.
b.	Strengthen the hierarchy of the roads (transit ways, local access roads) and at the same time strengthen the general need for accessibility, respect the areas' characteristics and protect those areas from transit traffic.
c.	Implement in these areas methods of one way traffic and of speed limitations (30 km areas).
d.	No increase and even sometimes decrease the capacity of roads giving access to the city in order to contain the morning congestion out of its limits.
e.	Regulate the capacity of roads leaving the city in order to limit the risks of congestion of its internal network during the evening peak hour.

4.5 Actions in Favour of Walking and Cycling

At present, in Brussels about 40% of the people who leave their home during the day only move about on foot. The regional movements policy must be more attentive to these two categories of users than it used to be.

On the other hand, very few people, about 1%, use a (motor)bike although they are much more numerous in cities such as Brussels where promotion campaigns in favour of their use were conducted. These two subjects constitute the last part of the action proposals for the Regional Movements Plan. They are listed in table 4.7.

Table 4.7. Actions in the Field of Pedestrian and two-wheel traffic

- Improve the comfort and safety of pavements and pedestrian crossings on roads by taking measures concerning parking.
- Develop cycle lanes across "non-transit" areas.
- Promote combined "bicycle-public transport" travel.

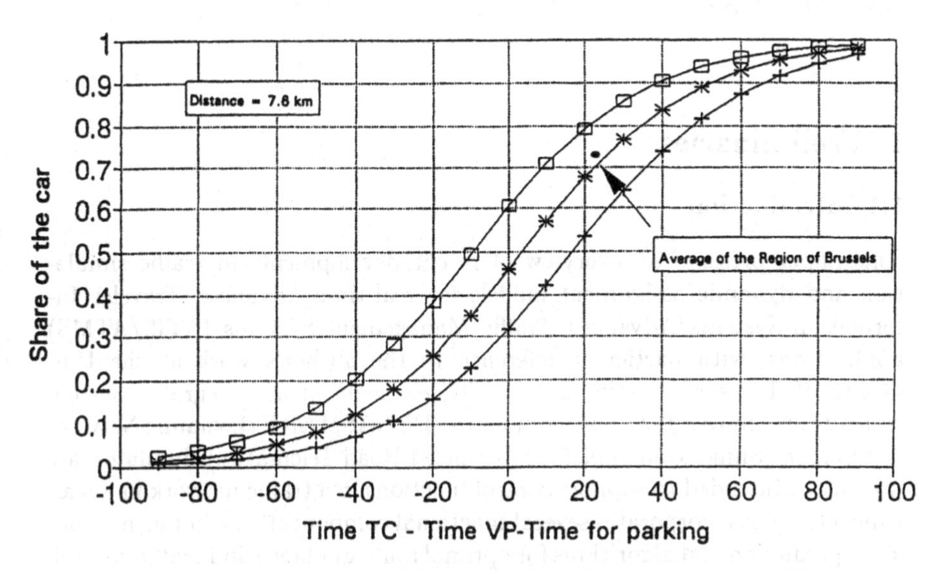

Fig. 4.1. Proportion of the use of the private car by working people as a function of the difference in travel duration between public and private transport, by motorization-level class.

Dynamic Traffic Simulation and Assignment: Models, Algorithms and Application to ATIS/ATMS Evaluation and Operation

Hani S. Mahmassani

The University of Texas at Austin,
Texas 78712, U.S.A.

1. Preliminaries

1.1 Introduction

This paper presents an overview of recent developments in traffic simulation and dynamic assignment models for real-time Advanced Traveler Information Systems/Advanced Traffic Management Systems (ATIS/ATMS) applications, with particular reference to the author's work at the University of Texas at Austin. Specifically, elements of an integrated system called DYNASMART-X, built around the DYNASMART (Dynamic Network Assignment-Simulation Model for Advanced Road Telematics) simulator, are presented. Intended to support control functions in a traffic network in a real-time setting, it incorporates several functional components, including network state prediction and algorithms for optimal route guidance and traffic control.

The paper is structured as follows. First, in the remainder of this section, we explain the principal functions of various dynamic traffic assignment models that arise in the context of ITS (Intelligent Transportation Systems) applications to road traffic networks. This is followed, in Section 2, by a description of the conceptual framework and structure of the DYNASMART simulation-assignment model, which is the core of the DTA system. In Section 3, we describe an algorithm to solve a time-dependent System Optimum path assignment problem, as well as modification to solve a User Equilibrium version of the problem. In Section 4, the model is extended to consider multiple user classes, defined on the basis of information supply strategy and user response to information, for the provision of real-time route guidance instructions. The algorithm and the simulation-assignment framework are illustrated in Section 5 through a set of numerical experiments on a test network intended to evaluate the effectiveness of different control strategies, including dynamic route guidance through real-time information dissemination. In Section 6, we discuss the real-time application of the above algorithm in a rolling horizon implementation of the DTA capabilities for on-line control, and the principal issues affecting the real-time execution of the DTA system functions. This centralized approach to real-time route assignment and guidance is contrasted to a decentralized approach which relies on heuristic local rules that react to observed measurements. We compare these approaches through numerical experiments on a test network. Concluding comments are presented in Section 7.

1.2 Descriptive and Normative Dynamic Traffic Assignment Capabilities in ATMS/ATIS Architecture

Advanced traffic management systems (ATMS) envision a traffic management center (TMC) in charge of monitoring operations and developing and implementing control actions over the traffic network in a metropolitan area. The TMC receives information in real-time about prevailing conditions from several sources, including various detection devices (loop detectors in the pavement, video imaging, Automatic Vehicle Identification devices, and others) as well as from vehicle probes—vehicles equipped with location and navigation devices and two-way communication capabilities, that relay information about prevailing speed and location to the control center. This serves as a basis for traffic control actions, including incident response, traffic signal setting, variable message signs to inform users and influence their route choices, as well as the provision of route guidance instructions to vehicles equipped with two-way communication capabilities and in-vehicle display units (Catling [2]; Whelan [21]; Branscomb and Keller [1]).

To support the above ATMS/ATIS functions of the TMC, several methodological capabilities are required, to process the volumes of incoming information, analyze network operations, and determine control actions that optimize network performance. Central among these methodologies are dynamic traffic assignment techniques, along with several associated support functions, which are conveniently integrated into a dynamic traffic assignment (DTA) system. Two essential capabilities are required of the DTA system. The first is descriptive, and consists of describing how flow patterns develop spatially and temporally in a traffic network, typically given a set of desired trips between origins and destinations. This descriptive capability allows estimation of current state of the network especially when the latter is only partially observable, as well as prediction of future network states over time. At its core is the ability to model the outcomes of tripmaker decisions, primarily the decision of which path to take between origin and destination, but also possibly the decision of when to depart, and what mode to use. To the extent that these decisions are predicated on network conditions, which in turn depend on the users' decisions, network states have to be determined simultaneously with the tripmaker choices, generally in an iterative scheme. The estimated state of the network and predicted future states, in terms of flows, travel times and other time-varying performance characteristics on the various components of the network, are used in the on-line generation and real-time evaluation of a wide range of ATMS measures and ATIS messages. The core of the descriptive DTA capability is a traffic simulation model, which seeks to capture the dynamics of traffic flow movement in the network. Research at the University of Texas at Austin has resulted in the DYNASMART (Dynamic Network Assignment-Simulation Model for Advanced Road Telematics) simulation model, which is described in the next section (Mahmassani and Jayakrishnan [11]; Jayakrishnan et al. [7]).

To the extent that the actual traffic network is also monitored continuously via a variety of sensing devices and probe vehicles capable of two-way communication with the control center, an important external support function for on-line DTA consists of ensuring consistency of the simulation-assignment model results with actual observations, and updating the estimated state of the system accordingly. Another external support function is intended to perform the estimation and prediction of the origin-destination (OD) trip desires that form the load onto the traffic network, and are as such an essential input to the simulation-assignment core.

The second major DTA capability required for ATMS/ATIS operation is normative; it aims to provide route guidance information to tripmakers, generally to attain some systemwide objectives, taking into account the individual welfare of tripmakers and the longer-term credibility and acceptance of the information system. In this sense, the provision of route guidance information is viewed as an integral element of traffic network operations management, working in tandem with the traffic control systems and incident response management systems to optimize overall network performance and productivity. There are different ways of seeking to achieve this capability. The most natural is to search for path assignments that are in some way optimal from the system's perspective, subject to certain reasonableness and acceptability requirements from the standpoint of individual users. This would again require joint determination of paths and associated network conditions. This approach is in fact required for any ATIS information supply strategy for which the supplied information is to be accurate and/or achieves the intended objectives when users' reactions to the information are taken into account. Another approach would be to guide individual users in a more "local" manner, link to link, with the actual path followed resulting from the succession of links along which the user would have been guided. The latter, appropriate for a decentralized control architecture, though also implementable in centralized architecture, could be entirely based on measured network conditions, and as such be highly responsive to sudden changes in traffic conditions due to supply shocks such as incidents. It may however grossly underperform a system optimal routing strategy when traffic conditions are relatively stable and predictable.

Within the ATMS/ATIS control architecture, both descriptive and normative DTA capabilities must be provided; both are required in two distinct operational settings: real-time on-line operation (tactical), and off-line planning (strategic). Of course, the descriptive capability (e.g. the DYNASMART simulator) is also an essential component of most algorithms and procedures designed to provide the normative capability, in order for the latter to achieve consistency between the information supplied and the users' responses to the information. On the other hand, the route guidance instructions and/or other ATIS information produced as output of the normative capability, in a control center architecture, will be an input to the network state prediction performed by the descriptive DTA, to the same extent that the output of the adaptive signal controllers and other control elements are continuously fed to the descriptive DTA.

The above two capabilities (descriptive and normative), along with their support functions, have been integrated in the DYNASMART-X DTA System, intended to provide, in real time, the four primary functions, shown in Figure 1.1: estimation, prediction, consistency checking and route guidance. Note that DTA-SAM in this figure refers to the simulation-assignment modeling (SAM) descriptive capability. The DYNASMART-X system is configured to eventually meet varying needs and requirements of different communities, as well as different deployment scenarios and ITS architectures that are likely to vary across different cities and regions.

The next section describes the basic components of the simulator that lies at the core of the simulation-assignment framework, and provides the primary descriptive capability.

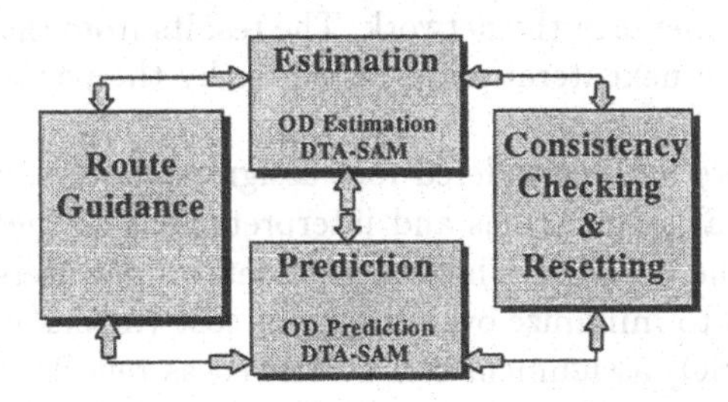

Fig. 1.1. DYNASMART-X DTA System Functional Diagram

2. Simulation-Assignment Modeling Framework

2.1 Problem Definition and User Assignment Rules

Consider a network $G(N, A)$ consisting of a set of nodes N connected by the set of directed arcs A. The time horizon of interest (e.g. the peak traffic period) is discretized into T small time slices, referred to as assignment intervals. Let r_{ij}^{τ} denote the number of users who intend to go from origin node i to destination node j during time interval $\tau, i, j \in N$ and $\tau = l, \ldots, T$. Suppose the TMC can provide information to users according to a set of information supply strategies S. Denote by R the set of possible response rules followed by network users (note that an element of R may itself be a collection of responses rules followed by different classes of users). The time-dependent assignment problem is to distribute trip desires $\{r_{ij}^{\tau}, \forall i, j, \tau \}$ to the network according to an assignment rule, I, where $I \in (S \times R)$ in this context, in a manner that is consistent with the temporal and spatial traffic processes that take place in the network (Mahmassani, Hu and Peeta [8]).

A conceptual framework of the simulation-based approach is elaborated in Figure 2.1. In this framework, vehicles are generated according to a time-dependent OD matrix, and assigned to routes specified by the assignment rules; depending on the particular rules adopted, the paths may be obtained either through direct application of individual path choice models, or through algorithmic steps intended to satisfy certain conditions in the network. The time-dependent flow pattern can be simulated by loading vehicles and representing their movements in the network. The results from the simulation may then be used in the next iteration, if called for by the particular assignment rule.

In this work, we have considered four assignment rules, corresponding to different behavioral assumptions and interpretations of the time-dependent flow patterns in the network. The first rule determines users' paths through the network so as to minimize overall system cost (in this case travel time). System optimal (SO) assignment can be viewed as reflecting an information supply strategy where vehicles are guided to use individual paths that are optimal for the system as a whole (i.e., normative route guidance), and users fully comply with the routing instructions. A SO assignment pattern does not usually correspond to an equilibrium situation as some users may be able to improve their individual trip times at the expense of greater cost to the system. However, a SO pattern provides a benchmark against which other assignments can be gauged.

The second assignment rule corresponds to a time-dependent User Equilibrium (UE), under which no user can improve his/her individual cost by unilateral route switching. Such a state could result from the long-term evolution of the system, as users somehow learn and adjust under the supplied information. However, there is no theoretical nor empirical justification to expect convergence to a UE pattern under inherently dynamic conditions.

The third assignment rule corresponds to a family of response rules to an information supply strategy under which users receive descriptive information on prevailing link trip times. The family of response rules consists of boundedly-rational path switching and selection rules, which include a myopic switching rule (always select the shortest path based on current conditions) as a special case.

The fourth assignment rule can be viewed as a special case of the preceding one, in which a vehicle is assigned to its current best path from the trip origin. Such an assignment would arise if a departing tripmaker could consult an origin-based information system (e.g. television or telephone) and select the shortest path to the destination under current traffic conditions without considering possible future congestion.

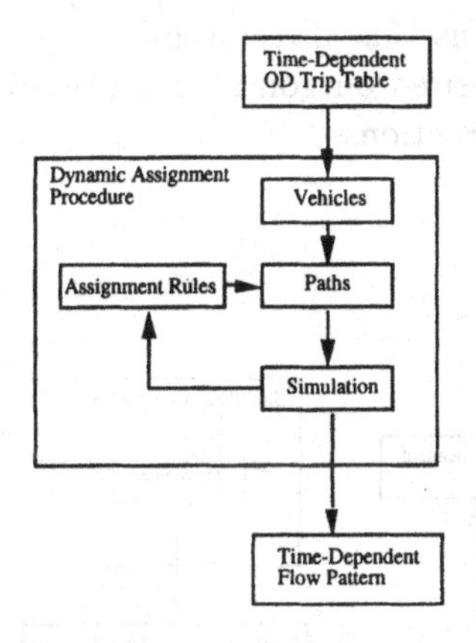

Fig. 2.1. Conceptual Framework of the Simulation-Assignment Procedure

The solution algorithms for these assignment rules are discussed in section 3. The next section describes the various components of the DYNASMART simulation assignment model, which is the principal methodology used here to represent the complex interactions taking place in the traffic network.

2.2 DYNASMART Simulation-Assignment Model

2.2.1 Summary of Features. In its present form, DYNASMART is primarily a descriptive analysis tool for the evaluation of information supply strategies, traffic control measures, and route assignment rules at the network level. The model is designed around a flexible structure that provides sensitivity to a wide range of traffic control measures for both intersections and freeways, capability to model traffic disruptions due to incidents and other occurrences, and representation of several user classes corresponding to different vehicle performance characteristics (e.g. cars vs. trucks), access to physical facilities (e.g. HOV lanes), different information availability status, and different behavioral rules.

2.2.2 Model Structure. The structure of DYNASMART is shown in Figure 2.2. The approach integrates traffic flow models, path processing methodologies, behavioral rules, and information supply strategies into a single simulation-assignment framework. The input data include a time-dependent origin-destination matrix (or a schedule of individual departures) and network data. Given the network representation, which includes link characteristics as well as control parameters, the simulation component will take a time-dependent loading pattern and process the movement of vehicles on links and the transfers between links according to specified control parameters. These transfers, which are determined by path processing and path selection rules, require instructions that direct vehicles approaching the downstream node of a link to the desired outgoing link. The user behavior component is the source of these instructions.

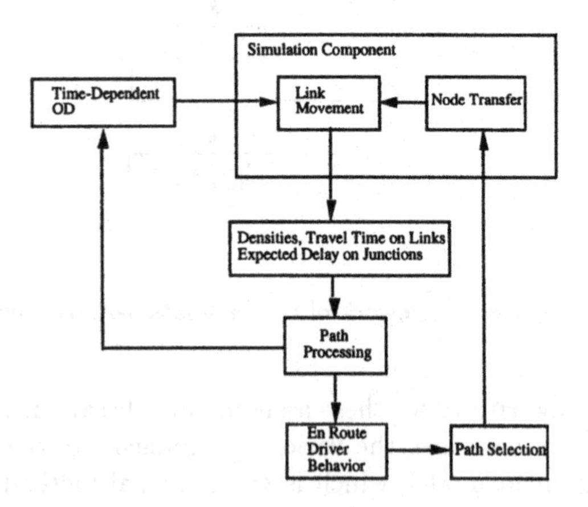

Fig. 2.2. Structure of DYNASMART Simulation-Assignment Model

2.2.3 Traffic Simulation Component. DYNASMART uses established macroscopic traffic flow models and relationships to model the flow of vehicles through a network. Whereas macroscopic simulation models do not keep track of individual vehicles, DYNASMART moves vehicles individually or in packets, thereby keeping a record of the locations and itineraries of the individual particles. This level of representation has also been referred to as "mesoscopic". Multiple user classes of different vehicle performance characteristics are modeled as packets, consisting of one or more passenger car units; for instance, a bus is represented by a packet with two (or other user-specified values) passenger car units. The traffic simulation consists of two principal modules: link movement and node transfer, as described hereafter.

Link Movement: The link movement is a process for moving vehicles on links during each scanning time interval in the simulation (time step). Note that the network links are subdivided into smaller sections or segments for traffic simulation purposes. The vehicle concentration prevailing on a section over a simulation time step is determined from the solution of the finite difference form of the usual continuity equation, given the concentration as well as inflows and outflows over the previous time-step (Jayakrishnan et al. [7]). Using the current concentration, the corresponding section's speeds are calculated according to a modified Greenshield speed-density relationship, namely:

$$V_i^t = (V_f - V_0)(1 - K_i^t/K_0)^\alpha + V_0$$

where,

V_i^t, K_i^t = mean speed and concentration in section i during the t-th time step,
V_f, V_0 = mean free speed and the minimum speed, respectively,
K_0 = jam concentration, and
α = a parameter used to capture the sensitivity of speed to the concentration.

Other traffic stream models may also be incorporated in DYNASMART based on field investigation.

Node Transfer: The node transfer module performs the link to link or section to section transfer of vehicles at nodes. For interrupted link flow, it appropriately allocates the right of way according to the prevailing control strategy. The output of the node transfer includes the number of vehicles that remain in queue and the number added to and subtracted from each link section for each simulation time step. A wide range of traffic control measures for both intersections and freeways are reflected in the outflow and inflow capacity constraints that govern the node transfer (Mahmassani et al. [9]).

2.2.4 User Behavior Component. One of the principal features of DYNASMART that allows it to interface with activity-based behavioral models is its explicit representation of individual tripmaking decisions, particularly for path selection decisions, both at the trip origin and en-route. Behavioral rules governing route-choice decisions are incorporated, including the special case in which drivers are assumed (required) to follow specific route guidance instructions. Experimental evidence presented by Mahmassani and Stephan [16] suggested that commuter route choice behavior exhibits a boundedly-rational character. This means that drivers look for gains only outside a threshold, within which the results are satisfying and sufficing for them. This can be translated to the following route switching model (Mahmassani and Jayakrishnan [12]):

$$\delta_j(k) = \begin{cases} 1 & \text{if} \quad TTC_j(k) - TTB_j(k) > \max(\eta_j . TTC_j(k), \tau_j) \\ 0 & \text{otherwise} \end{cases}$$

where $\delta_j(k)$ is a binary indicator variable equal to 1 when user j switches from the current path to the best alternate, and 0 if the current path is maintained; $TTC_j(k)$ and $TTB_j(k)$ are the trip times along the current path and along the best path from node k to the destination on current path, respectively; η_j is a relative indifference threshold, and τ_j is an absolute minimum travel time improvement needed for a switch.

The threshold level may reflect perceptual factors, preferential indifference, or persistence and aversion to switching. The quantity η_j governs users' responses to the supplied information and their propensity to switch. The minimum improvement τ_j is currently taken to be identical across users according to user defined values. Results of laboratory experiments indicate that τ_j is on average equal to one minute, while η_j is about 0.2 for typical urban commutes (Mahmassani and Liu [13]).

2.2.5 Path Processing. The path processing component of DYNASMART determines the route-level attributes (e.g. travel time), for use in the user behavior component, given the link-level attributes obtained from the simulator. For this purpose, a multiple user class K-shortest path algorithm with movement penalties is interfaced with the simulation model to calculate K different paths for every OD pair. However, in order to improve the model's computational performance, the K paths are not re-calculated every simulation time step, but at pre-specified intervals. In the interim, the travel times on the set of K current paths are updated using the prevailing link travel times at each simulation time step (Mahmassani et al. [8]) .

2.2.6 Traffic Control Elements. An integrated urban road system consists of a surface street network, a freeway system, and freeway corridors. DYNASMART provides the ability to explicitly model an array of ATMS control elements. The major element for surface streets is signal control, which includes pretimed control and actuated control, as well as signal coordination along arterials. Ramp metering, variable message signs, and High-Occupancy Vehicle (HOV) lanes are the major controls for the freeway system. DYNASMART uses a fixed time increment simulation approach, all the signal operations can be readily advanced according to a time clock. Phase number, offset, green time, red time and amber time, need to be defined for every phase in order to advance the operation.

For actuated signal control, DYNASMART evaluates the possible number of vehicles that can reach the stop line in the active phase, and compares it with the capacity under the current green time period. If the green time is insufficient for the active phase, the green time is extended for a suitable length that is determined by the number of vehicles in queue. The green time is bounded by the assumed minimum and maximum green times. These calculations are performed for each simulation interval (Mahmassani et al [8]).

3. System Optimal and User Equilibrium Assignment

Assignment under either rule one (SO) or rule two (UE) stated in section 2.1 requires the joint determination of time-varying path flows and mutually consistent performance measures (travel times) on the network components. We describe an iterative solution procedure, in which a traffic simulator (such as DYNASMART) is used to represent the traffic interactions in the network. First, we consider the SO problem.

3.1 Problem Statement

The system optimal (SO) time-dependent traffic assignment problem is encountered by a central controller seeking to route vehicles optimally through the network, given knowledge of the vehicles' origin-destination (O-D) desires over the period of interest. This problem is central to the ATIS/ATMS context, and forms the basic building block for all other formulations considered, namely the UE formulation, the Multiple User Classes (MUC) formulation, and its Rolling Horizon (RH) implementation. Formulation issues of dynamic assignment problems in the context of ATIS/ATMS are discussed in detail in Mahmassani et al. [9] and Mahmassani and Peeta [15].

Following the formulation of section 2.2, we also consider a traffic network represented by a directed graph $G(N, A)$, a set of time-dependent O-D vehicle trip desires known for the entire duration of interest (peak period), expressed as the number of vehicle trips r_{ij}^{τ} leaving node i for node j in time slice (assignment interval) $\tau, \forall i, j \in N$ and $\tau = 1, \ldots, T$. The assignment interval corresponds to a period over which O-D demands are not expected to vary much; the decision variables (number of vehicles r_{ijk}^{τ} assigned to alternative paths) are defined for the assignment intervals, which are expected to be of the order of minutes, say 3 to 5 minutes, in this formulation. The problem is to determine a time-dependent assignment of vehicles to network paths and corresponding arcs, i.e. to find the number of vehicles r_{ijk}^{τ} that follow path $k = 1, \ldots, K_{ij}$ between i and j at time $\tau, \forall i, j \in N$ and $\tau = 1, \ldots, T$, as well as the associated numbers of vehicles on each arc $a \in A$ over time, so as to minimize total travel time in the system, as follows.

3.1.1 Definition of Variables and Notation. The following variables and notation are used in the various formulations:

$i =$ subscript for origin node
$j =$ subscript for destination node
$n =$ node in the network, $n \in N$
$a =$ arc (or link) in the network, $a \in A$
$k =$ subscript for a path in the network
$\tau =$ subscript denoting the time interval in which assignment is made (i.e. departure time from trip origin)
$t =$ subscript denoting current time interval
$\Delta =$ length of a time interval
$T' =$ total duration (peak period) for which assignment is to be made
$r_{ij}^{\tau} =$ number of vehicles who wish to depart from i to j in period τ
$r_{ijk}^{\tau} =$ number of vehicles who wish to depart from i to j in period τ assigned to path k
$\delta_{ijk}^{\tau ta} =$ dynamic arc-path incidence indicator, equal to 1 if vehicles going from i to j assigned to path k at time τ are on link a at beginning of period t, i.e.
$\delta_{ijk}^{\tau ta} = 1$, if r_{ijk}^{τ} is on arc a at beginning of period t
$\delta_{ijk}^{\tau ta} = 0$, if arc a does not belong to path k
$\delta_{ijk}^{\tau ta} = 0$, if $\tau > t$
$\delta_{ijk}^{\tau ta} = 0$, if r_{ijk}^{τ} is not on arc a at beginning of period t
$T_{ijk}^{\tau} =$ path travel time for vehicles going from i to j assigned to path k at time τ
$x_{ijk}^{\tau ta} =$ number of vehicles (i to j) assigned to path k in period τ which are on link a at the beginning of period t
$d_{ijk}^{\tau ta} =$ number of vehicles (i to j) assigned to path k in period τ which enter arc a in period t
$m_{ijk}^{\tau ta} =$ number of vehicles (i to j) assigned to path k in period τ which exit link a in period t
$x^{ta} =$ total number of vehicles on link a at the beginning of period t
$d^{ta} =$ total number of vehicles which enter link a in period t
$m^{ta} =$ total number of vehicles which exit link a in period t
$C(n) =$ set of links directed towards node n
$B(n) =$ set of links directed away from node n

3.1.2 Mathematical Formulation. Given:

$$r_{ij}^{\tau}, \quad \forall i, j \text{ and } \tau = 1, \ldots, T'$$

Objective function:

$$\text{Min} \sum_{\tau} \sum_{i} \sum_{j} \sum_{k} (r_{ijk}^{\tau} T_{ijk}^{\tau})$$

or

$$\text{Min}[T(r_{ijk}^{\tau}), \quad \forall i,j,k,\tau]$$

Subject to:

$$r_{ij}^{\tau} = \sum_{k} (r_{ijk}^{\tau}), \quad \forall i,j,\tau \tag{3.1}$$

$$\sum_{c} m^{tc} = \sum_{b} (d^{tb}), \forall t, c \in C(n), b \in B(n), n = i \text{ or } j \tag{3.2}$$

$$x^{ta} = x^{t-1,a} + d^{t-1,a} - m^{t-1,a}, \quad \forall t,a \tag{3.3}$$

$$x^{ta} = \sum_{k}\sum_{\tau}\sum_{i}\sum_{j}(r_{ijk}^{\tau}\delta_{ijk}^{\tau ta}), \quad \forall t,a \tag{3.4}$$

$$T_{ijk}^{\tau} = \sum_{t}\sum_{a}[\delta_{ijk}^{\tau ta}.\Delta], \quad \forall i,j,k,\tau \tag{3.5}$$

$$\delta_{ijk}^{\tau ta} = f(r_{ijk}^{\tau}), \quad \forall i,j,k,\tau,t,a \tag{3.6}$$

$$d^{ta} = \sum_{k}\sum_{\tau}\sum_{i}\sum_{j}\delta_{ijk}^{\tau ta}, \quad \forall t,a \tag{3.7}$$

$$m^{ta} = \sum_{k}\sum_{\tau}\sum_{i}\sum_{j}m_{ijk}^{\tau ta}, \quad \forall t,a \tag{3.8}$$

$$\tau \leq t \tag{3.9}$$

$$\delta_{ijk}^{\tau ta} = 0 \text{ or } 1 \tag{3.10}$$

$$\text{All variables (other than } \delta_{ijk}^{\tau ta}) \geq 0 \tag{3.11}$$

The objective function is stated in two alternative forms above. The first states that the total travel time of the assigned vehicles in the system is aggregate of the product of the number of vehicles assigned to a particular path (from a given origin to a given destination at a particular time) and the corresponding path travel time. This assumption is realistic when assignment intervals are reasonably small (in which case there are not more than two or three vehicles to a particular path from an origin to a destination). The second form of the objective function simply states that the total travel time of all vehicles assigned to the various paths during the period of interest is some function of the assignment. This objective function can be evaluated by any available means. We do it through simulation.

Constraint (3.1) is a definitional constraint stating that O-D desires assigned to the various paths should sum up to the demand (conservation at the origin). Constraint (3.2) states that vehicles cannot be stored at intermediate nodes, that is, the number of vehicles exiting from all links incident on an intermediate node should equal the number of vehicles entering all links incident from that node at any given time. Constraint (3.3) represents the conservation of vehicles on a link and states that the total number of vehicles on any link at the end of the current time interval is the net algebraic sum of vehicles on that link at the end of the previous time period, vehicles entering that link during the current period and vehicles exiting that link during the current period. Constraints (3.4), (3.5) and (3.6) represent the time-dependent link-path incidence relationships which fundamentally characterize the dynamic assignment problem. Constraint (3.4) represents the dynamic relationship between the number of vehicles assigned to various paths and their aggregation on links. Constraint (3.5) calculate the path travel times using the dynamic link-path incidence variables. The number of time steps in which the dynamic incidence variable takes a value 1 implies the number of discrete time steps that a vehicle (or a group of vehicles) spent in the system, and multiplying with D gives the actual travel time in the system. Constraint (3.6) states that the dynamic link-path incidence variables are a function of the assignment, giving rise to a fixed point problem which defines the essence of the dynamic assignment problem.

Constraints (3.7) and (3.8) are definitional constraints for the number of vehicles entering and exiting links in the various time intervals. Constraint (3.9) defines temporal correctness. Constraint (3.10) restricts the dynamic incidence variables to take values of 0 or 1. Constraint (3.11) represents the non-negativity requirement.

The formulation incorporates dynamic link-path incidence variables which relate path flows to link flows. The fundamental difficulty in solving dynamic assignment problems (SO or otherwise) is that the dynamic incidence variables are themselves a function of the assignment, giving rise to a complicated fixed-point problem. Essentially, the resulting formulation, which involves nonlinearities in the objective function as well as in the constraints, yields generally undesirable mathematical properties that preclude the guarantee of global optimality. Additional discussion of this formulation and its properties is given in Peeta and Mahmassani [18].

An alternative scenario is that the controller has O-D trip desires only for the present period, and future O-D desires are treated as random variables with known probability distributions (based on historical data), giving rise to a stochastic programming formulation of the problem. This formulation is discussed in Mahmassani et al. [9].

3.2 Solution Algorithm

A simulation based algorithm is used to solve the above SO DTA problem. It consists of a heuristic iterative procedure in which the DYNASMART model described in section 2.2 is used to represent the traffic interactions in the network, thereby evaluating the performance of the system under a given assignment. As discussed in Mahmassani et al. [9], the use of a traffic simulator to evaluate the SO objective function and model system performance circumvents the need for link performance functions and link exit functions, as well as ensures FIFO, captures link interactions and precludes unintended holding of traffic, ensuring consistency with realistic traffic behavior (Mahmassani and Peeta [15]. The procedure assigns vehicles to various paths directly, obviating the need to infer a path assignment from the solution to a link-based formulation.

Figure 3.1 illustrates the general solution framework. The algorithm is an extension of well-known solution methods for the static assignment problem, with key differences in each component of the algorithm and significant additional implementation challenges. The vehicular trip times from the current simulation provide the basis for a direction finding mechanism in the search process. The complexity of the interactions captured by the simulator when evaluating the objective function generally preclude the kind of well-behaved properties required to guarantee a descent direction in each iteration and/or convergence of the search process to the minimum. The time-dependent least marginal travel time paths are obtained using the time-dependent least cost path algorithm described in Ziliaskopoulos and Mahmassani [22]. The solution methodology avoids complete path enumeration between O-D pairs.

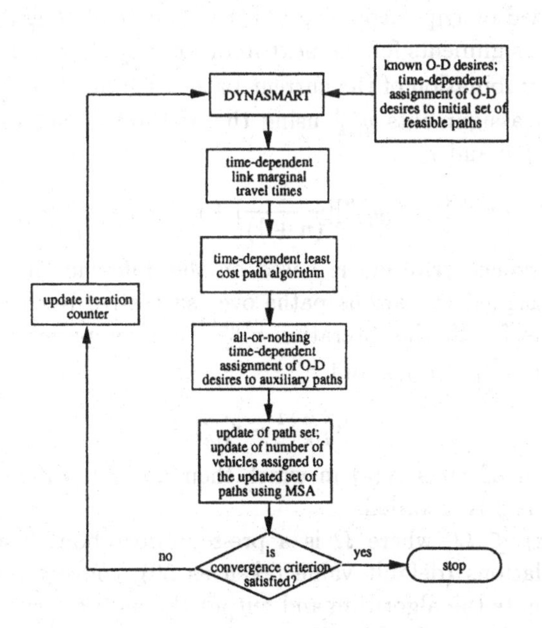

Fig. 3.1. Solution Algorithm for the SO DTA Problem

The algorithmic steps of the procedure are as follows:

1. Set the iteration counter $n = 0$. Assign the given O-D desires r_{ij}^{τ} $\forall i, j$ and $\tau = 1, \ldots, T$, to a time-dependent initial set of feasible paths K(I). Hence, at the start of the iterative search process, the initial solution is given by the assignment $r_{ijk}^{\tau,0}$, $\forall i, j, \tau = 1, \ldots, T$, and $k \in K(I)$.
2. The traffic network under the set of path assignments $r_{ijk}^{\tau,n}$ for the entire duration of interest is simulated using DYNASMART, which evaluates the total system travel time objective function. In addition, a number of time-dependent link level performance measures are obtained as simulation output including the link travel times T^{ta}, and the number of vehicles on links $x^{ta}, \forall t, a$.
3. Compute link marginal travel times $t^{ta}, \forall t, a$. In the present implementation, approximate marginal travel times are calculated using the link travel times T^{ta} and the numbers of vehicles on links x^{ta} in the neighborhood of each link (the procedure is described in detail in Mahmassani et al.[10].
4. Compute the time-dependent least marginal travel time paths $k^*, \forall i, j$ and τ.
5. Perform an all-or-nothing assignment of all O-D desires r_{ij}^{τ} for a given i, j and τ to the corresponding least cost marginal path k^*. This gives the auxiliary number of vehicles on paths, $y_{ijk^*}^{\tau,n}$ $\forall i, j$ and τ.
6. Path update is done by checking if $k^* \in K_{ij}^{\tau}$, the set of paths between i and j utilized by trips departing at time τ, and including it if it does not. The path assignments for the next iteration $r_{ijk}^{\tau,n+1}$ are obtained through a convex combination of the current path assignments $r_{ijk}^{\tau,n}$ and the auxiliary path assignments $y_{ijk}^{\tau,n}$ using the Method of Successive Averages (MSA) $\forall i, j, k$ and τ:

$$r_{ijk}^{\tau,n+1} = [y_{ijk}^{\tau,n}][\frac{1}{(n+1)}] + [r_{ijk}^{\tau,n}][1 - \frac{1}{(n+1)}]$$

7. The convergence criterion is based on the difference in the number of vehicles assigned to various paths over successive iterations. The path assignments for the last iteration $r_{ijk}^{\tau,n+1}$ are compared to current path assignments $r_{ijk}^{\tau,n}$ $\forall i, j, k$ and τ,

$$|r_{ijk}^{\tau,n+1} - r_{ijk}^{\tau,n}| \leq \varepsilon$$

The number of cases $N(\varepsilon)$ in which their absolute difference is greater than a value ε is recorded.

8. a) If $N(\varepsilon) \leq \Omega$, where Ω is a pre-set upper bound on the number of violations (default value taken as 50), convergence is assumed. Terminate the algorithm and output the path assignments $r_{ijk}^{\tau,n+1}$ as the solution to the SO DTA problem.

   b) If $N(\varepsilon) > \Omega$, the convergence criterion is not satisfied. Update $n = n + 1$. Go to Step 2 with the new current path assignments $r_{ijk}^{\tau,n+1}$.

3.3 Modification to Obtain User Equilibrium Solution

As previously discussed, the solution to the time-dependent UE problem is obtained by assigning vehicles to the shortest average travel time paths instead of the least marginal paths in the direction finding step (step 5). The algorithmic steps for UE assignment are virtually identical to those for the SO solution except for the specification of the appropriate arc costs and the resulting path processing component of the methodology. Specifically, use the (time-dependent) average travel times on links instead of the marginal travel times in the shortest path calculations. In the above solution procedure, this simplifies step 3 and modifies step 4 accordingly.

4. Time-Dependent Multiple User Classes Algorithm

This problem extends the previous single class formulations by recognizing that only a fraction of all users in a network are likely to be equipped, and that users may differ in terms of information received and/or response behavior. The central controller now seeks to optimize overall network performance through the provision of real-time routing information to equipped motorists, taking into account different user classes in terms of information availability, information supply strategy, and driver response behavior. The problem statement is discussed in section 4.1, followed by the solution algorithm.

4.1 Problem Statement

The problem assumes that the TMC has complete a priori information in the form of time-dependent O-D desires for users in each class and seeks a time-dependent traffic assignment that provides the number of vehicles of each class on the network links and paths satisfying system-wide objectives as well as the conditions corresponding to the behavioral characteristics of each user class. This problem therefore addresses the normative DTA capability discussed in Section 2, coupled with a descriptive capability to represent actual user behavior and system dynamics in conjunction with route guidance generation.

As before, consider the traffic network represented by a directed graph $G(N, A)$. The analysis period of interest, taken here as the peak period, is discretized into small equal intervals $t = 1, \ldots, T$. Given a set of time-dependent O-D vehicle trip desires for the entire duration of the peak period, expressed as the number of vehicle trips $r_{ij}^{\tau u}$ of user class u leaving node i for node j in time slice $\tau, \forall i, j \in N, \tau = 1, \ldots, T$, and $u = 1, \ldots, U$, determine a time-dependent assignment of vehicles to network paths and corresponding arcs. In other words, find the number of vehicles $r_{ijk(u)}^{\tau u}$ of user class u that follow path $k(u) = 1, \ldots, K_{ij}^{u}$ between i and j beginning at time $\tau, \forall i, j \in$

$N, \tau = 1, \ldots, T$, and $u = 1, \ldots, U$, as well as the associated numbers of vehicles on each arc $a \in A$ over time. Here, $u = 1, \ldots, 4$ corresponds to the following four user classes:

Class 1: equipped drivers who follow prescribed system-optimal (SO) paths. The solution will assign these users to paths that impose the least marginal cost (time) on the system from the origin to the destination.

Class 2: equipped drivers who follow user optimum routes. The solution will assign these users to paths that minimize their own average cost (time) from the origin to the destination, so that no member of this class could improve his/her travel time by unilaterally changing paths.

Class 3: equipped drivers who follow a boundedly-rational switching rule in response to descriptive information on prevailing conditions. This emulates the behavior of users who receive real-time information similar to that offered by AUTOGUIDE namely best paths based on link travel times that may not recognize future conditions (that would prevail at the time of actual traversal). The behavioral rules applicable to this class are described in section 2.2.4.

Class 4: non-equipped drivers who follow externally specified paths, such that $r_{ijk(4)}^{\tau 4}$ are known for all i, j, k and τ.

A mathematical statement of the problem is presented in Mahmassani et al. [9] and Peeta and Mahmassani [19].

4.2 Solution Algorithm

Figure 4.1 illustrates the simulation-based solution algorithm for the above problem, obtained by extending the corresponding single class procedure. It consists of an inner loop that incorporates a direction finding mechanism for the SO and UE classes based on the simulation results of the current iteration, namely the experienced trip times and associated marginal trip times. Convergence is sought by obtaining search directions for the SO (user class 1) and UE (user class 2) classes. User class 3 (BR), which follows behavioral rules in response to descriptive current traffic information is not directly involved in the search process. The paths of these users are obtained based on the traffic pattern that evolves in the network for the current path assignment (and the behavioral rules assumed), unlike classes 1 and 2 which obtain their paths based on search directions from the experience of previous iterations. Hence, from an algorithmic standpoint there is no direct guiding mechanism involved in obtaining the paths for user class 3, other than their being predicated on the assignment strategy for the SO and UE class vehicles. As illustrated in the figure, they form the outer loop of the iterative procedure. The unequipped users (class 4 or PS) are exogenous to the search and represent constant background information (for each iteration) as their paths remain unchanged.

The DYNASMART simulator captures the interactions taking place among the four user classes in the traffic network. It allows evaluation of the resulting network flow patterns and associated system performance for a given solution to the multi-class assignment problem (i.e. for a given set of path assignments); it provides the basis for extracting the information that guides the search process of the algorithm, and actually determines the assignment solution for user class 3 internally to the simulation through the embedded behavioral rules for these users. Further detail can be found in Mahmassani et al. [9] and Peeta and Mahmassani [18], [19].

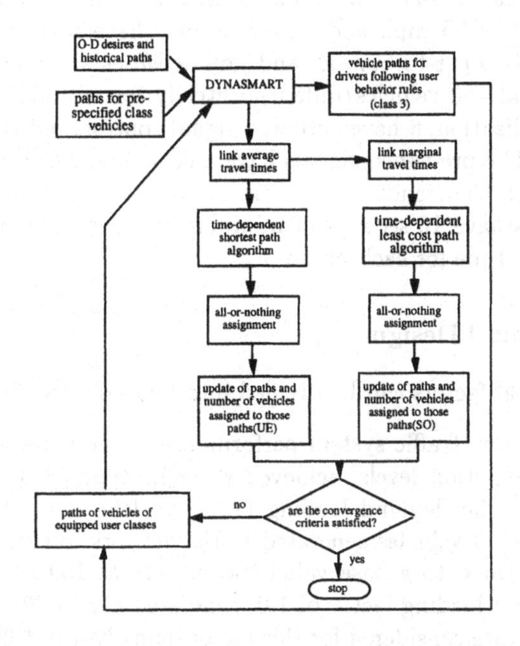

Fig. 4.1. Multiple User Classes Solution Algorithm

5. Illustrative Results

The purpose of the numerical experiments is to illustrate both descriptive and normative capabilities of the above modeling framework, and highlight some important substantive findings regarding the provision of real-time information to users as a traffic management technique. Specifically, we compare system performance under the different assignment rules and information strategies discussed previously. We first describe the structure and traffic characteristics of the test network, followed by an overview of the experimental design and related assumptions. The simulation results are discussed in section 5.3.

5.1 Description of the Test Network

Figure 5.1 depicts the test network, which includes 50 nodes and 168 links and consists of a freeway surrounded by a street grid. There are 10 demand zones with 32 origins and 10 destinations. The time-varying origin-destination matrix is given in terms of number of vehicles for each 5-minute (assignment) interval from zone to zone. All arcs are directed and have two lanes except the entrance and exit ramps that connect the street network to the freeway, which have one lane. Each link is 0.25 miles long. The freeway links have a mean free speed of 55 mph and all other links have a 30 mph free speed. The maximum bumper-to-bumper and jam densities are assumed to be 260 vehicles/mile and 160 vehicles/mile respectively for all links; 26 nodes have pre-timed signalization, 8 have actuated signal control, and the rest have no signal control. The pre-timed signals have a 60 second cycle length with two phases, each with 26 seconds of green time and 4 seconds of amber time. The actuated signals have 10 seconds of minimum green time and 26 seconds of maximum green time for each phase.

5.2 Experimental Design

The experimental factors can be divided into two primary categories:

1. Demand levels: Traffic system performance is compared under different network congestion levels, achieved through appropriate demand loading factors. The demand loading factor is defined as the ratio of the total number of vehicles generated in the network during the simulation period compared to a base value (taken here as 15,032 vehicles, corresponding to a loading factor of 1.0, generated over a 30-minute period). Three levels are considered for this factor, namely, 1.0, 1.25, and 1.5, corresponding to low, medium, and high network congestion, respectively. The temporal loading profile exhibits a typical peaking pattern associated with the morning commute. In all the loading patterns, the first five minutes is taken as start-up time, as vehicles generated in this period are not "tagged" for statistical accumulation. All other vehicles are tagged vehicles and contribute to the calculated performance measures.
2. Information availability: Vehicles are first categorized into two information availability groups: equipped and non-equipped. Equipped vehicles can be further differentiated into three classes: System Optimal (SO) class, UE class, and BR class (classes 1, 2, and 3, respectively, in the previous section). The non-equipped vehicles constitute class 4, as they follow pre-specified paths from the standpoint of the methodology. For

users in class 3 (BR), the path switching rule followed is that described in section 2.2.4. The parameters used for these rules are assumed to be fixed throughout the experiments. The mean relative indifference band (η) has a value of 0.2; the absolute minimum threshold bound is assumed to be one minute for all users. The quantity η_j is treated as a random variable and is assumed to follow a triangular distribution, with mean η and a range $\eta/2$. Users in class 4 are assumed to follow the initial paths prescribed before entering the network. As a result, their paths are assumed fixed during the assignment procedure. In this case, class 4 users were assigned to their respective current best path in the initial step of the procedure (equivalent to receiving pre-trip descriptive information at the origin, e.g. via computer or telephone). The assumptions about assignment rules and user class mix for the numerical experiments are summarized in Table 5.1.

It can be noted that cases I through III require a single simulation of the peak-period using a single run of DYNASMART, whereas cases IV through VI involve the application of the iterative algorithms described in sections 3 and 4. The algorithm was applied until convergence was reached, for each demand level under these three cases.

5.3 Discussion of Results

Overall system performance under each of the scenarios considered is captured by the average travel time (over all tripmakers), taken from the time of departure to the time of arrival at the destination of each trip. Other measures are typically used in such evaluations, both at the overall network level as well as for specific network components; these can be displayed through the graphical user interface.

The average travel time is shown graphically in Figure 5.2 for all six cases listed in Table 5.1, for the three loading levels considered. In case IV, all users are assumed to follow route guidance instructions intended to optimize overall system performance. As such, that case should result in the lowest total travel time or cost in the network. For clarity of presentation, the average trip times are shown as a percentage of that minimum, attained under case IV (SO) under the lowest congestion level (loading factor of 1.0); the value of 100% therefore corresponds to the SO trip time in the base case.

As expected from basic traffic theory, travel time increases (and the corresponding average speed decreases) as the overall demand, in the form of the loading factor, increases. This pattern is revealed consistently in Figure 5.2, for each of the six cases considered. As noted previously, case IV (the SO case) provides an absolute lower bound on the performance of the given network for any given loading pattern. The difference in overall travel time between the all-SO and UE cases (IV and V) is initially small at the lowest demand, but increases considerably for the 1.25 and 1.5 loading factors, confirming previous results (Mahmassani and Peeta [14]).

The results shown here have important implications for the provision of real-time information to motorists in congested traffic networks. First, it can be noted that cases I, II and III generally perform much worse than cases IV, V and VI. The main difference is that the former correspond to uncoordinated strategies, where users make decisions independently in response to the same currently prevailing information, whereas the latter corresponds to coordinated strategies, where at least a fraction of the users are acting according to instructions that recognize the interactions taking place in the network among all users. In fact, the latter were obtained as the convergence of an iterative procedure (described in section 4) seeking consistency between user actions and network conditions, to achieve certain systemwide conditions. The conclusion from this is that providing all users with current information and letting them make independent decisions may be inefficient for the system. Within these uncoordinated strategies, we can further see that, especially at the higher loading levels, some improvement in travel time is obtained by allowing users to switch en-route (cases II and III, vs. case I where users select their path at the origin and cannot switch).

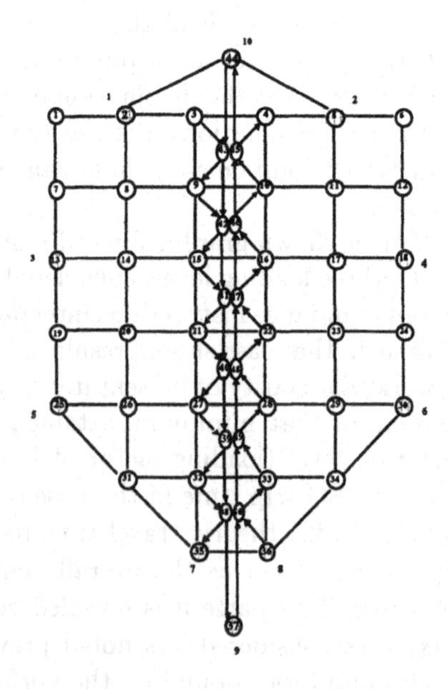

Fig. 5.1. Network Structure

Among the coordinated strategies, SO is of course always best, by definition. What is notable, however, is the extent to which it outperforms the UE case (V), as the congestion level increases in the network (at loading factors 1.25 and 1.5). Under low demand, UE and SO are close; however the difference rapidly increases, and reaches about 20% under high demand, again consistently with previous results (Mahmassani and Peeta [14]). In other words, there are significant benefits to be attained from a systemwide standpoint by seeking to route drivers optimally in the network; the state that would result from all users seeking to minimize their own trip time, even if they could reach an equilibrium in the system, would still be vastly inferior to the system optimum.

Table 5.1. Experimental Setup

CASE	SO Class	UE Class	BR Class	PS Class
I (PS)	0%	0%	0%	100%
II (BR)	0%	0%	100%[1]	0%
III (BR)	0%	0%	100%[2]	0%
IV (SO)	100%	0%	0%	0%
V (UE)	0%	100%	0%	0%
VI (MUC)	25%	25%	25%	25%

[1] In this case, myopic switching is assumed for all vehicles.
[2] In this case, the boundedly-rational en-route switching rule with a bound of one minute and a threshold of 0.2 is used.

The final implication of the results is the most powerful, and most significant from a practical standpoint. It is obtained by examining the MUC results relative to the SO and UE cases. Specifically, we see that the MUC results, with only 25% of the users following SO guidance, in a state that is much closer to the 100% SO case than to the UE case. When one recognizes that it is unlikely to ever have 100% compliance with any type of instructions, it is very significant to see that the optimum is approached with only a fraction of the users (in this case only 25%) of the users following SO route guidance.

6. Real-Time DTA Strategies and Implementation

The primary strategy considered for the implementation of dynamic assignment capabilities in real-time is a rolling horizon framework, described in section 6.1. It exemplifies a centralized predictive approach to DTA. In contrast, we also present, in section 6.2, a decentralized reactive approach, which relies on heuristic rules used by a spatially distributed system of local controllers, with varying degrees of inter-communication, operating on limited

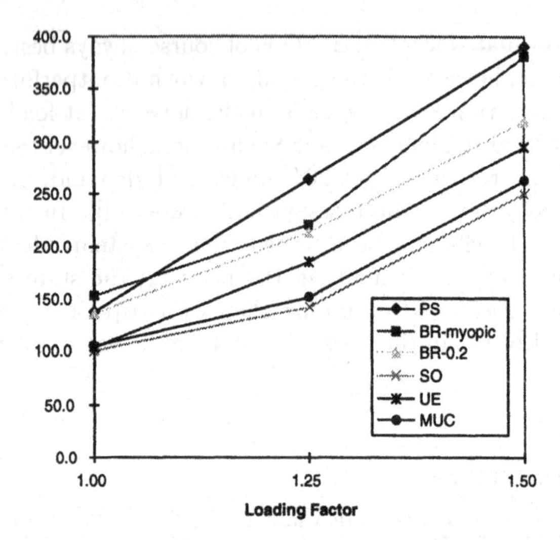

Fig. 5.2. Average Travel Time (expressed as a percentage of the SO base case under loading factor of 1.0) vs. Overall Demand (loading factor) Under Different Assignment Rules and Information Strategies

local information obtained from sensors to provide route guidance to vehicles in the network.

6.1 Rolling Horizon (RH) Approach for Real-Time Implementation

The principal mechanism proposed for implementing the above DTA capabilities in real-time is the rolling horizon approach, used previously for production-inventory control (Wagner [20]), and in transportation systems for on-line demand-responsive traffic signal control (Gartner [3], [4]). The underlying philosophy behind the RH approach is that current events will not be influenced by events "far" into the future, i.e. that assignment of current vehicles may be performed with only limited consideration of vehicles to be assigned "far" into the future, as currently assigned vehicles would probably be out of the system by that time. The stage length h in Figure 6.1 depicts the time horizon considered when making current assignments (its value in actual problems is network specific). For a given stage, the problem encountered is analogous to the complete information availability scenario, albeit, only for the duration h of the stage. The system is solved for optimality only for this duration, and O-D desires for the roll period are assigned to the paths determined. The path assignments in each stage are determined for the entire stage, but implemented only for the roll period. The time frame is now "rolled" forward by the roll period, and the above process is repeated till the end of the duration of system operation, possibly on a continuing basis. Hence, a series of optimizations are performed in quasi real-time. The estimation and prediction functions of the DTA system shown in Figure 1.1 need to be exercised, to produce O-D desires for the entire stage. The O-D desires beyond the stage length h are assumed to be zero. From a simulation standpoint, it is necessary to ensure proper initial conditions as one advances from one stage to the next.

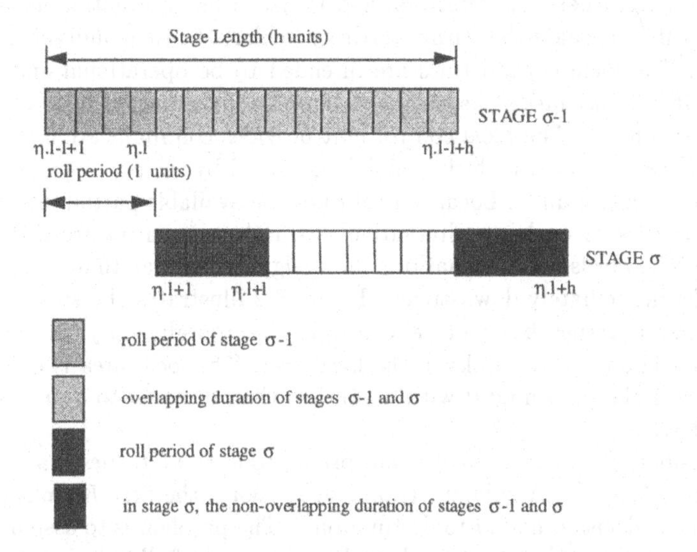

Fig. 6.1. Rolling Horizon Implementation Framework

The selection of the values of l and h represents a compromise that reflects quality of OD prediction, computational requirements, rate of change in the traffic system, and solution quality. Simulation experiments have suggested roll periods of the order of 10 to 15 minutes, with stages of about 20 to 30 minutes in duration for DTA implementation (Mahmassani et al. [10]).

In addition to the simulation-assignment model and normative route guidance modules to generate a solution for the next roll period, the support routines generating the OD predictions and performing consistency checking and resetting must be executing, possibly several times per roll period, in a real-time implementation of the overall DYNASMART-X DTA system shown in Figure 1.1. The time frame for executing the suite of modules involved in generating a solution (route guidance/state prediction) is referred to as an execution cycle.

6.2 Decentralized Reactive Route Control Using Local Rules

In contrast to the above centralized control architecture, with its heavy requirements in terms of input information which may be available only with varying degrees of accuracy and confidence, algorithmic complexity and computational intensiveness, hierarchical distributed architectures provide for 'locally-oriented' real-time strategies for vehicle routing that rely on limited available information. The decentralized approach envisions a set of local controllers scattered or distributed in the network, where every controller can only extract limited "raw" information (speed, travel time, concentration, etc.) from network detectors, and utilizes this information using local control rules to guide the within-territory vehicles to their individual destinations. The local control rules are intended to be operational and robust under different scenarios of spatial and temporal information availability in a real-time context. The local control unit provides communication to drivers in a territory the size of which is mainly governed by the processing capabilities of the control units. Local control rules use available partial information and heuristics to evaluate alternative sub paths emanating from the decision node towards the destination, and assign vehicles at that node among the links immediately downstream. Figure 6.2 illustrates the spatial extent of the area governed by one local controller. Temporally, only current travel times are known for all links in the local area. The local area is defined by the set of links (and nodes) with depth less than or equal to a pre-specified number K.

Consider a vehicle v going from origin node $O(v)$ to destination node $D(v)$, $v = 1, \ldots, V$. A subpath (i, j, m, K) denotes the first K links of path m from the decision node i to destination j. The problem is to assign vehicle v to an outgoing link $a \in B(i)$, where $B(i)$ is the set of all links incident from I; such decisions are made repeatedly upon reaching the next decision node, until v reaches $D(v)$. Assignment decisions are reached by control units after considering the relative merit or disutility of alternative subpaths, as captured by local and non-local state variables. The latter describe the expected state beyond the knowledge level K. The logic of the proposed local control is analogous to that of the A^* graph search algorithm that uses heuristic information to select the next node to be scanned. The A^* algorithm relies on an evaluation function, F, which has two components: the cost of reaching the node from the start node, G, and the cost of reaching the goal from the node, H. The node chosen for expansion is the one for which $F = G + H$ is minimum (Pearl [17]).

Let $P(O(v), D(v))$ denote the path actually followed by vehicle v over its trajectory, and $TT(O(v), D(v))$ the corresponding trip time. In making the nodal assignments of individual vehicles to an emanating link, the desired underlying objective is to minimize $TT(O(v), D(v))$, where the summation is taken over all vehicles, $v = 1, \ldots, V$, seeking to depart from anywhere in the network over the duration of a given planning horizon of interest. Of course, the kind of local heuristic assignment rules considered here will not in general achieve the underlying global objective. The interest is to explore the performance of such rules, for different functional specifications and parameter values, relative to what might be achievable by more global algorithms, recognizing the imperfect knowledge (e.g. of future O-D demands) under which the latter might operate.

A penalty function $G(l)$, specified as a function of various local state variables, is defined to capture the performance over the subpath l (consisting of K links) under currently prevailing conditions. The desirability of the remaining portion of the path, from the end of the K-th link to $D(v)$, is estimated according to a heuristic function H(l). Thus, $[G(l) + H(l)]$ provides the basis for evaluating alternative subpaths from the current node to the destination. The specification of the heuristic function may reflect varying degrees of "knowledge" or "intelligence", with varying corresponding effort in terms of computation, data acquisition, data processing and/or prediction.

A family of rules have been developed on the basis of the criteria by which subpaths are evaluated and the assignment process is performed (Hawas [5]). We highlight here the specification of one of these rules which distributes vehicles among several subpaths using a splitting model operationalized using the logit form. A generalized subpath disutility or penalty function is developed. It comprises local state variables (travel-time and concentration), and non-local variables (expected travel time). The proposed penalty function can be specified as the approximate current marginal travel time along the subpath. The marginal time the system incurs by adding one vehicle to a subpath may be approximated by the vehicle's own travel time, from i to j, plus the marginal effect on all other subpath vehicles (assuming that the vehicle affects the subpath vehicles only). This can be expressed as:

$$G_{ij}^{t}(l) = T_{ij}^{Lt}(l) + M.K_{ij}^{Lt}(l).S_{ij}^{Lt}(l)$$

The state variable, $T_{ij}^{Lt}(l)$, refers to the travel time of subpath l at time t, where (L) superscript is used to indicate that this variable is local (can be actually measured). The term $(K_{ij}^{Lt}(l).S_{ij}^{Lt}(l))$ expresses the total number of vehicles along subpath l. The coefficient M is the average marginal effect of the added vehicle at i on any of the $(K_{ij}^{Lt}(l).S_{ij}^{Lt}(l))$ vehicles; M is expected to decrease with higher knowledge levels to account for the diminishing marginal effect on the subpath vehicles as they move further away from the decision point. The heuristic function can be specified as:

$$H_{ij}^{t}(l) = AT_{ij}^{NLt}(l)$$

The state variable $AT_{ij}^{NLt}(l)$ is an approximation of the anticipated non-local travel time from the end of subpath to destination j. It can be calculated by extrapolating the local travel time, historical information, or it may be replaced by corresponding information exchanged from adjacent controllers. The superscript (NL) indicates that this variable is a non-local variable (and cannot be measured directly according to the problem assumptions). This variable is calculated using local speed estimates and heuristics (Hawas [5]).

Denote by F^t_{ij*} the disutility of the most promising subpath l^*, which is the minimum value of $F^t_{ij}(l) = G^t_{ij}(l) + H^t_{ij}(l)$ for all feasible subpaths. The term "feasible subpaths" refers to the set of subpaths of apparent acceptable performance (Hawas [5]). The share of any feasible subpath l is inversely proportional to the penalty value, $F^t_{ij}(l)$, and is given by:

$$p^t_{ij}(l) = \frac{EXP[\theta(F^t_{ij}(l) - F^t_{ij*})]}{\sum_f EXP[\theta(F^t_{ij}(f) - F^t_{ij*})]}$$

The above equation allocates vehicles to feasible subpath l based on the difference between $F^t_{ij}(l)$ and F^t_{ij*}. The θ parameter, or dispersion factor, affects the share of each subpath by allocating higher flows to subpaths of lesser disutility.

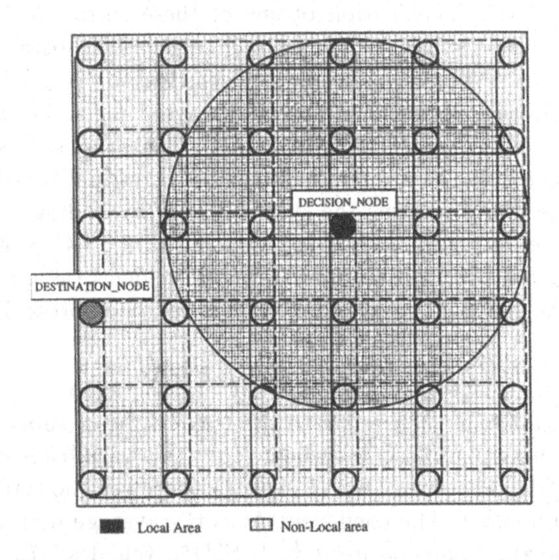

Fig. 6.2. Local and non-local areas in a sample network

6.3 Comparative Experiments

A comparative analysis of network performance is conducted under the predictive centralized RH and the decentralized reactive strategies (Mahmassani and Hawas [6]). The test network configuration is identical to the network of Figure 5.1. The spatial distribution of O-D pattern is approximately uniform. The experiments were conducted using a total of about 21500 vehicles generated over a period of 30 minutes.

The effect of the stage and roll lengths in the rolling horizon implementation described in Section 6.1 on overall performance is summarized in Table 2. All the results are presented as percentages of the benchmark performance (total travel time under SO with full a priori knowledge of the time varying OD demand). Note that in these experiments, no rerouting of vehicles is permitted once they are assigned to their respective paths. Under these conditions, a degradation of performance relative to the full-information SO solution appears to be of the order of 30%. The performance of the rolling horizon procedure relative to the benchmark scenario is better under longer roll periods and stage lengths. In addition, the marginal improvement in performance (relative to hindsight) appears to increase with the increase in duration of the roll period and stage length. Note however that increasing the roll period but keeping the stage length constant, does not necessarily lead to improvement in performance.

Table 6.1. Rolling Horizon Procedure Results

Roll (min)	Stage (min)	Total Travel Time as % of the Benchmark (Full-Information SO)Travel Time
5	20	127%
5	15	139%
10	20	128%

Experiments with the reactive decentralized strategy are conducted using different operating scenarios with various loading factors and knowledge levels (Hawas [5]). Table 6.2 shows the distributed system performance results, which indicate that the higher the knowledge level (up to a specific level), the better the performance. This could be attributed to the greater accuracy of $F_{ij}^t(l)$. Results of the simulation experiments indicated that overall performance is not particularly sensitive to non-local travel time at high knowledge levels. Current concentration measurements are highly beneficial in all but low loading situations. Under low loading conditions, while travel time information is highly significant, concentration has a slight effect on overall performance. At such low concentration levels, the system might incur higher marginal cost if a subpath other than the one of minimal travel time is used. Under higher loading conditions, it is favorable to assign vehicles to less congested subpaths that may not be of minimum travel time.

Most notable in the results of Table 6.2, compared to those in Table 6.1, is that the decentralized strategy, based on rather simple rules that react to current measurements, appears to be quite competitive in terms of overall performance with the predictive centralized RH approach, under the scenarios considered in these experiments. Naturally, the latter are too limited to offer a basis for generalization; however, they reveal a rather promising approach that offers considerable simplicity of implementation.

The simulation-assignment model was also utilized to assess the relative robustness of the distributed scheme and the centralized approach in dealing with incident traffic conditions. Different incident scenarios were considered in terms of incident location, duration, and severity (Hawas and Mahmassani [6]). These experiments indicated that the distributed scheme is more robust under incident conditions due to its greater ability to rapidly respond to changes in network conditions.

Table 6.2. Decentralized System Performance

Knowledge Level (Depth)	Total Travel Time as % of the Benchmark (Full-Information SO) Travel Time
K=1	142%
K=3	131%
K=5	129%

7. Concluding Comments

Dynamic traffic assignment (DTA) is an essential capability for the operation of intelligent transportation systems, and the successful deployment of telecommunications and information technologies for traffic network management and control. This paper has described several DTA problem formulations that correspond to different functional capabilities that arise in conjunction with on-line operation of ATMS/ATIS, as well as off-line assessment of operational measures and information supply strategies. These capabilities have been integrated in a DTA system for real-time operation that also incorporates procedures for ensuring consistency between model values and actual measurements from sensors, as well as various state estimation and prediction capabilities.

The representation of the dynamics of traffic flow and user decisions (in response to information and control actions) in networks is of critical importance in conjunction with the analysis and operation of intelligent transportation systems. Proper description of flow propagation in networks, including queueing at junctions, under time-varying loading patterns, is essential; yet no satisfactory analytic representation that satisfies the laws of physics and traffic science, while at the same time yielding a mathematically tractable and well-behaved mathematical formulation, is at present available, and most attempts have to date been unsuccessful. For this reason, computer simulation of tripmaker decisions and traffic processes plays a central role in the solution methodologies for the various problem formulations discussed in this paper. Simulation allows representation of a complex array of entities and description of their interaction in a large-scale traffic network. The DYNASMART simulation-assignment logic is noteworthy in that it combines a microscopic level of representation of individual tripmakers and drivers, with a macroscopic description of some of the interactions taking place in the traffic stream. This allows the attainment of robust solutions with acceptable accuracy at a fraction of the computational cost that would have been required with a completely microscopic representation of traffic maneuvers.

From a substantive standpoint, simulation experiments performed to evaluate system performance have provided important insights into the effectiveness of intelligent transportation system technologies and operational concepts. For example, the results presented in Section 5 strongly suggest that benefits attainable through coordinated real-time route guidance strategies that seek a system optimum are quite robust vis-à-vis driver compliance with the supplied instructions, as a substantial portion of the potential benefits may be attained with only a moderate fraction of complying users. Also notable are the results obtained with the real-time decentralized reactive strategies, compared with the considerably more elaborate centralized rolling horizon solution approach with its heavy dependence on O-D trip information.

The deployment of intelligent transportation systems will continue to give rise to increasingly challenging problems that require careful methodological development. It would be a misconception to assume that these problems and their solution approaches represent a minor incremental modification of existing methodologies, developed primarily for static equilibrium conditions. Significant challenges and opportunities exist for developments ranging from the representation of the dynamic traffic and behavioral processes, to the formulation of the assignment and route control problems jointly with other control dimensions, such as signalization and road pricing, in addition to the development of efficient solution of these problems both off-line, for evaluation purposes, as well as on-line for real-time control and operation purposes. The latter will undoubtedly proivide ample opportunities for exciting and significant developments in the basic and fundamental aspects of these problems.

Finally, it is important to recognize that many of the issues encountered in the real-time execution of simulation-based frameworks for large scale complex systems evaluation and control are not limited to the intelligent transportation systems arena. These are also encountered in a growing class of telecommunications and information-driven systems across a wide spectrum of service sectors that call for real-time decision-making and resource allocation under real-time information (such as finance, retailing, logistics and distribution). As such, it is desirable to seek unifying frameworks and strategic constructs that cross disciplinary and domain boundaries to address common fundamental issues.

Acknowledgments

The author wishes to acknowledge the contribution of current and former members of his research team at the University of Texas at Austin. All work described in the paper is joint work, and draws liberally from papers and reports co-authored with several colleagues and former students, including Drs G.-L. Chang, Y. Hawas, T.-Y. Hu, R. Jayakrishnan, S. Peeta, and A. Ziliaskopoulos. The author is particularly indebted to Dr. Yaser Hawas for his help in preparing Section 6 of this paper. Dr Russ Taylor's contribution to the software engineering aspects of DYNASMART-X is gratefully acknowledged. This paper is based in part on work supported by the U.S. Federal Highway Administration and administered through Martin Lockheed's Oak Ridge National Laboratory. Additional support was provided by the U.S. Department of Transportation through the Southwest Region University Transportation Center. Of course, the author is solely responsible for the contents of this paper.

References

1. L.M. Branscomb, J.H. and Keller. Converging Infrastructures: Intelligent Transportation and the National Information Infrastructure. MIT Press, Cambridge, MA, 1996.
2. I. Catling (Ed.) Advanced Technology for Road Transport: IVHS and ATT. Artech House, Boston, 1994.
3. N.H. Gartner. OPAC: A Demand-Responsive Strategy for Traffic Signal Control. *Transportation Research Record*, 906:75–81, 1982.
4. N.H. Gartner. Simulation Study of OPAC: A Demand-Responsive Strategy for Traffic Signal Control. Gartner and Wilson (eds.) Transportation and Traffic Theory, Elsevier Science Publishing Company, 1983.
5. Y. Hawas. A Decentralized Approach and Local Search Procedures for Real-Time Route Guidance in Congested Vehicular Traffic Networks. Ph.D. Dissertation, Department of Civil Engineering, The University of Texas at Austin, 1996.
6. Y. Hawas and H.S. Mahmassani. Real-Time Dynamic Traffic Assignment For Route Guidance: Comparison Of Global Predictive Vs. Local Reactive Strategies Under Stochastic Demands. *Proceedings of 8th IFAC Symposium on Transportation Systems*, Chania, Greece, pp. 557-562, 1997.

7. R. Jayakrishnan, H.S. Mahmassani, and T.Y. Hu. An Evaluation Tool for Advanced Traffic Information and Management Systems in Urban Networks. *Transportation Research C*, 2C(3):129-147, 1994.
8. H.S. Mahmassani, T.Y. Hu and S. Peeta. Microsimulation Based Procedures for Dynamic Network Traffic Assignment. Proceedings of the 22nd European Transport Forum (The PTRC Summer Annual Meeting), Nottingham, UK, 1994.
9. H.S. Mahmassani, T.Y. Hu, S. Peeta and A. Ziliaskopoulos. Development and testing of Dynamic Traffic Assignment and Simulation Procedures for ATIS/ATMS Applications. Technical Report DTFH61-90-R00074-FG, Center for Transportation Research, The University of Texas at Austin, 1994.
10. H.S. Mahmassani, Y. Hawas, H.H. Lin, A. Abdelfatah, E. Miller and T.Y. Hu. Sensitivity Tests and Guidelines for the System Optimal Dynamic Traffic Assignment Procedure for ATIS/ATMS. Technical Report DTFH61-90-C-00074-BX, Center for Transportation Research, University of Texas, Austin, TX, 1994.
11. H.S. Mahmassani and R. Jayakrishnan. Dynamic Simulation-Assignment Methodology to Evaluate In-vehicle Information Strategies in Urban Traffic Networks. Proceedings of Winter Simulation Conference WSC'90, New Orleans, LA, pp.763-769, 1990.
12. H.S. Mahmassani and R. Jayakrishnan. System Performance and User Response under Real-Time Information in a Congested Traffic Corridor. *Transportation Research A*, 25A(5):293-307, 1991.
13. H.S. Mahmassani and Y.H. Liu, Y.-H. Models of User Pre-trip and En-route Switching Decisions in Response to Real-time Information. Proceedings of the Eighth IFAC/IFIP/IFORS Symposium on Transportation Systems, Chania, Crete, pp. 1427-1432, 1997.
14. H.S. Mahmassani and S. Peeta. Network Performance under System Optimal and User Equilibrium Dynamic Assignments: Implications for ATIS. *Transportation Research Record*, 1408:83-93, 1993.
15. H.S. Mahmassani and S. Peeta. System Optimal Dynamic Assignment for Electronic Route Guidance in a Congested Traffic Network. Gartner and Improta (eds.), Urban Traffic Networks: Dynamic Flow Modelling and Control, Springer-Verlag, Berlin, pp. 2-27, 1995.
16. H.S. Mahmassani and D.G. Stephan, D.G. Experimental Investigation of Route and Departure Time Dynamics of Urban Commuters. *Transportation Research Record*, 1203:69-84, 1988.
17. J. Pearl. Heuristics: Intelligent Search Strategies for Computer Problem Solving. Addison-Wesley, Reading, MA, 1984.
18. S. Peeta and H.S. Mahmassani. System Optimal and User Equilibrium Time-Dependent Traffic Assignment in Congested Networks. *Annals of Operations Research*, 60:81-113, 1995.
19. S. Peeta, H.S. Mahmassani. Multiple user Classes Real-time Traffic Assignment for On-line operations: A Rolling Horizon Solution Framework. *Transportation Research C*, 3C(2):83-98, 1995.
20. H.M. Wagner. Principles of Operations Research. Prentice-Hall, Englewood Cliffs, NJ, 1977.
21. R. Whelan. Smart Highways, Smart Cars. Artech House, Boston, 1995.
22. A. Ziliaskopoulos, H.S. Mahmassani. A Time-Dependent Shortest Path Algorithm for Real-Time Intelligent Vehicle/Highway Systems Applications. *Transportation Research Record*, 1408:94-100, 1993.

Traffic Modeling and Variational Inequalities Using GAMS

Steven P. Dirkse[1] and Michael C. Ferris[2]

[1] GAMS Development Corporation,
1217 Potomac Street NW,
Washington, D.C. 20007, USA

[2] University of Wisconsin – Madison,
1210 West Dayton Street, Madison,
Wisconsin 53706, USA.

Summary. We describe how several traffic assignment and design problems can be formulated within the GAMS modeling language using newly developed modeling and interface tools. The fundamental problem is user equilibrium, where multiple drivers compete noncooperatively for the resources of the traffic network. A description of how these models can be written as complementarity problems, variational inequalities, mathematical programs with equilibrium constraints, or stochastic linear programs is given. At least one general purpose solution technique for each model format is briefly outlined. Some observations relating to particular model solutions are drawn.

1. Introduction

Models that postulate ways to assign traffic within a transportation network for a given demand have been used in planning and analysis for many years, see [41] and references therein. A popular technique for such assignment is to use the shortest path between the origin and destination points of a given journey. Of course, such a path depends not only on the physical distance between these two points, but also on the mode of transport and the congestion experienced during the trip.

To account for congestion, traffic assignment models use the notion of user equilibrium or Wardropian equilibrium [42]. In this context, the travel time used to define the "distance" between the origin and destination is a function of the length and capacity of each arc of the path, and the total flow on that path. Thus, the fact that many users can travel along an arc will affect the time it takes any particular user to traverse the arc. User equilibrium occurs when all users travel along their shortest path, where distance is measured using the above definition for time. Underlying the model is the notion of noncooperative behavior - everyone is out for themselves. This should be contrasted with the notion of system equilibrium when a traffic controller assigns every vehicle to a particular path to minimize the total distance traveled. Note that these kinds of odels are typically used to predict the steady-state volume of traffic on a network, not to look at the dynamic behavior.

Due to the absence of an overall objective in the user equilibrium, this problem is more easily cast in the context of a variational inequality. In this paper, we describe how to formulate, solve, and extend models for traffic assignment using the notion of a mixed complementarity problem, a specialization of the variational inequality. Many papers have discussed the formulation and solution of user equilibrium problems using complementarity and variational inequality models, as well as the applications of these problems to urban planning; see [18, 19, 21, 30, 40].

The first two sections of the paper show how the user equilibrium problem is cast as a mixed complementarity problem (MCP), formulated within the GAMS modeling language, and solved using the PATH algorithm. Some examples of problems formulated using these tools are described in [6].

There are many cases when a modeler wishes to optimize an objective function subject to the system being in equilibrium. These problems are commonly called Mathematical Programs with Equilibrium Constraints (MPEC's). The recent monograph [29] describes the current state of optimality theory and algorithms related to such problems. In Section 4., we describe the basic structure of an MPEC and give some examples of traffic design problems that can be formulated as such.

Section 5. describes extensions to the GAMS modeling language that allow MPEC's to be formulated within the language. Furthermore, an outline of new tools for large scale implementation is given. These tools allow algorithm developers direct access to relevant function and derivative values via subroutine library calls. It is intended that this suite of routines, MPECLIB, will foster the development of new applications and test problems in the MPEC format. In the ensuing section, we show how these tools are used in an analysis of a tolling problem over a network representing Sioux-Falls. The problem is cast as an MPEC and two algorithms based on an implicit programming approach, namely DFO [5] and the bundle-trust region algorithm [39], are used to demonstrate the ability of these tools and the power of the modeling format.

The final section of the paper treats some modeling issues related to traffic assignment and path choice in networks subject to failure. Here again, we show how recently developed modeling tools are effective for investigating complex issues in transportation design and analysis.

2. MCP models: User equilibrium

The mixed complementarity problem (MCP) is defined in terms of some lower and upper bounds $\ell \in \mathbf{R^n}$ and $u \in \mathbf{R^n}$ satisfying $-\infty \leq \ell < u \leq +\infty$ and a nonlinear function $F: \mathbf{B} \to \mathbf{R^n}$. Here $\mathbf{B}$ represents the box $\mathbf{B} := [\boldsymbol{\ell}, \mathbf{u}] = \{\mathbf{z} \in \mathbf{R^n}: \ell_\mathbf{i} \leq \mathbf{z_i} \leq \mathbf{u_i}\}$. The variables $z \in \mathbf{R^n}$ solve $MCP(F, \ell, u)$ if for some variables $w \in \mathbf{R^n}$ and $v \in \mathbf{R^n}$

$$w_i \geq 0,\ v_i \geq 0,\ \ell_i \leq z_i \leq u_i, \quad i = 1, \ldots, n \tag{2.1}$$

$$F_i(z) = w_i - v_i, \quad i = 1, \ldots, n \tag{2.2}$$

and

$$w_i(z_i - \ell_i) = 0 \text{ and } v_i(u_i - z_i) = 0, \quad i = 1, \ldots, n. \tag{2.3}$$

Note that any solution z of MCP trivially satisfies the following implications

$$z_i = \ell_i \Rightarrow z_i \neq u_i \Rightarrow v_i = 0 \Rightarrow F_i(z) \geq 0 \tag{2.4}$$

$$z_i = u_i \Rightarrow z_i \neq \ell_i \Rightarrow w_i = 0 \Rightarrow F_i(z) \leq 0 \tag{2.5}$$

and

$$\ell_i < z_i < u_i \Rightarrow w_i = v_i = 0 \Rightarrow F_i(z) = 0. \tag{2.6}$$

Several interesting special cases of MCP exist for particular choices of ℓ and u. When $\ell \equiv -\infty$ and $u \equiv +\infty$, it follows from (2.6) that $w = v = 0$ and (2.1) is satisfied for any $z \in \mathbf{R^n}$. Hence, the problem becomes the classical square system of nonlinear equations

$$F(z) = 0. \tag{2.7}$$

Many techniques used to solve MCP are inspired by techniques used for such systems. Another special case is when $\ell \equiv 0$ and $u \equiv +\infty$, whereupon it can be easily seen that the problem (2.1)-(2.3) becomes

$$0 \leq z \perp F(z) \geq 0. \tag{2.8}$$

Here we have introduced the notation "$\perp$" that signifies the two adjacent quantities are orthogonal, that is, in addition to the explicit inequalities $0 \leq z$ and $F(z) \geq 0$ we have

$$z^T F(z) = 0.$$

In effect this enables us to rewrite (2.1) and (2.3) more succinctly as

$$0 \leq z - \ell \perp w \geq 0 \tag{2.9}$$

$$0 \leq u - z \perp v \geq 0. \tag{2.10}$$

Problem (2.8) is commonly called the nonlinear complementarity problem (NCP) and has been the subject of much research over the past three decades. A plethora of applications can be found in [14, 15]; this paper is concerned with applications arising from traffic and transportation management.

The general variational inequality VI(F, C) is defined using an arbitrary convex set $C \subseteq \mathbf{R^n}$ as the following system of inequalities

$$z \in C,\ \langle F(z), y - z\rangle \geq 0 \quad \forall y \in C. \tag{2.11}$$

It can be reformulated as an MCP using a transformation involving multipliers. We consider two cases separately. If the feasible set C is a box, then it is elementary to show that (2.11) and the MCP (2.1)-(2.3) defined by F

and C are completely equivalent, as their solution sets are identical. When C is polyhedral rather than rectangular, (2.11) can be reduced to an MCP by explicitly including the dual variables to the constraints defining C. Thus, given a box $\mathbf{B}$ and a set $X := \{z: Az \leq b\}$, where $A \in \mathbf{R}^{\mathbf{m}\times\mathbf{n}}$, it can be shown that (2.11) with $C = \mathbf{B} \bigcap \mathbf{X}$ is equivalent to $\mathrm{VI}(H, \mathbf{B} \times \mathbf{R}^{\mathbf{m}}_{+})$, where

$$H(z,u) = \begin{bmatrix} F(z) + A^T u \\ -Az + b \end{bmatrix}.$$

When equality constraints are used to define X, the associated dual variables u are free. Two advantages to using the MCP formulation as opposed to the NCP are the explicit treatment of simple bounds on the variables z and the availability of free variables, which enable the explicit representation of equality constraints. This is more efficient than introducing extra variables and equations to deal with bounds and equality constraints.

We now proceed to show how to model the user equilibrium problem as an MCP. In all models used for the analysis of traffic congestion, there is a transportation network given by a set of nodes $\mathcal{N}$ and a set of arcs $\mathcal{A}$. In the equilibrium setting, it is usually assumed that drivers compete noncooperatively for the resources of the network in an attempt to minimize their costs, where the cost of traveling along a given arc $a \in \mathcal{A}$ is a nonlinear function $c_a(f)$ of the total flow vector f with components f_b, $b \in \mathcal{A}$. Let $c(f)$ denote the vector with components $c_a(f)$, $a \in \mathcal{A}$. There are two subsets of $\mathcal{N}$ that represent the set of origin nodes $\mathcal{O}$ and destination nodes $\mathcal{D}$ respectively. The set of origin-destination (O-D) pairs is a given subset $\mathcal{W}$ of $\mathcal{O} \times \mathcal{D}$; associated with each such pair is a travel demand that represents the required flow from the origin node to the destination node.

There are at least two equilibrium techniques used for generating models of traffic congestion on such a network. The first model is based on considering all the paths between the origin-destination pairs, and the second uses a multicommodity formulation, representing each origin or destination node as a different commodity. Both of these formulations use the Wardropian characterizations of equilibria [42], a special case of a Nash equilibrium (see [15, 23]).

2.1 A path based formulation

The given path based formulation follows [20]. For each $w \in \mathcal{W}$, let $\mathcal{P}_w$ represent the set of paths connecting the O-D pair w and $\mathcal{P}$ represent the set of all paths joining all O-D pairs of the network. Let ξ_p denote the flow on path $p \in \mathcal{P}$; let $\gamma_p(\xi)$ be the cost of flow on this path which is a function of the path flow vector ξ. Let Δ be the arc-path incidence matrix with entries

$$\delta_{ap} \equiv \begin{cases} 1 & \text{if path } p \in \mathcal{P} \text{ traverses arc } a \in \mathcal{A} \\ 0 & \text{otherwise.} \end{cases}$$

It is clear that f and ξ are related by

$$f = \Delta \xi.$$

When the path cost $\gamma_p(\xi)$ on each path p is assumed to be the sum of the arc costs on all the arcs traversed by p, that is, if

$$\gamma(\xi) \equiv \Delta^T c(f),$$

the model is called *additive*. Finally, we introduce variables τ_w that depict the minimum transportation cost (or time) between O-D pair $w \in \mathcal{W}$. The travel demand between O-D pair w is assumed to be a function $d_w(\tau)$ of the vector τ in the path formulation. The model is called a *fixed-demand model* if each $d_w(\tau)$ is a constant function; the general model is often called the *elastic demand model.*

The Wardrop equilibrium principle [42] states that each driver will choose the minimum cost path between every origin destination pair and through this process, the paths used will all have equal cost; paths with costs higher than the minimum will have no flow. Mathematically, this principle can be phrased succinctly as

$$0 \le \gamma_p(\xi) - \tau_w \perp \xi_p \ge 0, \quad \forall w \in \mathcal{W}, p \in \mathcal{P}_w. \tag{2.12}$$

The demand is satisfied if

$$\sum_{p \in \mathcal{P}_w} \xi_p \ge d_w(\tau), \quad \forall w \in \mathcal{W},$$

and the equilibrium condition of zero excess demand can be stated as follows,

$$0 \le \sum_{p \in \mathcal{P}_w} \xi_p - d_w(\tau) \perp \tau_w \ge 0, \quad \forall w \in \mathcal{W}. \tag{2.13}$$

Conditions (2.12) and (2.13) clearly define a nonlinear complementarity problem with (ξ, τ) as the variables.

For networks of reasonable size with many O-D pairs, the enumeration of all paths connecting elements of $\mathcal{W}$ is prohibitive. Thus, the above path-flow formulation is not suitable for a generic complementarity code. Nevertheless, there are path-generation schemes [24, 33] that utilize this formulation and generate the paths only if they are needed. The alternative multicommodity formulation to be discussed below completely removes the necessity of enumerating the paths.

2.2 A multicommodity formulation

In this alternative formulation of the traffic equilibrium problem, a commodity is associated with each destination node. For simplicity, we assume that each node in $\mathcal{D}$ is a destination. Let K be the cardinality of $\mathcal{D}$. The variable $x = (x^1, x^2, \ldots, x^K)$ represents the flows of the commodities $1, 2, \ldots, K$ with x_{ij}^k denoting the flow of commodity k on arc $(i, j) \in \mathcal{A}$. The variable $t = (t^1, t^2, \ldots, t^K)$ is composed of components t_i^k that represent the minimum cost (or time) to deliver commodity k (i.e. to reach destination k) from node i. Associated with each pair (k, i), $k \in \mathcal{D}$ and $i \in \mathcal{N}$, is the travel demand d_i^k, which is a function of the minimum cost vector t. There are two sets of equilibrium conditions. The first represents conservation of flow of commodity k at node i and is given by

$$\sum_{j:(i,j)\in\mathcal{A}} x_{ij}^k - \sum_{j:(j,i)\in\mathcal{A}} x_{ij}^k = d_i^k, \qquad \forall i \in \mathcal{N}, k \in \mathcal{D}.$$

In terms of the standard node-arc incidence matrix $\mathcal{I}$ of the network and the demand function $d^k(t)$, these constraints can be rewritten as

$$\mathcal{I}x^k = d^k(t), \qquad \forall k \in \mathcal{D}. \tag{2.14}$$

The second condition ensures that if there is positive flow of commodity k along arc (i, j), then the corresponding time to deliver that commodity is minimized:

$$0 \leq c_{ij}(f) + t_j^k - t_i^k \perp x_{ij}^k \geq 0 \qquad \forall (i, j) \in \mathcal{A},\ k \in \mathcal{D}, \tag{2.15}$$

where the arc flow vector f is given by

$$f \equiv \sum_{k\in\mathcal{D}} x^k.$$

This is typically termed "Wardrop's second principle", although it appears first in his article [42]. It is clear that given an enumeration of the paths, the solution generated from a path based formulation can easily yield a solution to the multicommodity formulation, and vice versa.

Eliminating the flow vector f, conditions (2.14) and (2.15) define an MCP in the arc flow vector x and minimum travel cost vector t. Early studies of the traffic equilibrium problem have focused on the case of a separable cost function and constant demand function; that is, for all $a \in \mathcal{A}$ and $k \in \mathcal{D}$,

$$c_a(f) = c_a(f_a), \text{ and } d^k(t) = d^k \text{ (a constant)}.$$

A specific example of a separable cost function c_a is given by

$$c_a(f_a) = A_a + B_a \left[\frac{f_a}{\gamma_a}\right]^4 \qquad \forall a \in \mathcal{A},$$

for particular data A_a, B_a and γ_a. Cost functions of this kind have been used extensively in transportation research. Nonseparable and nonintegrable cost functions $c(f)$ have also been used in the literature; see for example [21].

It is shown in [13] how to implement the multicommodity formulation as an MCP within the GAMS modeling language. For more discussion of the complementarity approach to traffic equilibrium, see [22].

3. MCP algorithm: PATH

There has been a great deal of research into algorithms for solving MCP and NCP. Some of these recent developments are surveyed in [12]; much of this work has investigated nonsmooth analysis and algorithms based on the systems of equations

$$\min\{z_i,\ F_i(z)\} = 0 \quad i = 1, \ldots, n \tag{3.1}$$

and

$$z_i + F_i(z) - \sqrt{z_i^2 + F_i^2(z)} = 0,\ i = 1, \ldots, n. \tag{3.2}$$

However, we will outline here only the basic ideas behind the PATH algorithm [7, 8] for MCP since this is currently the most widely used code for solving such problems due to the fact that it is available as a GAMS subsystem. This code is intended for large scale applications and uses sparse matrix technology in its implementation. Since our aim is to look at traffic applications which we formulate as nonlinear complementarity problems, we will describe the algorithm only in this context.

Many of the ideas relating to complementarity theory can be thought of as simple generalizations of equation theory. To understand this, we consider the notion of a "normal cone" to a given convex set C at some point $z \in C$, a generalization of the notion of a normal to a smooth surface. This set consists of all vectors c which make an obtuse angle with every feasible direction in C emanating from z, that is

$$N_C(x) := \{c \colon \langle c, y - z\rangle \leq 0, \text{for all } y \in C\}.$$

When $C = \mathbf{R}^n_+$, the nonnegative orthant in $\mathbf{R}^n$, we label this set as $N_+(z)$. A little thought enables one to see that the nonlinear complementarity problem is just the set-valued inclusion,

$$0 \in F(z) + N_+(z),\ z \in \mathbf{R}^n_+.$$

It is possible to look at NCP from this geometric standpoint. Firstly, it is well-known that $N_+(x_+)$ is characterized by the rays $x - x_+$, where $(x_i)_+ := \max\{x_i, 0\}$ is the projection of x onto the nonnegative orthant. An equivalent formulation of (2.8) is therefore to find a zero of the nonsmooth equation

$$0 = F(x_+) + x - x_+$$

The map $F_+ := F(x_+) + x - x_+$ is sometimes referred to as the "normal map", see for example [11, 37, 38]. The equivalence is established by noting that if z solves NCP, then $x = z - F(z)$ is a zero of the normal map, and if x is a zero of the normal map, then $z = x_+$ is a solution of the NCP.

Under the assumption that F is smooth, it is easy to see that the normal map is a smooth map on each of the orthants of $\mathbf{R^n}$ and is continuous on $\mathbf{R^n}$. However, the normal map is not in general differentiable everywhere. It is an example of a piecewise smooth map and is intimately related to a manifold defined by the collection of faces of the set $\mathbf{R^n_+}$, called the normal manifold. It is well known that the collection of these faces precisely determine the pieces of smoothness of the normal map. For example, it was shown in [38] that these pieces are the (full dimensional) polyhedral sets $\mathcal{F} + N_{\mathcal{F}}$ that are indexed by the faces $\mathcal{F}$ of C; here $N_{\mathcal{F}}$ represents the normal cone to the face $\mathcal{F}$ at any point in its relative interior. When the set is $\mathbf{R^n_+}$, the faces are given by $\{x : x \geq 0, x_I = 0\}$, where I runs over the subsets of $\{1, 2, \ldots, n\}$; the pieces of the manifold in this case are precisely the orthants of $\mathbf{R^n}$.

In the context of nonlinear equations, Newton's method proceeds by linearizing the smooth function F. Since F_+ is nondifferentiable, the standard version of Newton's method for NCP approximates F_+ at $x^k \in \mathbf{R^n}$ with the piecewise affine map

$$L_k(x) := F({x^k}_+) + \nabla F({x^k}_+)(x_+ - {x^k}_+) + x - x_+.$$

Thus, the piecewise smooth map F_+ has been approximated by a piecewise affine map. The Newton point x^k_N (a zero of the approximation L_k) is found by generating a path p^k, parametrized by a variable t which starts at 0 and increases to 1 with the properties that $p^k(0) = x^k$ and $p^k(1) = x^k_N$. The values of $p^k(t)$ at intermediate points in the path satisfy

$$L_k(p^k(t)) = (1 - t)F_+(x^k).$$

The path is known to exist locally under fairly standard assumptions and can be generated using standard pivotal techniques to move from one piece of the piecewise affine map to another. Further details can be found in [36].

A Newton method for NCP would accept x^k_N as a new approximation to the solution and re-linearize at this point. However, as is well known even in the literature on smooth systems, this process is unlikely to converge for starting points that are not close to a solution. In a damped Newton method for smooth systems of nonlinear equations (2.7), the merit function $F(x)^T F(x)$ is typically used to restrict the step size and enlarge the domain of convergence. In the PATH algorithm[7], the piecewise linear path p^k is computed and searched using a non-monotone watchdog path-search of the merit function $\|F_+(x)\|^2$. The watchdog technique allows path-searches to be performed infrequently, while the non-monotone technique allows the Newton point to be accepted more frequently. The combination of these techniques helps avoid convergence to minimizers of the merit function that are not zeros of F_+, without detracting from the local convergence rates [7].

The extension of the above approach to MCP is straightforward; furthermore, computational enhancements are described in [9].

4. MPEC models: Tolling and inverse problems

An MPEC consists of two types of variables, namely design variables $x \in \mathbf{R}^n$ and state variables $y \in \mathbf{R}^m$. A function $\theta: \mathbf{R}^{n+m} \to \mathbf{R}$ is the overall objective function to be minimized, subject to two sets of constraints. The first set $(x, y) \in Z$ represents joint feasibility constraints for x and y, with $Z \subseteq \mathbf{R}^{n+m}$ representing a nonempty closed set. The second set of constraints are the equilibrium constraints, defined by the equilibrium function $F: \mathbf{R}^{n+m} \to \mathbf{R}^m$ and the closed convex set $C \subseteq \mathbf{R}^m$. These constraints force the state variables y to solve a VI parametrically defined by $F(x, \cdot)$ and C. The MPEC can be writtens succinctly in the following manner.

$$
\begin{array}{ll}
\text{minimize} & \theta(x, y) \\
\text{subject to} & (x, y) \in Z \\
\text{and} & y \text{ solves } \mathrm{VI}(F(x, \cdot), C).
\end{array}
\tag{4.1}
$$

A special case of the MPEC is the bilevel program in which the mapping $F(x, \cdot)$ is the partial gradient map (with respect to the second argument) of a real-valued C^1 function. In the next two sections, we describe some traffic models that can be formulated as MPEC's and show how to carry out this modeling within GAMS.

One example of an MPEC occurring in traffic network design is described in [31]. The idea is to determine values for some design variables, e.g. arc capacities, that minimize a weighted some of the investment cost and the system operating costs. Another example is described in [4]. This is an example of an inverse problem, where estimates of O-D demands are given and the network planner wishes to adjust these demands minimally in order to satisfy the equilibrium conditions.

A third example is present in tolling. The tolling model is based on an underlying assumption of user equilibrium. As we outlined in Section 2., user equilibrium assumes that each driver uses his shortest path, where distance is measured using a function for time of journey based on the total flows on the arcs.

When a particular arc is tolled, this increases the monetary cost of traversing that arc. We use a simple weighting to add this cost to the distance each user tries to minimize. Of course, complex human behavioral models could be used to generate realistic weightings and potentially add nonlinear cost effects. Our model currently ignores these facets of the problem and simply adds the toll p_a to the cost function c_a for each arc. If we view these tolls as parameters, we have not changed the structure of the user equilibrium problem at all. If we view the tolls as design variables, however, we now have quite a bit of flexibility in designing the tolling structure, and can do so with certain objectives in mind.

What is the objective of adding tolls to some of the arcs of the network? In some cases, it is to maximize system revenue, in which case the upper level objective function for the equilibrium problem defined by (2.14) and (2.15) is as follows:

$$\theta(p, f, t) = \sum_{a \in \mathcal{A}} p_a f_a \tag{4.2}$$

Note that the design variables in this problem are p_a (which are typically nonzero for a small subset of $\mathcal{A}$, and the state variables are f and t. Note that if tolls p_a are set too high, then drivers will use other arcs creating a decrease in f_a and possibly forcing total revenue to decrease.

In other cases, a traffic controller is attempting to impose tolls with the aim of reducing congestion. In these cases, a typical objective function is system cost

$$\theta(p, f, t) = \sum_{a \in \mathcal{A}} c_a(f_a)$$

Other objectives arise in different applications.

The GAMS model that we developed for testing MPECLIB and the solvers that we implemented arose from such a tolling problem. The model used in our example was based on the objective shown in (4.2). The GAMS source of the model is available via anonymous ftp from

ftp://www.cs.wisc.edu/math-prog/mcplib/traffic/.

Apart from the particular data and model equations used, the formulation in GAMS follows the structure we now outline for a simple model using the newly developed MPECLIB.

5. Interfacing models and algorithms: MPECLIB

We have developed a software interface between the GAMS modeling language and MPEC algorithms that allows users to model practical, large-scale MPEC's and algorithm developers to link in their solvers for such problems. We believe such an interface serves two purposes. Firstly, the data from realistic applications is most easily made accessible to researchers developing codes for these problems via interfaces similar to ours. This is an essential ingredient in algorithmic development, testing, and comparison. Secondly, it is only when efficient codes can be applied to real applications, such as tolling problems, and can show an improvement over the existing heuristic schemes that the modeling format of an MPEC becomes a serious alternative to such heuristics.

Unfortunately, developing an interface that easily allows both algorithm developers and application experts to perform their respective research is difficult. We have chosen to allow modelers to use the GAMS [3] modeling language and force algorithmic development to occur in Fortran or C, although a possible extension of this work to support algorithm development in Matlab [17] is possible. Comparable tools that enable algorithmic development for MCP have previously been described in [10], with the result that many more complementarity applications are now being developed.

From a modeling standpoint, the MPEC interface tool is very similar to the MCP interface in GAMS. This allows the application expert to move from an MCP to an MPEC formulation with very little difficulty. The modeling interface to MPEC is similar to the MCP one, consisting of the usual GAMS language features (e.g. sets, parameters, variables, equations, control structures) and extensions to the GAMS `model` and `solve` statements. These extensions are required to allow the modeler to specify the complementarity conditions (i.e. the equilibrium constraints) along with an objective variable and its defining equation. Just as with MCP's, the MPEC `model` statement includes a collection of equation-variable pairs, where each pair defines a complementarity relationship between the function determined by the equation and the variable in the pair. In case equations and variables are indexed by sets, functions and variables with matching indices are paired. In addition, the MPEC `model` may include an unpaired "objective" equation. While MCP must have an equal number of functions and variables, this is not the case with MPEC models. Each function in an MPEC must have a matching variable, but unmatched variables are now allowed; these are the design variables x in (4.1), while the "matched" variables are the state variables y.

The objective variable and the direction of optimization are specified in the GAMS `solve` statement, using the `MPEC` keyword as the model type. In order to define the objective, a scalar "objective" variable and an equation defining this variable are often used. In this case, the objective variable must appear only in the objective equation, and must appear linearly in it. As an example, consider the simple MPEC below (a reformulation of [32], Example 1):

$$\begin{array}{ll} \text{minimize} & \theta(x,y,u) := {x_1}^2 - 2x_1 + {x_2}^2 - 2x_2 + {y_1}^2 + {y_2}^2 \\ \text{subject to} & x_i \in [0,2] \\ \text{and} & (y,u) \text{ solves } \mathrm{MCP}(F(x,\cdot,\cdot),\mathbf{B}), \end{array} \tag{5.1}$$

where

$$F(x,y,u) := \left[\begin{array}{c} -2x_1 + 2y_1 + 2(y_1-1)u_1 \\ -2x_2 + 2y_2 + 2(y_2-1)u_2 \\ -(y_1-1)^2 + .25 \\ -(y_2-1)^2 + .25 \end{array} \right]$$

and $\mathbf{B} := \{(\mathbf{y},\mathbf{u}) \in \mathbf{R}^4 \colon \mathbf{u} \geq 0\}$. The GAMS model for this example is given in Figure 5.1. Readers familiar with GAMS will understand the sets, vari-

```
set I / 1 * 2 /;
alias (I,J);

variables
obj,
x(J)                'design variables',
y(I)                'state variables',
u(I)                'state vars, duals in MCP reformulation of VI';
x.lo(J) = 0;
x.up(J) = 2;
u.lo(I) = 0;

equations
objeq,
Fy(I),
Fu(I);

objeq .. obj =e= sum(J, sqr(x(J))) - 2 * sum (J, x(J))
                    + sum(I, sqr(y(I)));

Fy(I) .. (-2)*x(I) + 2*y(I) + 2*(y(I)-1)*u(I) =e= 0;

Fu(I) .. (.25) =g= sqr(y(I)-1);


model oz1 / objeq, Fy.y, Fu.u /;

option mpec=bundle;
solve oz1 using mpec minimizing obj;
```

Fig. 5.1. GAMS Model for (5.1)

ables, and equations declared in Figure 5.1 immediately, while those not familiar with GAMS will appreciate the concise yet descriptive style and should recognize the parallel to (5.1). In the `model` statement, we see the "objective" equation given first, while the remaining pairs define the equilibrium constraints. The variables `y` and `u` are paired with equations, and are state variables. The variable `x` is not paired, so it is a design variable.

While an indexed GAMS variable will often have components of only one type (i.e. design or state variables) this is not always the case. For example, it is possible to pair a variable `w(I)` declared and defined over the set `I` with an equation `g(I)` declared over the same set `I` but defined over a subset `II` of `I`. In this case, the components of `w` with indices in `II` will be state variables matched to `g`, while the components of `w` with indices in `I\ II` will be design variables.

When the GAMS `solve` statement is executed, the equation-variable pairs defining the MPEC model, together with information about the sets, variables, and equations themselves, are sent to the disk as a sequence of scratch files, and an MPEC solver is called. This solver uses the interface library MPECLIB to read and interpret the scratch files, evaluate functions and gradients, and write solution data.

Interface Initialization

```
int mpecInit (char *controlFileName, int indexStart,
              int diagRequired, int noObj,
              mpecRec **mpec);
void sparseInit (mpecRec *mpec, int colPtrx[], int cpxdim,
                 int rowIdxx[], int rixdim,
                 int colPtry[], int cpydim,
                 int rowIdxy[], int riydim);
```

The first task of the solver is to call the `mpecInit` routine to read in the scratch files and construct a problem of the form

$$\begin{array}{ll} \text{minimize} & \theta(x,y) \\ \text{subject to} & x \in \mathbf{B_x} \\ \text{and} & y \text{ solves } \mathrm{MCP}(F(x,y), \mathbf{B_y}). \end{array} \tag{5.2}$$

Note that the form of (5.2) is not as general as that specified in (4.1). In (5.2) we force the modeler to use $Z = \mathbf{B_x} \times \mathbf{R^m}$, although in practice there may be models that include side constraints of the form $h(x) = 0$ or $g(x, y) = 0$. Furthermore, we assume the equilibrium problem is written as an MCP. The motivation for this restriction is simply that there are currently no large scale implementations that allow for such side constraints. Such extensions to the model format would be easy to implement whenever appropriate solvers for these problems reach maturity.

If there is any inconsistency in the MPEC specification, it is detected during this initialization phase. It is here that the variables are partitioned into design and state variables, and the required maps are set up to support efficient function and gradient evaluation by other routines. Also, parameters to `mpecInit` exist allowing the user to specify if index ranges must begin with 0 (C style) or 1 (Fortran style), whether or not a dense Jacobian diagonal is required, and whether the model is an MCP instead of an MPEC (the library can be used for both model types). A pointer to a record containing all information specific to this model is passed back to the calling routine. This pointer will be passed on all subsequent MPECLIB calls.

In order to fully initialize the MPEC model, some space is required for the row indices and column pointers used to store the sparse Jacobians. Rather than allocating this space inside the library, the `mpecInit` routine returns estimates of the amount of space required for this to the calling routine.

A second routine, `sparseInit`, must then be called, which passes in these arrays, as well as the number of elements actually allocated for them. This routine completes the initialization, using the space provided it. The assumption here is that the user will not modify these arrays, as they are used by both solver and interface library. This assumption saves having to store two copies of the sparsity structure and copy it to the user's data structure at each derivative evaluation.

Variable Bounds and Level Values

```
void getxBounds (mpecRec *mpec, double lb[], double ub[]);
void getxLevels (mpecRec *mpec, double x[]);
void setxbar (mpecRec *mpec, double xbar[]);
void getyBounds (mpecRec *mpec, double lb[], double ub[]);
void getyLevels (mpecRec *mpec, double y[]);
```

The routines to obtain variable bounds and initial level values are for the most part self-explanatory. The `setxbar` routine is used to store a vector of design variables $\bar{x}$ in MPECLIB for use in subsequent calls to function and gradient routines that pass only state variables y. This is useful for solvers that implement a two-level solution scheme in which the inner solver (an MCP code) has no knowledge of the variables x in the outer optimization problem. In our current implementation, the box $\mathbf{B_y}$ does not depend on y, so the `getyBounds` routine would be called only once. A possible generalization is to allow $\mathbf{B_y}$ to depend on x, in which case a new function with the input parameter x would be required. This function would of course be called whenever x changed.

Function and Jacobian Evaluation

```
int getF ( mpecRec *mpec, double x[], double y[],
           double F[], double *theta);
int getdF (mpecRec *mpec, double x[], double y[],
           double F[], double *theta,
           double Jx[], int colPtrx[], int rowIdxx[],
           double Jy[], int colPtry[], int rowIdxy[],
           double dthetadx[], double dthetady[]);
int getFbar ( mpecRec *mpec, int n, double y[], double F[]);
int getdFbar (mpecRec *mpec, int n, int nnz, double y[],
              double F[],
              double Jy[], int colPtry[], int rowIdxy[]);
```

The routine `getF` takes the current point (x, y) as input and outputs the value of the functions F and θ at this point. The routine `getdF` does this also, but it computes the derivative of F and θ as well. The derivative of F w.r.t. x and y are returned in separate matrices, both stored sparsely in row index, column pointer fashion, while the derivative of θ w.r.t. x and y is returned as two dense vectors. The routines **getFbar** and **getdFbar** are similar, but in these routines, the input x is assumed to be the constant value $\bar{x}$ fixed in the previous call to **setxbar**. In this case, derivatives w.r.t. x and objective function values and derivatives are also not passed back. These routines are designed for use by an algorithm solving an inner (MCP) problem.

Solver Termination

```
void putxLevels (mpecRec *mpec, double x[]);
void putyLevels (mpecRec *mpec, double y[]);
void putObjVal (mpecRec *mpec, double theta);
void putStatus (mpecRec *mpec, int modelStat, int solverStat);
int mpecClose (mpecRec *mpec);
```

Once a solution has been found, the solver must pass this solution on to the interface library. This is done via the **putxLevels**, **putyLevels**, and **putObjVal** routines. The **putStatus** routine is used to report the model status (e.g. local optimum found, infeasible, unbounded, intermediate nonoptimal) and solver status (e.g. normal, iteration limit, out of memory, panic termination) via integer codes. All of this information is stored in the **mpec** data structure and written to disk when the **mpecClose** routine is called. When the solver terminates, these files are read by GAMS so that the solution information is available for reporting purposes, as data to formulate other models, etc.

6. MPEC algorithms: Implicit approaches

Algorithms for solving MPEC's are not nearly as well developed as those for MCP. Given the efficiency of the known methods for solving MCP and NLP, we describe here several techniques for solving MPEC's using an implicit approach that solves a sequence of MCP's.

We assume first that the problem has the form (5.2). For the implicit programming approach to work, a further assumption is needed, namely that there is a (locally) unique solution y of the equilibrium problem $\mathrm{MCP}(F(x,\cdot),C)$ for each value of x. We denote this solution by $y(x)$. Under these assumptions, the problem (5.2) is equivalent to the implicit program:

$$\begin{aligned} &\text{minimize} \quad \Theta(x) = \theta(x, y(x)) \\ &\text{subject to} \quad x \in \mathbf{B_x}. \end{aligned} \tag{6.1}$$

This implicit programming formulation has the advantage of simple constraints, but a rather complex, in fact nondifferentiable objective function Θ, even though the original functions f and F may be smooth. Nonsmoothness results from the underlying complementarity condition.

A promising solution strategy for the implicit program (6.1) is to apply a "bundle method", that is an algorithm specifically designed to solve non-smooth optimization problems. This idea is presented in [27, 28, 32]. The implementation of the bundle method we used, `btnc` [39], was developed for bound constrained problems and requires the user to provide a Fortran implementation to evaluate the objective function and a subgradient at a user supplied point. The formulas to generate these quantities were developed in [32] for the MPEC format described in (5.2).

The Fortran function was easy to code using the routines developed in Section 5.; essentially the only modification needed to the PATH solver was an option to allow the optimal basis of an MCP to be returned to the caller.

Derivative free optimization has a long history in mathematical programming. We used MPECLIB in conjunction with a prototype Matlab implementation of DFO [5] provided by Ph. Toint. This implementation was designed under the assumption that the dimension of the underlying optimization problem is small and the time to evaluate the objective function is large. Even though the underlying MCP is large dimensional, provided the number of tolled arcs is small, these properties are present in the implicit form of tolling model (6.1). Based on these assumptions, the DFO algorithm uses multivariate interpolation to develop good local models and ideas from the trust region literature to search the (small dimensional) space.

While both of these MPEC solvers succeeded in solving a large class of tolling problems of the form outlined in Section 4., their performance was not entirely satisfactory. Even though the number of variables in the underlying MCP can be large, preliminary numerical results demonstrate the fact that the current implementations of both these codes are limited to small numbers of design variables. We believe that future research and improvements in these and other algorithms will lead to substantially more robust and efficient implementations. The use of MPECLIB will remain critical in each such development.

7. Stochastic models in GAMS

In most practical instances, many of the variables in the previous formulations are not known with certainty, but are estimated from observed data, perhaps using an MPEC formulation of an inverse problem. Even when we are willing to believe these estimations as being truly representative of the actual data in the problem, the network may be subject to dynamic changes that mean its behavior over time varies considerably. Thus, when failures occur in a network (due for example to accidents or roadworks), the shortest paths that each user views in the network changes, and each driver tries to compensate for these changes by modifying their path choice.

In this stochastic setting, a typical formulation of the user equilibrium problem assumes that all users have complete information regarding the plans that every other user has under all possible scenarios or states of the network. We believe this to be an unreasonable assumption. Instead, we review below some recent work [16] on how a single user can modify their shortest path in order to develop a plan that is more robust to failures in the network. We show how to use some extensions of the GAMS modeling language to implement these models, and briefly describe the results of this work.

The model assumes that the network may be in one of finitely many states characterized by different travel times along the arcs, and allows transitions between the states according to a continuous-time Markov chain. The objective is to guide the vehicles in a manner minimizing the total expected travel time.

We describe the case when the only possible transitions are between state 0 (representing the *normal* operation mode) and states $\ell \neq 0$ (representing *failure* modes). The rate of transition from 0 to $\ell \neq 0$ will be denoted by λ^ℓ, and the transition rate back by μ^ℓ, see Figure 7.1. The general case is treated

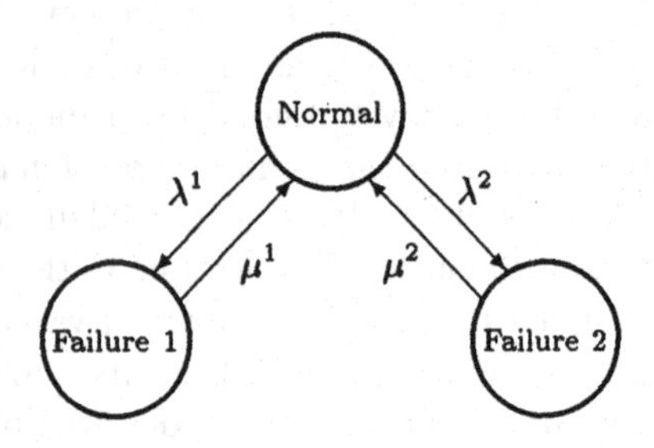

Fig. 7.1. Transition diagram of network's states.

in [16].

The problem is as follows. At each node $n \in \mathcal{N}$ there is a constant demand flow s_n that must be moved through the network to some destination node D at the minimum expected travel time. To facilitate the analysis and to provide ground for more general cases we make the following simplifying assumptions.

(A1) If the state of the system changes from k to ℓ when a vehicle is on arc (i, j) the travel time on (i, j) remains equal to c_{ij}^k for this vehicle; it experiences new travel times only after hitting j.

(A2) The products $\lambda^\ell c_{ij}^0$ and $\mu^\ell c_{ij}^\ell$ are much smaller than one for each ℓ and each $(i, j) \in \mathcal{A}$.

Condition (A1) amounts to assuming that failures occur at the initial sections of the arcs and do not affect those who have passed them. It is equally simple to consider other cases, except the notation then becomes more involved.

Condition (A2) implies that the probability of more than one state transition during the travel time of a vehicle on an arc (i,j) is negligibly small.

Whenever the arcs are uncapacitated, the problem can be solved as a stochastic shortest path problem [1, 2, 34]. We analyze the capacitated problem, in which the main issue is the interaction between vehicles that started at different times but reach a node at the same time, thus leading to jams. In this general case there are arc capacities u_{ij}^{ℓ}, associated with states $\ell = 0, \ldots, L$, so we cannot ignore the interactions between different flow subvectors, if they share the same arc at the same time.

We assume that all travel times are integer and let M be an upper bound on all of them. Suppose that a transition from state 0 to state ℓ takes place, and let $t = 0$ denote the time of this transition. Let $Y_{ij}^{\ell}(t)$ be the flow of re-routed vehicles entering arc (i,j) at time t. They satisfy the flow conservation equations

$$\sum_{(i,j)\in\mathcal{A}} Y_{ij}^{\ell}(t) - \sum_{\substack{(j,i)\in\mathcal{A} \\ c_{ji}^{\ell}\le t}} Y_{ji}^{\ell}(t-c_{ji}^{\ell}) = \sigma_i(t), \quad i\in\mathcal{N}\setminus D, \quad t=0,1,2,\ldots, \tag{7.1}$$

where $\sigma_i(t)$ is the inflow into i of the vehicles that experienced the state transition while traveling along the arcs leading to i:

$$\sigma_i(t) = \sum_{\substack{(j,i)\in\mathcal{A} \\ c_{ji}^{0}>t}} X_{ji}. \tag{7.2}$$

Since the supply (7.2) vanishes after a finite time (for which an upper bound M is known), we know that the flows Y^{ℓ} will vanish after a finite time, too, although this time may be much larger than M.

Since we have many sources, and the network is not layered, we cannot ignore the interactions of the rescheduled flow Y^{ℓ} with the flow $X^{\ell}(t)$ of vehicles that started *after* the state transition to ℓ. We make a simplifying assumption that further state transitions do not occur during the time that we are calculating X^{ℓ}. Even with this assumption, we cannot avoid modeling the initial non-stationary phase, when the re-routed flow $Y^{\ell}(t)$ and the new flow $X^{\ell}(t)$ interact. The policy that we develop under this assumption is termed a one-step lookahead policy.

Denoting by T the optimization horizon and by $Z^{\ell}(t) = Y^{\ell}(t) + X^{\ell}(t)$ the effective flow after the state transition, we obtain the problem

$$\min \sum_{t=0}^{T} \sum_{(i,j)\in\mathcal{A}} c_{ij}^{\ell} Z_{ij}^{\ell}(t) \tag{7.3}$$

$$\sum_{(i,j)\in\mathcal{A}} Z_{ij}^{\ell}(t) - \sum_{\substack{(j,i)\in\mathcal{A}\\ c_{ji}^{\ell}\le t}} Z_{ji}^{\ell}(t-c_{ji}^{\ell}) = s_i + \sigma_i(t), \quad i\in\mathcal{N}\setminus D, \quad t=0,1,\ldots,T, \tag{7.4}$$

$$0 \le Z_{ij}^{\ell}(t) \le u_{ij}^{\ell}, \quad (i,j)\in\mathcal{A}, \quad t=0,1,\ldots,T, \tag{7.5}$$

where the additional supply $\sigma_i(t)$ is given by (7.2). The optimal value $Q^{\ell}(X)$ of the above problem is the rescheduling cost for the plan X, when transition to state ℓ occurs.

There are reasons to believe that the value T does not matter for determining the robust plan X, provided T is large enough, and that the solution to (7.3)–(7.5) becomes, for large t, equal to a solution of the 'steady-state' problem associated with state ℓ:

$$\min \sum_{(i,j)\in\mathcal{A}} c_{ij}^{\ell}\overline{X}_{ij}^{\ell}$$

$$\sum_{(i,j)\in\mathcal{A}} \overline{X}_{ij}^{\ell} - \sum_{(j,i)\in\mathcal{A}} \overline{X}_{ji}^{\ell} = s_i, \quad i\in\mathcal{N}\setminus D,$$

$$0 \le \overline{X}_{ij}^{\ell} \le u_{ij}^{\ell}, \quad (i,j)\in\mathcal{A}.$$

To avoid some terminal effects associated with the fact that the vehicles that start late cannot make it to the destination anyway, and therefore choose short arcs, we may augment (7.3)–(7.5) with terminal conditions:

$$Z_{ij}^{\ell}(t) = \overline{X}_{ij}^{\ell}, \quad t=T-\tau, T-\tau+1,\ldots,T-1,T, \quad (i,j)\in\mathcal{A}, \tag{7.6}$$

where τ is some constant (for example, the maximum travel time on the arcs). In fact, by choosing T (or τ) one may change the allowed length of the transient period, before the flow settles on the new steady-state solution. This is easily done from within GAMS.

We are now ready to formulate the robust planning problem in the capacitated case:

$$\min\Big\{ \sum_{(i,j)\in\mathcal{A}} c_{ij}^{0}X_{ij} + \sum_{\ell=1}^{L}\lambda^{\ell}Q^{\ell}(x)\Big\} \tag{7.7}$$

$$\sum_{(i,j)\in\mathcal{A}} X_{ij} - \sum_{(j,i)\in\mathcal{A}} X_{ji} = s_i, \quad i\in\mathcal{N}\setminus D, \tag{7.8}$$

$$0 \le X_{ij} \le u_{ij}^{0}, \quad (i,j)\in\mathcal{A}. \tag{7.9}$$

The functions $Q^{\ell}(x)$ are the optimal values of the re-routing problems in scenarios $\ell = 1,\ldots,L$.

Problem (7.7)–(7.9) is similar to two-stage stochastic programming problems (see [25, 35, 43]). Much is known about these problems, and efficient solution techniques exist that exploit the structure of the model in question (see [26, 43] and the references therein). The simplest approach, however, is to include the linear programs defining $Q^\ell(X)$ into (7.7)–(7.9) and construct a giant linear programming problem with a dual block angular structure:

$$\min \sum_{(i,j)\in\mathcal{A}} \left(c_{ij}^0 X_{ij} + \sum_{\ell=1}^{L} \lambda^\ell \sum_{t=0}^{T} c_{ij}^\ell Z_{ij}^\ell(t) \right) \tag{7.10}$$

subject to (7.8)–(7.9) and (7.4)–(7.5), and (7.6). This problem, the deterministic equivalent to a two-stage stochastic program, can be solved by standard linear programming techniques, such as the simplex method or interior point methods. In addition, a GAMS link to the SPOSL solver [26] for stochastic programs is under development and has been used on this problem. This link accepts the deterministic equivalent, but passes it to SPOSL in stochastic form for more efficient solution.

We have investigated the effects of our modeling format on a simple example using the same Sioux Falls network as in the MPEC model. Our GAMS implementation of the model is shown in Figure 7.2 and Figure 7.3.

Ignoring the possibility of failure on the arcs, we first solved the shortest path problems to find the solution shown in Figure 7.4. In Figure 7.3, this is computed as a special case of the robust plan, with 0 probability of failure. We then considered the possibility of 2 failures in the network, on arcs (1,2) and arc (21,24). We incorporated capacities of 0.5 only on arcs (15,14) and (22,23) so that when a failure of arc (21,24) occurs, all the flow could not be rerouted through these arcs. The loop statement in Figure 7.3 is used to compute the steady state solution for each arc failure, required to enforce (7.6). The resulting robust solution plan obtained from minimizing (7.10) subject to (7.8)–(7.9), (7.4)–(7.5) and (7.6) is depicted in Figure 7.5. An interesting paradox can be observed. Arcs (15,14) and (22,21) that are not used at all in the shortest path plan have saturating flow sent across them in the robust plan. This paradox can be explained by the fact that the arcs are heavily used for rerouting. Thus, to avoid the major expense of rerouting large amounts of flow through arc (10,11) in the event that (21,24) fails, it is better to send as much flow as possible away from potential bottlenecks. Also, in the robust plan, flow is sent from node 8 to node 6, which is in direct contrast to the shortest path solution depicted in Figure 7.4 which sends flow from 6 to 8.

In order to demonstrate the effect of our robust plan on the transient behavior of the jam, we show the flows on two representative arcs after a failure occurs. These rerouting flows are calculated under two different plans. The charts on the top of Figure 7.6 depict the transient behavior of the flows on the arcs (15,10) and (10,11) under the plan that chooses shortest paths initially, and then reroutes the flow when a failure occurs. The charts on the bottom depict the rerouting flows that occur when we follow the robust plan; both of these rerouting procedures allow a period $T_p = 40$ to attain the steady state solution. Note that on both of these arcs, the amount of flow

```
set      i 'nodes in traffic network' /1*24/;
alias (j,i), (k,i), (l,i);

set      dest(j) identification of destination nodes,
         arcs(i,i) arcs;
$include sioux-falls.dat

set time / t0*t100 /
    fixed(time) 'steady state solution reached' ;
fixed(time) = yes$(ord(time) gt 41);

* nodes in the stochastic tree, node0 the root
set nodes 'nodes' /node0*node2/;
parameter cost(nodes,i,i), capacity(i,i), tranrate(nodes);

cost(nodes,arcs) = coef_a(arcs);
cost("node1","1","2") = 100; cost("node2","21","24") = 100;
capacity(arcs) = inf;
capacity("15","14") = 0.5; capacity("22","23") = 0.5;

variables        z(nodes,i,j,time), obj;

equations        balance(nodes,i),
                 robbalance(nodes,i,time), costdef;

balance(nodes,i)$(ord(nodes) eq 1 and not dest(i))..
  sum(arcs(i,j), z(nodes,arcs,"t0"))
  - sum(arcs(j,i), z(nodes,arcs,"t0"))
  =e= demand(i);

robbalance(nodes,i,time)$(ord(nodes) gt 1 and not dest(i))..
  sum(arcs(i,j), z(nodes,arcs,time))
  - sum(arcs(j,i), z(nodes,arcs,time-cost(nodes,arcs)))
  =e= demand(i) +
  sum(arcs(j,i)$(cost("node0",arcs) ge ord(time)),
      z("node0",arcs,"t0"));

costdef..
  obj =e= sum(arcs(i,j), sum(nodes, tranrate(nodes)*
                sum(time, cost(nodes,arcs)*z(nodes,arcs,time))));
```

Fig. 7.2. Robust Path Choice: GAMS model using SP/OSL

```
z.lo(nodes,arcs,time) = 0.0;
z.up(nodes,arcs,time) = capacity(arcs);
z.fx(nodes,i,j,time)$(not arcs(i,j)) = 0;
z.fx(nodes,dest,j,time)$(arcs(dest,j)) = 0;
z.fx("node0",arcs,time)$(ord(time) gt 1) = 0;

model robust / costdef, balance, robbalance /;

tranrate(nodes) = 0; tranrate("node0") = 1;
option lp = sposl;
solve robust using lp minimizing obj;

tranrate("node1") = 0.01; tranrate("node2") = 0.05;

* determine steady state flows after each scenario has occurred
set inc_node(nodes); inc_node(nodes) = no;
equation steadyobj(nodes), steadybal(nodes,i);

steadybal(inc_node,i)$(not dest(i))..
  sum(arcs(i,j), z(inc_node,arcs,"t0"))
  - sum(arcs(j,i), z(inc_node,arcs,"t0"))
  =e=  demand(i);

steadyobj(inc_node)..
  obj =e= sum(arcs, cost(inc_node,arcs)*z(inc_node,arcs,"t0"));

model steady / steadyobj, steadybal /;

option lp = osl;
loop(nodes$(ord(nodes) gt 1),
  inc_node(nodes) = yes;
  solve steady using lp minimizing obj;
  z.fx(nodes,arcs,fixed) = z.l(nodes,arcs,"t0");
  inc_node(nodes) = no;
);

option lp = sposl;
solve robust using lp minimizing obj;
```

Fig. 7.3. Robust Path Choice: GAMS model using SP/OSL (cont)

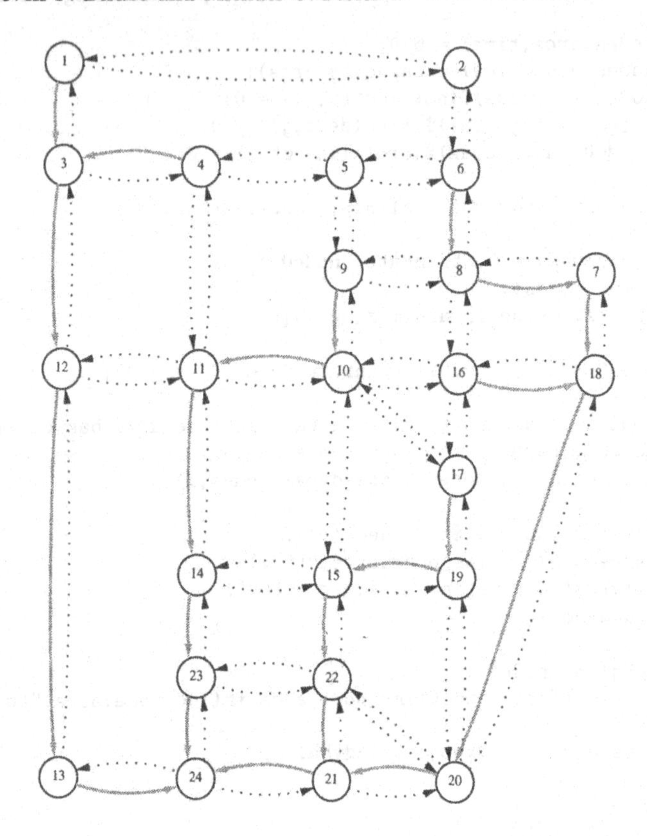

Fig. 7.4. Shortest path solution for Sioux Falls network.

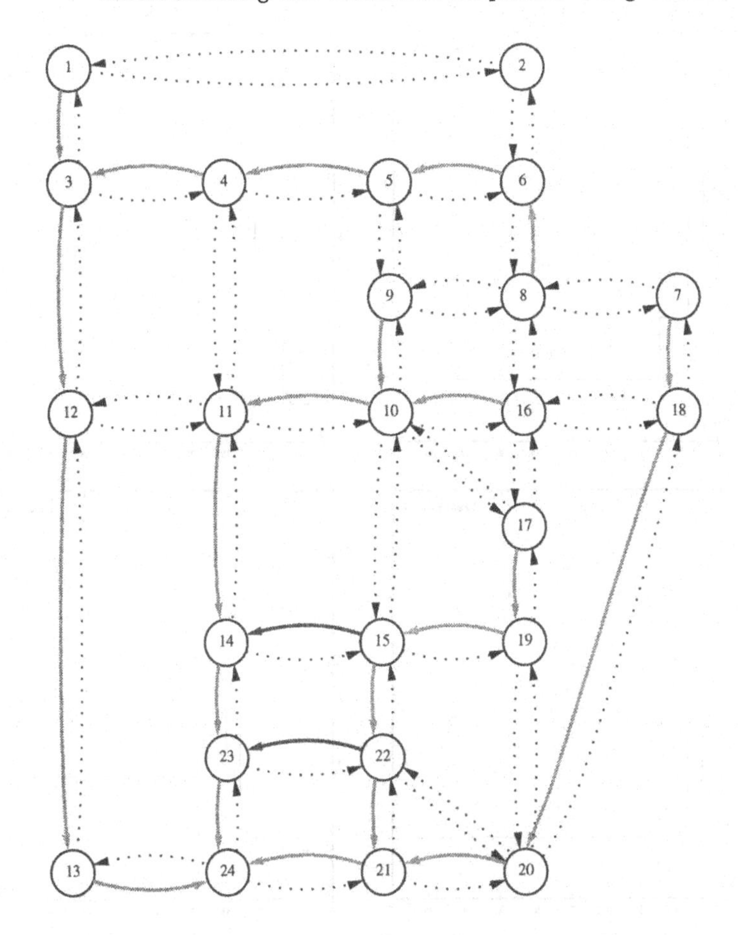

Fig. 7.5. Robust solution in capacitated case ($T = 40$).

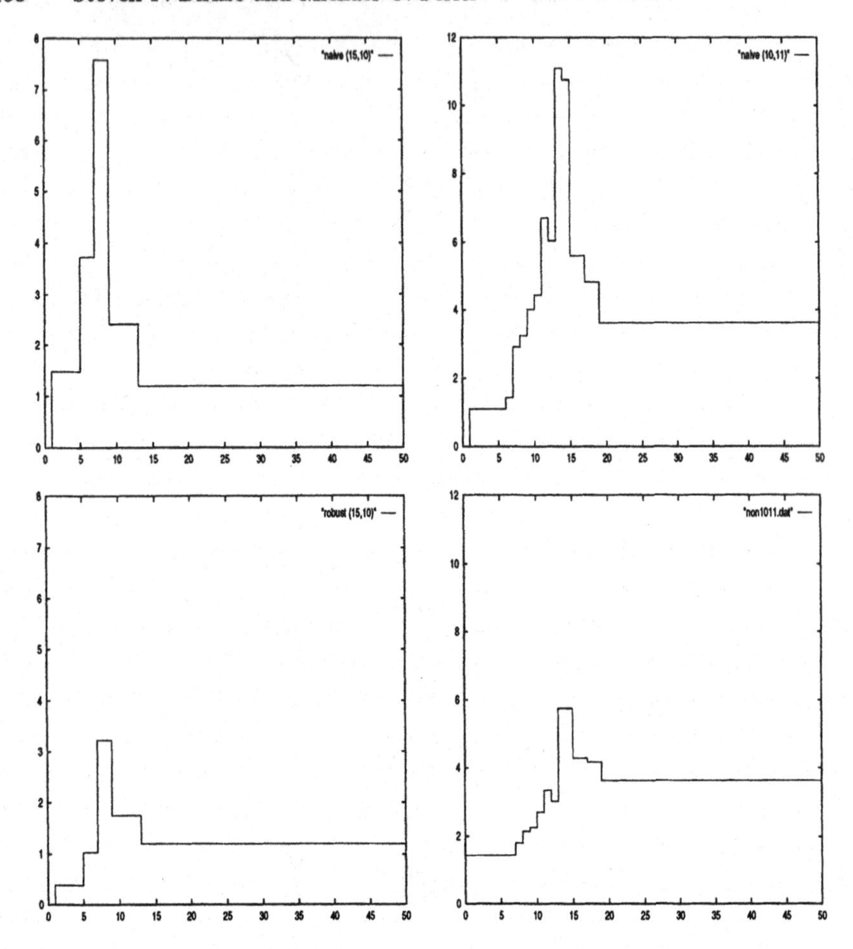

Fig. 7.6. Flow on (15,10) and (10,11) during transition from shortest path plan (top) or robust plan (bottom). Failure occurs at $t = 1$.

that has to be rerouted in the robust case is less than half that needed to be rerouted when the shortest path plan is followed.

8. Conclusions

In this paper, we have shown the connection between various forms of traffic design problems and modeling formats based on complementarity and variational inequalities. We have used the modeling language GAMS to develop all the models in this paper; extensions to other modeling languages such as AIMMS and AMPL could be similarly implemented.

Stochastic modeling and MPEC solution are active areas of current research. We believe that some emerging algorithms for MPEC and stochastic LP may prove to be suitably quick and robust to satisfactorily solve large classes of these models. The techniques we have described here allow for easy formulation of many MPEC's and interfacing with solvers. The class of problems for which the algorithms described herein are applicable is restricted by size of the problem and/or dealing with underlying nonconvexities. Future research extending both the models and tools outlined in this paper will remain critical for more reliable problem formulation and solution.

Acknowledgements

This material of this paper is based on research supported by National Science Foundation Grant CCR-9157632.

References

1. D. P. Bertsekas. *Dynamic Programming and Optimal Control.* Athena Scientific, Belmont MA, 1995.
2. D. P. Bertsekas and J. N. Tsitsiklis. An analysis of stochastic shortest path problems. *Mathematics of Operations Research*, 16:580–595, 1991.
3. A. Brooke, D. Kendrick, and A. Meeraus. *GAMS: A User's Guide.* The Scientific Press, South San Francisco, CA, 1988.
4. Y. Chen and M. Florian. O-D demand adjustment problem with congestion: Part I. model analysis and optimality conditions. Publication CRT-94-56, Centre de Recherche sur les Transports, Université de Montréal, Montréal, Canada, 1994.
5. A. R. Conn and Ph. L. Toint. An algorithm using quadratic interpolation for unconstrained derivative free optimization. In G. Di Pillo and F. Giannessi, editors, *Nonlinear Optimization and Applications.* Plenum Press, New York, 1996.
6. S. P. Dirkse and M. C. Ferris. MCPLIB: A collection of nonlinear mixed complementarity problems. *Optimization Methods and Software*, 5:319–345, 1995.

7. S. P. Dirkse and M. C. Ferris. The PATH solver: A non-monotone stabilization scheme for mixed complementarity problems. *Optimization Methods and Software*, 5:123–156, 1995.
8. S. P. Dirkse and M. C. Ferris. A pathsearch damped Newton method for computing general equilibria. *Annals of Operations Research*, 1996.
9. S. P. Dirkse and M. C. Ferris. Crash techniques for large-scale complementarity problems. In Ferris and Pang [14].
10. S. P. Dirkse, M. C. Ferris, P. V. Preckel, and T. Rutherford. The GAMS callable program library for variational and complementarity solvers. Mathematical Programming Technical Report 94-07, Computer Sciences Department, University of Wisconsin, Madison, Wisconsin, 1994.
11. B. C. Eaves. On the basic theorem of complementarity. *Mathematical Programming*, 1:68–87, 1971.
12. M. C. Ferris and C. Kanzow. Recent developments in the solution of nonlinear complementarity problems. Technical report, Computer Sciences Department, University of Wisconsin, 1997. In preparation.
13. M. C. Ferris, A. Meeraus, and T. F. Rutherford. Computing Wardropian equilibrium in a complementarity framework. Mathematical Programming Technical Report 95-03, Computer Sciences Department, University of Wisconsin, Madison, Wisconsin, 1995.
14. M. C. Ferris and J. S. Pang, editors. *Complementarity and Variational Problems: State of the Art*, Philadelphia, Pennsylvania, 1997. SIAM Publications.
15. M. C. Ferris and J. S. Pang. Engineering and economic applications of complementarity problems. *SIAM Review, forthcoming*, 1997.
16. M. C. Ferris and A. Ruszczyński. Robust path choice and vehicle guidance in networks with failures. Mathematical Programming Technical Report 97-04, Computer Sciences Department, University of Wisconsin, Madison, Wisconsin, 1997.
17. M. C. Ferris and T. F. Rutherford. Accessing realistic complementarity problems within Matlab. In G. Di Pillo and F. Giannessi, editors, *Nonlinear Optimization and Applications*. Plenum Press, New York, 1996.
18. M. Florian, editor. *Traffic Equilibrium Methods*, Berlin, 1976. Springer-Verlag.
19. M. Florian. Nonlinear cost network models in transportation analysis. *Mathematical Programming Study*, 26:167–196, 1986.
20. T. L. Friesz. Network equilibrium, design and aggregation. *Transportation Research*, 19A:413–427, 1985.
21. T. L. Friesz, R. L. Tobin, T. E. Smith, and P. T. Harker. A nonlinear complementarity formulation and solution procedure for the general derived demand network equilibrium problem. *Journal of Regional Science*, 23:337–359, 1983.
22. P. T. Harker. *Lectures on Computation of Equilibria with Equation–Based Methods*. CORE Lecture Series. CORE Foundation, Louvain-la-Neuve, Université Catholique de Louvain, 1993.
23. A. Haurie and P. Marcotte. On the relationship between Nash–Cournot and Wardrop equilibria. *Networks*, 15:295–308, 1985.
24. D. W. Hearn, S. Lawphongpanich, and J. A. Ventura. Restricted simplicial decomposition: Computation and extensions. *Mathematical Programming Study*, 31:99–118, 1987.
25. P. Kall and S. W. Wallace. *Stochastic Programming*. John Wiley & Sons, Chichester, 1994.
26. A. J. King. SP/OSL version 1.0 stochastic programming interface library user's guide. Technical Report RC 19757, IBM T.J. Watson Research Center, Yorktown Heights, New York, 1994.

27. M. Kočvara and J. V. Outrata. On optimization of systems governed by implicit complementarity problems. Technical Report 513, Institute of Applied Mathematics, University of Jena, Leutragraben, 1994.
28. M. Kočvara and J. V. Outrata. On the solution of optimum design problems with variational inequalities. In *Recent Advances in Nonsmooth Optimization*, pages 172–192. World Scientific Publishers, Singapore, 1995.
29. Z.-Q. Luo, J. S. Pang, and D. Ralph. *Mathematical Programs with Equilibrium Constraints*. Cambridge University Press, 1996.
30. T. L. Magnanti. Models and algorithms for predicting urban traffic equilibria. In M. Florian, editor, *Transportation Planning Models*, pages 153–186. North Holland, 1984.
31. P. Marcotte. Network design problem with congestion effects: A case of bilevel programming. *Mathematical Programming*, 34:142–162, 1986.
32. J. V. Outrata and J. Zowe. A numerical approach to optimization problems with variational inequality constraints. *Mathematical Programming*, 68:105–130, 1995.
33. J. S. Pang and C. S. Yu. Linearized simplicial decomposition methods for computing traffic equilibria on networks. *Networks*, 14:427–438, 1984.
34. G. H. Polychronopoulos and J. N. Tsitsiklis. Stochastic shortest path problems with recourse. *Networks*, 27:133–143, 1996.
35. A. Prékopa. *Stochastic Programming*. Kluwer Academic Publishers, Dordrecht, 1995.
36. D. Ralph. Global convergence of damped Newton's method for nonsmooth equations, via the path search. *Mathematics of Operations Research*, 19:352–389, 1994.
37. S. M. Robinson. Mathematical foundations of nonsmooth embedding methods. *Mathematical Programming*, 48:221–229, 1990.
38. S. M. Robinson. Normal maps induced by linear transformations. *Mathematics of Operations Research*, 17:691–714, 1992.
39. H. Schramm and J. Zowe. A version of the bundle idea for minimizing a nonsmooth function: Conceptual idea, convergence analysis, numerical results. *SIAM Journal on Optimization*, 2:121–152, 1992.
40. M. J. Smith. The existence, uniqueness and stability of traffic equilibria. *Transportation Research*, 13B:295–304, 1979.
41. Ph. L. Toint. Transportation modelling and operations research: A fruitful connection. In Ph. L. Toint, M. Labbe, K. Tanczos, and G. Laporte, editors, *Operations Research and Decision Aid Methodologies in Traffic and Transportation Management*. Springer-Verlag, 1997.
42. J. G. Wardrop. Some theoretical aspects of road traffic research. *Proceeding of the Institute of Civil Engineers, Part II*, pages 325–378, 1952.
43. R. J.-B. Wets. Large scale linear programming techniques in stochastic programming. In Yu. M. Ermoliev and R. J. B. Wets, editors, *Numerical Techniques for Stochastic Optimization*, pages 65–93. Springer-Verlag, Berlin, 1988.

Multicriteria Evaluation Methods and Group Decision Systems for Transport Infrastructure Development Projects

Katalin Tanczos

Technical University of Budapest,
Department of Transport Economics,
H-1111 Budapest, Hungary

Summary. In this paper the author presents an approach and a set of techniques to support the multicriteria evaluation and group decision used for the selection and ranking of transport infrastructure development projects. First the author shows the way from the traditional monocriterion to the multicriteria decision making, then she gives an overview of the stages of multicriteria analysis, describing the basic terms and the theoretical considerations of the multi-attribute evaluation process. Considering the main purposes of the countries in Central and Eastern Europe with respect to the joining process to the European Union the author shows the complexity of the transport systems. She identifies the main elements of the transport systems, describes the policy frameworks which are effective at the European scale and therefore they have to be taken account at the group decisions related to large size transport infrastructure development projects in the countries with economies in transition. Finally the author presents an extension of multiple criteria decision aid by considering the ranking problem of a finite set of transport infrastructure investment projects subject to a preference relation pair set generated within the framework of multicriteria analysis process and a quasi-equal financial expenses rhythm-constraint.

1. From traditional monocriterion approach to multicriteria decision making

Until 1970, managers had to formalise decision-making problems as follows. One is given a well-defined set A of feasible alternatives a. In general, there are two possible forms for A: the analytic form where a feasible alternative $a = (x_1, \ldots, x_m)$, which leads to A being a subset of R^m, and the enumerative form, in which A is defined by a list of alternatives without any explicit links to mathematical formulations of constraints. The decision maker (D) has a real valued function g (a unique criterion) defined on A precisely reflecting his or her preferences in that "D prefers a' over a if and only if $g(a') > g(a)$ and D is indifferent between a' and a if and only if $g(a') = g(a)$".

Within the analytic form of A, we obtain that $g(a) = g(x_1, \ldots, x_m)$. The decision maker then had to distinguish between the deterministic case, where $g(a)$ is computed without any reference to random variables, and the probabilistic case, where one or more random variables Y intervene (by the means of some characteristics of probabilistic distributions and utility functions) in the computation of $g(a)$. The decision maker then had a well-formulated mathematical problem: he or she had to find (meaning "to discover") $a* \in A$ such that $g(a*) \geq g(a)$ for all $a \in A$.

Many decision problems, specially, those arising in the infrastructure development of the transport sector today are complicated by the need to consider a range of issues, such as those relating to environment, quality of life, sustainability of development, and by the participation of divergent interest groups. To reflect this, the majority of the transport infrastructure development problems has to deal with multiple objectives and methods which are designed to assist groups of decision makers. The evaluation processes have to integrate the quantitative and qualitative aspects of transport infrastructure development. These types of decision problems involve multiple objectives where the decision makers have to consider the trade-offs between the benefits offered by the various development alternatives. Considering the main features of multicriteria decision making, the general framework can be described as follows.

A well-defined set A of feasible alternatives a (in either analytic form or enumerative form) being given, the decision maker(s) has a well-defined set of attributes or consequences to which he or she refers to judge if:

- he or she prefers a' to a: $a'Pa$;
- he or she prefers a to a' : aPa';
- he or she is indifferent between a' and a: $a'Ia$.

Moreover, the decision maker's preferences are supposed to be completely shaped in his or her mind, meaning that P is asymmetric and transitive, and that I is reflexive, symmetric and transitive. Consequently, everything occurs as if there was a real valued function U defined on A such that

$$a'Pa \text{ iff } U(a') > U(a) \text{ and } a'Ia \text{ iff } U(a') = U(a).$$

The utility $U(a)$ can again be envisaged in a deterministic or in probabilistic context. Within these definitions, the decision maker's problem has again a well-formulated mathematical formulation: by referring to U, explicitly known or implicitly present in his mind, the task is to find (which again means "to discover") $a* \in A$ such that $U(a*) \geq U(a)$ for all $a \in A$.

Considering the two possible contexts, two fundamental distinctions have to be mentioned with respect to the nature of the utility U. In the deterministic case, attributes or consequences are supposed to be exactly known for each a, in such a way that, after a more or less complete sub-aggregation, it is possible to synthesise them by the means of n criteria $g_k(a)$ reflecting partial preferences. In other words, one assumes that $g_k(a)$ exist such that, if a' and a are such that $g_j(a') = g_j(a)$ for all $j \neq k$, then D prefers a' to a if and only if $g_k(a') > g_k(a)$, and D is indifferent between a' and a if and only if $g_k(a') > g_k(a)$. Consequently, we can write $U(a) = V[g_1(a), \ldots, g_n(a)]$, where, if A is known analytic form, $g_k(a) = g_k(x_1, \ldots, x_m)$. In the probabilistic case, attributes or consequences are viewed as random variables Y_k

characterized, for each $a \in A$, by a probabilistic distribution $\delta^a(y_1, \ldots, y_n)$. $U(a)$ is then a function of this distribution.

Before dealing with the problem of multicriteria analysis of the transport infrastructure development projects, we review the basic mathematical difficulties arising in multicriteria decision making. The first difficulty occurs in connection with the standard aggregation formula. Indeed, the question is to know what are the assumptions or axioms that justify relations of the form

$$V[g_1, \ldots, g_n] = \sum_{j=1}^{n} w_j[g_j] \quad \text{(additive form)},$$

$$V[g_1, \ldots, g_n] = \sum_{j=1}^{n} k_j.G_j \quad \text{with } k_j \geq 0 \quad \text{(weighted sum)},$$

$$U(a) = \sum_{y_1, \ldots, y_n} u(y_1, \ldots, y_n)\delta^a(y_1, \ldots, y_n) \quad \text{(general form of expected utility value}$$

where $u(y_1, \ldots, y_n)$ is a multi-attribute utility function, or

$$U(a) = V[g_1(a), \ldots, g_n(a)] \quad \text{with } g_j(a) = \sum_{y_j} \delta_j^a(y_j)$$

where $u(y_j)$ is a partial utility function attached to the j-th attribute and $\delta^a j(y_j)$ is the marginal probabilistic distribution of y_j? The second question deals with the nature of the set E of strongly efficient alternatives, where we say that $a \in A$ is a strongly efficient iff, whatever $b \in A$ verifying $g_j(b) > g_j(a)$, there exists at least one criterion g_k such that $g_k(b) < g_k(a)$, this property being true whatever the initial criterion g_j chosen. Does a given alternative a belong to E? How can E be generated? Is a given property (convexity) verified by E? Finally the convergence of choice procedures is also a problem. In particular, when we only know some properties of $U(a)$ and answers given by the decision-maker in conformity with $U(a)$, can we prove that a given organisation of questions automatically leads to the final choice $a*$?

There are two crucial preoccupations within this framework of multicriteria decision making. The main objective is to describe or discover something which is viewed as a fixed and ever present entity, either a utility function $U(a)$, or an optimum $a* \in A$. Consequently, effort of researchers are directed towards concepts, axioms and theorems liable to be used so as to define conditions under which the existence of the entity which must be discovered is guaranteed; and to help the decision-maker to reach the right solution.

To find the way from multicriteria decision making to multicriteria decision aid it is useful to identify some limitations on objectivity. The practice of operations research and multicriteria decision making had shed light on five major aspects. First, the borderline between what is and what is not feasible is often fuzzy and frequently modified in the light of what is found through the study itself. Second, in many real world problems, the decision maker, as a person truly able to make the decision, does not really exist: usually, several actors take part in the decision process and there is a confusion between the one who ratifies the decision and what is called the decision maker. The third aspect is that, even when decision maker is not a mythical person, his or her preferences are very seldom well shaped: in and among areas of firm convictions lie zones of uncertainty, half-held belief or, indeed, conflicts and contradictions. Of course, data is, in many cases, imprecise, and/or defi ned in an arbitrary way, and, finally, it is in general impossible to say that a decision is a good or a bad one by referring only to a mathematical model: organisational and cultural aspects of the whole decision process which leads to a given decision also contribute to its quality and success. If we want to avoid leaving aside these limitations on objectivity, a truly scientific foundation for an optimal decision seems impossible.

The main objective of multicriteria decision aid is to construct or create something which is viewed as liable to help an actor taking part in a decision process, either to shape, and/or to argue, and/or to transform his or her preferences, or to make a decision in conformity with his or her goals. Consequently, researchers' efforts are directed towards concepts, properties, and procedures liable to be used in order: to extract from the available information what appears as really meaningful in the perspective of what needs to be built; to help to shed light on decision maker's behaviour by bringing to him arguments able to strengthen or weaken his or her own convictions.

In contrast with multicriteria decison making, the general framework of multicriteria decision aid may be given as follows.

- A not necessarily stable set A of potential actions a is given; two potential actions can be put jointly into operation (unlike alternatives).
- Comparisons can be conducted between these potential actions, based on n criteria (or pseudo-criteria) $g_j(a)$ reflecting, with a certain fuzziness, the preferences of one or several actors on behalf of whom decision aid is provided. The family F of the n criteria is built in reference to what is taken into consideration by actors for shaping, and/or arguing, and/or transforming their preferences. Aside from probabilistic distributions, thresholds can be used to take into account imprecision, and/or uncertainty, and/or inaccurate determination.
- We therefore obtain an ill-defined mathematical problem: on the basis of the family F and additional inter-criteria information, elaborate: either a mathematical model allowing comprehensive (unlike partial comparisons for which only one criterion comes into play, the $n-1$ others keeping the same value) comparisons of potential actions, or a procedure helping to reflect upon and to progress in the formulation of comprehensive comparisons of potential actions.

Finally, the aid can be provided by the selection of a better action (optimum), the assignment of each action to an appropriate predefined category according to what we want it to become afterwards (for instance acceptance, rejection, delay for additional information); and the ranking of those actions which seem to be the most satisfactory (subset $A_0 \subset A$) according to a total or partial pre-order.

In this framework, the basic theoretical and methodological preoccupations in multicriteria decision aid are concerned with

- the nature, the generation, and the formal definition of potential actions to be considered;
- the nature and the formal definition of criteria: type of consequences and attributes, sources of imprecision, uncertainty and inaccurate determination, personality and number of actors;
- the conditions to be satisfied by the family of criteria in order to be appropriate to play the role devoted to it (organisational and cultural aspects);
- the nature and the quantification of the inter-criteria information which is required to formulate comprehensive comparisons: especially, how to characterise the specific role of each criterion due to its own importance;
- the logic and the properties of the aggregation models by which comprehensive comparisons are totally or partially formalised: such mathematical models may use a synthesising single criterion or not (one or more outranking relations, fuzzy or not for example);
- the logic and the properties of the procedures (interactive or not) by which the final selection, assignment or ranking is made. Since such the latter can be viewed as predefined, it becomes impossible to found the validity of a procedure primarily on a purely mathematical property of convergence. Furthermore, since the final decision, assignment or ranking is more like a creation than a discovery, the validity of a procedure depends on some mathematical properties which make it conform to given requirements, as well as on the way it is used and integrated into a decision process.

Overviewing some theoretical considerations we now provide some general conclusions. The objective of multicriteria approaches is clearly to help managers to make "better" decisions, but what is the meaning of "better"? This meaning is not independent from the process by which the decision is made and implemented, which then implies that our concepts, tools and procedures cannot be conceived from the perspective of discovering (more or less approximately) pre-existing truths which can be universally imposed. Nevertheless, methodical decision aid based upon appropriate concepts and procedures can play a significant and beneficial role in guiding a decision making process. Therefore, solutions obtained by solving multicriteria decision making well-formulated problems constitute a fundamental background for multicriteria decision aid. We also stress that the aim of multicriteria decision aid is, above all, to enhance the degree of conformity and coherence between the evolution of a decision making process, and the value system and the objectives of those involved in this process. For that purpose, concepts, tools and procedures must be helpful in making our way in the presence of ambiguity, uncertainty and an abundance of bifurcation.

2. A general overview of the multicriteria analysis

2.1 The frame of the multicriteria analysis

In the context of transport infrastructure development projects, the general frame of the multicriteria analysis consists of the following successive steps:

- identification of
  - the decision maker(s):
    - publicly elected officials,
    - transportation agency managers,
    - private sector managers,
    - corporate officials,
    - elected or appointed government officials,
    - representatives of the local inhabitants or authorities,
    - experts of financial institutions;
  - the decision level:
    - international (multinational),
    - government (national),
    - regional (local),
    - local (company level);
  - the time horizon of decision:
    - operative,
    - strategic,
    - political;
  - the purpose of decision:
    - to find the "best" solution,
    - ranking,
    - resource allocation;
- identification of the alternative courses of action (variants for development);
- identification of the attributes which are relevant to the decision problem (attributes directly measured for the assessment):
  - land use:
    - community and neighbourhood for proximity to city centre,
    - proportion of mixed land use,
    - proportion of undeveloped land area,
    - density of population,
    - age of dwellings in area,
    - location of social institutions,
    - location of neighbourhood boundaries;

- economic impacts:
  - employment,
  - income,
  - business activity,
  - residential activity,
  - effects on property,
  - regional and community plans,
  - resource consumption;
- social impacts:
  - displacement of people,
  - accessibility of facilities and services,
  - effects of terminals on neighbourhoods,
  - special user groups;
- physical impacts:
  - aesthetics and historic value,
  - infrastructure;
- impacts on the ecosystems:
  - air quality (CO, HC, NO, Sulphur oxides, particulate),
  - noise,
  - vibration,
  - disruption or damage to adjacent properties,
  - used land;
- public safety:
  - dead,
  - seriously injured,
  - slightly injured;
- energy
- assignment of values for each attribute to measure the performance of the alternatives on that attribute;
- determination of a weight for each attribute;
- taking a weighted average of the values assigned to that alternative for each alternative,
- making a (provisional) decision;
- performing sensitivity analysis to see how robust the decision is to changes in the figures supplied by the decision maker.

Regularly, with the repetition of the process described above it is possible to improve the accuracy and "goodness" of the decision making procedure.

Overviewing the multicriteria decision process it is also useful to describe a few basic definition and theoretical considerations. In the analysis we implicitly make a number of assumptions about the decision maker's preferences. These assumptions can be regarded as the axioms of the procedure, in that they represent a set of postulates which may be regarded as reasonable. If the decision maker accepts the axioms, and if he or she is rational, then he or she should also accept the preference rankings. The generally considered axioms are:

- decideability: ability to decide which of two options is to be preferred;
- transitivity means if $a > b$ and $b > c$, then $a > c$;
- summation: if $a > b$ and $b > c$ then the strength of preference of a over c must be greater than the strength of a over b (or b over c);
- solvability means to obtain a value function;
- finite upper and lower bounds for value: in assessing values we assume that the best option and the worst option are not infinite.

In the multicriteria evaluation model the decision making problem can be described as follows: there are n alternatives with m criteria. This type of decision situations contains (one or) more decision makers who are to evaluate and rank a finite number of alternatives with respect to a finite number of criteria. Evaluation criteria have to fit to the next requirements:

- completeness: all the attributes have to be included;
- operationality: specific enough to evaluate;
- decomposability: the attribute can be judged independently;
- absence of redundancy: to avoid double-counting
- minimum size: eliminating attributes which do not distinguish between the options;
- mutual preference independence.

Let $A_1, A_2, \ldots, A_n$ denote the alternatives and $C_1, C_2, \ldots, C_m$ the criteria. Assume that the data related to the alternatives are known. Let $a_{ji} \geq 0, i = 1, \ldots, m, j = 1, \ldots, n$ denote the value of the j-th alternative with respect to the i-th criterion. The data of the decision problem may then be written in a matrix form:

	C_1	C_2	$\ldots$	C_i	$\ldots$	C_m
A_1	a_{11}	a_{12}	$\cdot$	a_{1i}		a_{1m}
A_2	a_{21}	a_{22}	$\cdot$	a_{2i}		a_{2m}
$\cdot$	$\ldots$	$\ldots$	$\ldots$	$\ldots$	$\ldots$	$\ldots$
A_j	a_{j1}	a_{j2}	$\cdot$	a_{ji}		a_{jm}
$\cdot$	$\ldots$	$\ldots$	$\ldots$	$\ldots$	$\ldots$	$\ldots$
A_n	a_{n1}	a_{n2}		a_{nj}		a_{nm}

In order to build this matrix, it is necessary to derive, or each course of action (alternative) facing the decision maker, a numerical score to measure its attractiveness to him or her. If the decision does not involve risk and uncertainty then this score is the value of the course of action. If the decision involves risk and uncertainty then this score is the utility of the course of action. Elements of the decision matrix can be given in different scales (nominal, ordinal, interval, cardinal).

Before the practical illustration of a multicriteria decision making example the main elements and characteristics of the transport systems will be described to identify those criteria which may be considered as the most relevant for the evaluation and ranking of transport infrastructure development projects.

2.2 The complexity of the transport systems

Any assessment of transport infrastructure to be developed calls for a whole range of criteria, the weighing and ranking of which can only be carried out by means of a process of essentially policy trade-offs. Different categories of evaluation criteria had been proposed by an ECMT study (Simon,1994 [11]). These criteria may be derived from the analysis of a complex transport system.

The multimodal transport systems comprise a set of basic elements, like:

- infrastructure network (mode-specific),
- interface (intermodal terminals, stations, ports),
- ancillary (for operation and maintenance),
- rolling stock, vehicles, fuels,
- human capital,
- information (information system and telematics including passenger information, bookings, reservation planning, scheduling, trip, vehicle monitoring),
- finance (availability revenue, subsidisation).

To identify the measurable outcomes and the main evaluation criteria of the transport systems, some new specific terms related to European conformity have to be determined. One of the most important requirements, in the prespective of the integration of the Central and Eastern European countries in the European Union, is the interoperability of the transport systems. Interoperability is not a one-off, absolute state, rather it is a dynamic set of circumstances, which change through time. It can be defined as the ability of two, or more transport systems to operate effectively and efficiency together to fulfill consumer's requirements of a transport system. Different dimensions of interoperability can be distinguished:

- technical interoperability is achieved when different transport systems are linked in ways which effectively and efficiently extend the network of services;
- interconnectivity is achieved when different transport systems, of either the same or different modes, are physically and operationally linked to facilitate transfers across the boundaries between different systems. It necessitates the completion of missing links in the physical infrastructure, information, ticketing systems and transport services;

- organisational introperability occurs when different organisations are willing and able to cooperate to provide transport services;
- juridical interoperability means the harmonisation between European and national legislation;
- full interoperability is achieved when transport systems are deemed to have technical, organisational and juridical interoperability.

Policies to overcome impediments to interoperability may be analysed with reference to the general model of system organisation; especially the technological, corporate and regulatory policy frameworks. The delivery of transport service leads to a set of measurable outcomes from the activities of any transport system. The workings of any transport system are conducted in the context of a number of policy frameworks. The elements and outcomes are described and analysed only in so far as they act as impediments to the interoperability. The defined elements are neither mode- nor function-specific in order that they are effective as a model for the definition of transport systems in general. The nature and mix of the elements vary between different mode- and function-specific transport systems, and between different countries within the European Union. These elements are key aspects of system organisation.

As an example, the impediments and the measurable outputs to interoperability in the element's of an inter urban bus and coach system are presented below:

The example presented above illustrates the difficulties of the full and precise selection and identification of the evaluation criteria with reference to the transport infrastructure development.

2.3 Application of a multicriteria evaluation procedure in a simple example

Given five different development alternatives A_1, A_2, A_3, A_4, A_5 evaluated according to four different criterion C_1, C_2, C_3, C_4. The weight of the criterion i is w_i, where

$$\sum_{i=1}^{m} w_i = 1 \quad w_i \geq 0 \quad \forall i$$

The scores for a 5-categories evaluation process are given below:

The qualified values (V) of the five alternatives according to the four criteria and the scores (S) assigned to the class of evaluation are given in the next table:

Indicators of preference (c_{ij}) and disqualification (d_{ij}) are computed

$$c_{ij}\% = \sum_{\substack{i \to j \\ i \leftrightarrow j}} w_i 100 \quad \text{and} \quad d_{ij}\% = \frac{\max_i |h_j - h_i|}{H} 100$$

Elements within a common framework of system organisation	Impediment to operability in an inter urban and coach system
infrastructure network	motorways and trunk roads
	secondary and minor roads
	bridges and tunnels
interface	stations and interchanges
ancillary	breakdown and recovery services
	refueling facilities
	garaging and parking
rolling stock vehicles	size of vehicles
	number of seats per vehicles
	maximum weight of vehicles
	performance
fuels	availability of fuel
	price of fuel
	quality of fuel
human capital, levels of skills, training, working conditions	driver training
	route knowledge
	drivers hours
	shift patterns
information system, telematics	through ticketing
	reservation systems
	service control systems
	passenger information
finance, prices, fares, tariffs, revenue, subsidisation	investment on infrastructure
	vehicles (new)
	expenditure on maintenance
	fare box revenue

Measurable output in a general transport system	Impediment to interoperability in an interurban bus and coach system
availability of transport services	proportion of motorway standard roads
capacity of transport services	speed limits
	restrictions in the size of vehicles
	capacity of stations
distribution of services	extent of the network
safety	maximum speed
	requirement for seatbelts
	lane-restrictions on motor ways
quality of service	comfort of vehicles
	frequency of service
environmental impacts	limits on exhaust emissions

	C_1 $w_1 = 0,4$	C_2 $w_2 = 0,3$	C_3 $w_3 = 0,2$	C_4 $w_4 = 0,1$
very good(vg)	90	80	70	60
good(g)	70	65	60	55
medium (m)	50	50	50	50
satisfactory(s)	30	35	40	45
bad (b)	10	20	30	40

	C_1 $w_1 = 0,4$		C_2 $w_2 = 0,3$		C_3 $w_3 = 0,2$		C_4 $w_4 = 0,1$	
	V	S	V	S	V	S	V	S
A_1	vg	90	m	50	g	60	m	50
A_2	m	50	vg	80	s	40	v	60
A_3	s	30	g	65	vg	70	m	40
A_4	g	70	s	35	m	50	m	50
A_5	b	10	m	50	g	60	m	50

where

$$\max_i |h_j - h_i|$$

is the maximal difference of scores between the alternatives j and i, considering all criteria and $H = 90 - 10 = 80 =$ constant. Their values are given by the following table.

	A_1	A_2	A_3	A_4	A_5
A_1	-	$c_{12} = 60\%$ $d_{12} = 37.5\%$	$c_{13} = 50\%$ $d_{13} = 18.7\%$	$c_{14} = 100\%$ $d_{14} = 0\%$	$c_{15} = 100\%$ $d_{15} = 0\%$
A_2	$c_{21} = 40\%$ $d_{21} = 50\%$	-	$c_{23} = 80\%$ $d_{23} = 37.5\%$	$c_{24} = 40\%$ $d_{24} = 25\%$	$c_{25} = 80\%$ $d_{25} = 25\%$
A_3	$c_{31} = 50\%$ $d_{31} = 75\%$	$c_{32} = 20\%$ $d_{32} = 25\%$	-	$c_{34} = 50\%$ $d_{34} = 50\%$	$c_{35} = 90\%$ $d_{35} = 12.5\%$
A_4	$c_{41} = 10\%$ $d_{41} = 25\%$	$c_{42} = 60\%$ $d_{42} = 56.2\%$	$c_{43} = 50\%$ $d_{43} = 37.5\%$	-	$c_{45} = 50\%$ $d_{45} = 18.7\%$
A_5	$c_{51} = 60\%$ $d_{51} = 100\%$	$c_{52} = 20\%$ $d_{52} = 50\%$	$c_{53} = 10\%$ $d_{53} = 25\%$	$c_{54} = 60\%$ $d_{54} = 75\%$	-

The assortation graph for ranking the alternatives can be calculated starting from 100 % value of preference (P) and 0 % value of disqualification (Q):

$$\frac{P = 100\%}{c_{14} = c_{15} = 100\%} \quad \frac{Q = 0\%}{d_{14} = d_{15} = 0\%}$$

At this level the order of the alternatives can be identified between $A_1 \to A_4$ and $A_1 \to A_5$.

Decreasing the level of preference to the next discrete value of $c_{ij}(P = 90\%)$ and increasing the level of disqualification to the next discrete level of $d_{ij}(Q = 12.5\%)$ gives

$$\frac{P = 90\%}{c_{35} = 90\%} \quad \frac{Q = 12.5\%}{d_{35} = 12.5\%}$$

and the order of the alternatives can be identified between $A_3 \rightarrow A_5$. The next level for P and Q can be chosen like below:

$$\frac{P = 80\%}{c_{25} = 80\%} \quad \frac{Q = 25\%}{d_{25} = 25\%}$$

where the order of the alternatives can be identified between $A_2 \rightarrow A_5$, then following with

$$\frac{P = 80\%}{c_{23} = 80\%} \quad \frac{Q = 37.5\%}{d_{23} = 37.5\%}$$

the order of the alternatives can be identified between $A_2 \rightarrow A_3$. To stop the ranking procedure at preference level of 60 % and disqualification level of 40%:

$$\frac{P \geq 60\%}{c_{12} = 60\%} \quad \frac{Q \leq 40\%}{d_{12} = 37.5\%}$$

the order of the alternatives can be identified between $A_1 \rightarrow A_2$. The final assortation graph can be drawn as follows.

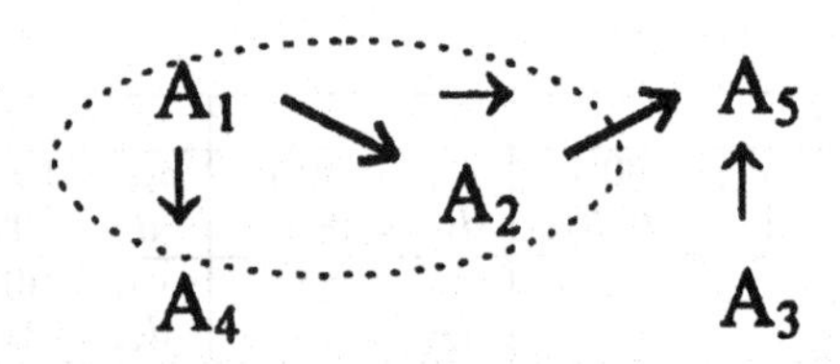

This graph means that only A_1 and A_2 alternatives are included in the final rank with the order: first is A_1 the second is A_2. Others can not be ranked at this level of preference and disqualification.

3. Group decision support system

3.1 From problem modelling to group ranking

As it was mentioned earlier the procedure of selection and ranking of the transport infrastructure development projects requires such a decision support tool which is able to consider not only multi attributes but also handle the interests of the different groups of the society. In this type of decision problems the decision makers from different fields (governmental or local authorities, traffic engineers, financial experts, ecologists, representatives of the travellers and citizens ...) with a common interest have the task of ranking certain transport infrastructure development projects that have been previously given and characterized by a finite set of criteria or attributes. The frame of a flexible and complex group decision support system for personal computers in Microsoft Windows environment which has been developed in Hungary (Csaki et al, 1995 [3]) will be presented as a tool meeting these requirements (Martel and Kiss, 1994 [7]).

The WINGDSS provides a final evaluation for every alternative ensuring a ranking according to the final scores. Moreover, the module for sensitivity analysis assists to achieve the desired ranking of the alternatives. The general structure of the group decision support system can be described as follows. First the hierarchy of criteria is determined with a gradual decomposition, resulting in a criterion tree (the nondecomposable criteria are the leaves of the tree).

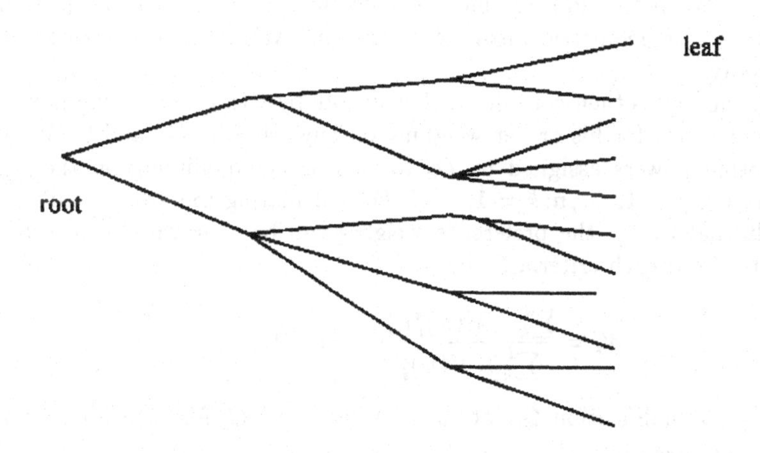

The second step is to define functions (procedures) to evaluate an alternative at each leaf criterion; the result is the score belonging to a leaf criterion. After the identification of all the decision makers, they first express the importance of any leaf criterion by assigning preference weights to them. Then starting from the lowest level of criterion tree, the combination of the preference weights and scores at the leaves results in scores at higher level nodes. The calculation proceeds toward the root where the final score to each alternative is determined. During this process, it is possible to assess voting powers to each decision maker at each criterion for weighing and qualifying.

3.2 The mathematical model

Let denote by $w_i^k \geq 0$ the weight assigned by the k-th decision maker to the i-th criterion, by $A_1, \ldots, A_n$ the n alternatives, by $C_1, \ldots, C_m$ the m criteria and by $D_1, \ldots, D_l$ the l groupmembers, i.e. decision makers. The procedure then includes the following steps:

- Individual ranking.
  The value a_{ij}^k given by the k-th decision maker, D_k, for alternative A_j, on the criterion C_i is determined.
- The normalised linear combination is calculated at each simple subtree N'

$$\mu_j^k = \frac{\sum_{i \in N''} w_i^k a_{ij}^k}{\sum_{i \in N'} w_i^k} \qquad j = 1, \ldots, n; \; k = 1, \ldots, l$$

- Proceeding on the tree toward the root, weights on the higher level criteria are combined with values obtained from one level below.
- The final individual score by the k-th decision maker, D_k for A_j will be the value assigned to the root, and the alternative will be ranked in descending order. (This additive multi-attribute model is applicable only to decision problems when the additive independence of the criteria can be proved. Arithmetic or geometric mean of scores and weights can be calculated optionally.)
- Group ranking is then considered. Let denote by $V(w)_i^k$ the voting powers assigned to D_k for his or her weighing on any criterion C_i, and by $V(q)_i^k$ the voting powers assigned to D_k for his or her qualifying on the leaf criteria C_i, $j = 1, \ldots, n; k = 1, \ldots, l$. For calculating the group utility on the alternative A_j, the preference weights will be aggregated into group weights W_i at each criteria C_i by

$$W_i = \frac{\sum_{k=1}^{l} V(w)_i^k w_i^k}{\sum_{k=1}^{l} V(w)_i^k} \qquad i = 1, \ldots, m.$$

  The group qualification Q_{ij} at each leaf criterion C_i for each alternative A_j is then given by

$$Q_{ij} = \frac{\sum_{k=1}^{l} V(q)_i^k a_{ij}^k}{\sum_{k=1}^{l} V(q)_i^k} \qquad i \in N'; \; j = 1, \ldots, n.$$

  The group utility of A_j is the result of the normalised linear combination of the aggregated qualification values with the aggregated weights, that is

$$U_j = \frac{\sum_i W_i Q_{ij}}{\sum_i W_i} \qquad j = 1, \ldots, n$$

  The best group alternative is the one associated with the highest group utility.

4. Multiple criteria ranking of transport infrastructure investment projects with budget constraints

4.1 Extension of multiple criteria decision aid

In countries with economies in transition, the infrastructure development of transportation is, for evident reasons, one of the most important preoccupations. But only limited financial sources are available for the construction of these projects. Government authorities, in collaboration with professional institutions, different political and civil organisations must therefore first define basic needs and constraints at the regional and national levels so as to arrive at a list of economically viable projects.

The identified transport infrastructure projects, i.e. those with a potential to increase welfare, have to undergo the classic feasibility analysis (Tanczos, 1985 [12] and Monigl, 1993 [8]). Following this analysis, a preliminary ranking is done under a more or less large cardinal set of relevant criteria and other constraints, within the framework of multiple criteria decision process (Brans and Vincke, 1985 [2]).

During the ranking process of a finite set of transport infrastructure investment projects (TIIP) subject to a preference relation pair set (PRS) generated within the framework of multicriterion analysis process (Kiss and Martel, 1994 [5]), a quasi-equal financial expenses' rhythm-constraint has to be taken into consideration with respect to the budget constraint of the countries being in transition. An extension of multiple criteria decision aid (Kiss and Tanczos, 1995 [6]) can assist to solve this problem.

4.2 Multicriteria ranking with budget constraint

The ranking problem of transport infrastructure investment projects (TIIPs) can be structured around the fundamental elements of ELECTRE II (Roy and Bertier, 1973 [9]). Let us denote by $Y = (Y_1, \ldots, Y_i, \ldots, Y_m)$ a family of criteria with regard to which each alternative (transport infrastructure investment project) is evaluated. A coherent set of criteria (not discussed in details here) must contain at least the following aggregated attributes:

- social impacts,
- traffic impacts,
- economic impacts,
- technical impacts,
- environmental impacts,
- security/safety,
- financing considerations (including the investment costs).

Let us also denote by $(X_1, \ldots, X_j, \ldots, X_n)$ a finite set of alternative projects to be ranked (i.e. the TIIPs set). An inter-ministerial committee, responsible for the quasi-equal spending of the limited financial sources for the most efficient development of transport infrastructure, continuously registers the incoming project proposals in a computerised information system. This list of projects is updated in every half year and only includes proposals after a careful pre-selection based on feasibility studies. Let us denote by $M_{ij} = \{M_{ij} = Y_i(X_j); i = 1, \ldots, m; j = 1, \ldots, n\}$ a performance matrix of the alternatives evaluated objectively and/or subjectively, according to each of the criteria. Finally, let $w_i = (w_1, \ldots, w_i, \ldots, w_m)$ be a set of weights (i.e. relative importance) associated with each criterion, where $\sum w_i = 1$ and $0 \leq w_i \leq 1, i = 1, \ldots, m$.

We also consider the preference-relation pair set (PRS) generated in the framework of a previously described multicriterion analysis process and eventually revised by the decision makers in the framework of pairwise comparisons in relation to the set of TIIP. When a PRS is used to obtain a ranking of TIIPs, it is crucially important to use a procedure to detect possible circuits in the oriented graph associated with this PRS. A simple and easy programmable detection procedure (Kiss, Martel and Nadeau, 1994 [5]) can be used for this purpose which is based on the fact that the equivalence class of a vertex X is defined by the intersections between the set of the vertexes which can be arrived at, starting from X (i.e. direct transitive closing) and the set of the vertexes from which X can be arrived at (i.e. inverse transitive closing). A circuit is detected if this intersection is not empty. By applying an aggregate decomposition procedure (Martel and Kiss, 1994 [7]) to the PRS, we obtain a partition of TIIP's elements as a n-part graph (i.e. direct decomposition). This aggregate decomposition algorithm is based on the following principle: convert a circuit-free oriented graph to an ordered multi-part graph. The transformation is done through a search and sequential separation of vertices whose external semi-degrees are zero. For various reasons the generated PRS is generally incomplete. From an incomplete PRS, many different hierarchical structures may be derived (and, conversely, several PRS may stem from a single hierarchical structure). It is precisely this non-mutual congruity and the many decomposition possibilities which are wished to be exploited by adding a quasi-equal financial expenses' rhythm-constraint to the generated PRS constraints.

In order to help the decision makers facing such a particular ranking problem, two decision support system oriented algorithms have been simultaneously developed (Kiss and Tanczos, 1995 [6]). The first is based on a heuristic computational approach, whereas the second represents a pure mathematical model based on a large scale linear programming approach. The obtained compromise hierarchical structure minimises the summarised value of the absolute deviations between the median value and the investment value of each project identified in the given hierarchical level providing quasi-equal spending for the transport infrastructure investment projects. The experimental results with real data exhibit the expected aspirations and suggest further possible applications, most probably, in employment policy and in achieving other macro-economic targets.

Taking into account the special circumstances of the countries in transition, not only the funding available for transport infrastructure investments is strictly limited and should be used in a quasi-equal schedule, but some special elements of the constructing capacity should have been used in a "smoothed" way as well. The most important components of these constructing capacity elements are the number of workers and the building machines with special functions. As regard the first one, the quasi-equal number of workers used during the projects' construction phases, is a very important requirement of the national economy to keep the unemployment average's level in the country below an acceptable value and to avoid the fluctuation of employed persons. The second type of capacity, which is important to use at a quasi-equal level is the kind of building machines which have special functions in the construction phase and which therefore may lead to bottlenecks. These machines must be utilised in quasi-equal level as well, due to the high costs involved.

These arguments stimulate our research effort to perfect the algorithm services for a flexible management of construction manpower and high value technological engines as well. This is an other important extension where the concept of creating compromise hierarchical project structures (aided by multicriterion analysis for transport investment decision support) can be adopted.

References

1. J.-P. Brans, B. Mareschal. The PROMCALC and GAIA decision support system for multicriteria decision aid. *Decision Support Systems*, 12:297–310, 1994.
2. J.-P. Brans, Ph. Vincke. A preference ranking organisation method. *Management Science*, 31:647–656, 1985.
3. P. Csáki, T. Rapcsák, P. Turchányi, M. Vermes. R and D for group decision aid in Hungary by WINGDSS, a Microsoft Windows based group decision support system. *Decision Support Systems*, 14:205–217, 1995.
4. P. Goodwin, G. Wright. Decision analysis for management judgement. John Wiley and Sons, New York, 1991.
5. L.N. Kiss, J.-M. Martel, R. Nadeau. ELECCALC - an interactive software for modelling the decision maker's preferences. *Decision Support Systems*, 12:311–326, 1994.
6. L.N. Kiss, K. Tanczos. Compromise Hierarchical structures in transport infrastructure investment. TIMS Conference, Singapore, 1995.
7. J.-M. Martel, L.N. Kiss. A Support to Consensus Reaching in Group Decision. *Group Decision and Negotiation*, 3:93–119, Kluwer Academic Publishers, 1994.
8. J. Monigl. Development of public transport system in Budapest (in Hungarian, with summary in English). *Urban Transport*, 5:261–276, Budapest, 1993.

9. B. Roy, P. Bertier. La methode ELECTRE II - Une application au media-planning. *Proceeding of O.R.'72*, M. Ross (Ed.), p. 291–302, North-Holland, 1973.
10. T.L. Saaty. The analytic hierarchy process. University of Pittsburgh, Pittsburgh, 1990.
11. J. Simons. Benefits of different transport modes. ECMT Round Table '93, Paris, 1994.
12. K. Tanczos. Multicriteria evaluation of transport system development (in Hungarian, with summary in English). Thesis, Hungarian Academy of Sciences, Budapest, 1985.

Recent Advances in Routing Algorithms

Gilbert Laporte

GERAD, Groupe d'études et de recherche en analyse des décisions,
3000 chemin de la Côte-Sainte-Catherine, Montréal, CANADA H3T 2A7
Centre de recherche sur les transports, Université de Montréal,
C.P. 6128, succursale Centre-ville, Montréal, CANADA H3C 3J7

Summary. In recent years, there have been several important algorithmic developments for the traveling salesman problem and the vehicle routing problem. These include some polyhedral results and related branch and cut algorithms, new relaxations, generalized insertion algorithms, and tabu search methods. Some of the most significant developments will be presented, together with indications on their computational value.

Key words: Traveling Salesman Problem, Vehicle routing problem.

1. Introduction

In recent years, there have been several important algorithmic developments for the *Traveling Salesman Problem* (TSP) and the *Vehicle Routing Problem* (VRP). The purpose of this article is to report on these advances. It extends an earlier survey by Laporte [54].

Both problems are defined on a graph $G = (V, A)$, where $V = \{v_1, \ldots, v_n\}$ is a vertex set representing *cities* or *customers*, and

$$A = \left\{(v_i, v_j) : i \neq j, v_i, v_j \in V\right\}$$

is an arc set. With every arc (v_i, v_j) is associated at non-negative cost c_{ij} representing a distance, a travel cost or a travel time. We will use these terms interchangeably. If $c_{ij} = c_{ji}$ for all $v_i, v_j \in V$, then $C = (c_{ij})$ and the problem are said to be *symmetrical*. In this case it is common to work with an edge set

$$E = \left\{(v_i, v_j) : i < j, v_i, v_j \in V\right\}$$

instead of A. If $c_{ik} + c_{kj} \geq c_{ij}$ for all $v_i, v_j, v_k \in V$, C is then said to satisfy the triangle inequality. The TSP consists of determining a minimum cost Hamiltonian circuit (if C is asymmetrical) or cycle (if C is symmetrical) on G. In the VRP, vertex v_1 represents a *depot* at which are based m identical vehicles. The value of m is either *fixed* at some constant or *bounded* above by $\overline{m}$. The VRP consists of designing a set of least cost vehicle routes in such a way that

1. every route starts and ends at the depot;
2. every city of $V \setminus \{v_1\}$ is visited exactly once by exactly one vehicle, and
3. some side constaints are satisfied.

We consider the following side constraints:

1. with every city is associated a non-negative demand q_i ($q_1 = 0$) and the total demand carried by any vehicle may not exceed the vehicle capacity Q;
2. every city v_i requires a non-negative service time δ_i ($\delta_1 = 0$).

The total length of any route (travel plus service times) may not exceed a preset upper bound L.

Both the TSP and the VRP play a central role in distribution planning and have been studied extensively over the past four decades. For the TSP, we refer to the book by Lawler, Lenstra, Rinnooy Kan and Shmoys [60], to Laporte [52] and to Jünger, Reinelt and Rinaldi [48]. For the VRP, see Magnanti [64], Bodin, Golden, Assad and Ball [9], Christofides [15], Golden and Assad [36, 37], Bodin [8], Laporte [53] and Fisher [26] . See also the recent bibliographies by Laporte and Osman [59] and by Laporte [55]. Our purpose is not to provide a full survey of these two problems, but rather to concentrate on some recent key results.

The remainder of this paper is organized as follows. Recent research on the design of exact TSP algorithms is extensive and we devote two sections to this topic: the asymmetrical and symmetrical cases are covered in Sections 2. and 3., respectively. Section 4. describes approximate algorithms for the TSP. Exact and approximate algorithms for the VRP are presented in Sections 5. and 6., respectively.

2. Exact algorithms for the asymmetrical traveling salesman problem

One of earliest references to the TSP is a paper by Menger [67], but the first formulation of this problem appears in Dantzig, Fulkerson and Johnson [20]. It defines a binary variable $x_{ij} (i \neq j)$ equal to 1 if and only if arc (v_i, v_j) is used in the optimal solution. This formulation is

$$\text{(TSP1) Minimize} \quad \sum_{i \neq j} c_{ij} x_{ij} \tag{2.1}$$

$$\text{subject to} \quad \sum_{j=1}^{n} x_{ij} = 1 \quad (i = 1, \ldots, n) \tag{2.2}$$

$$\sum_{i=1}^{n} x_{ij} = 1 \quad (j = 1, \ldots, n) \tag{2.3}$$

$$\sum_{v_i, v_j \in S} x_{ij} \leq |S| - 1 (S \subset V; 2 \leq |S| \leq n - 2) \tag{2.4}$$

$$x_{ij} \in \{0, 1\} \ (i, j = 1, \ldots, n; i \neq j) \, . \tag{2.5}$$

In this formulation, constraints (2.2) and (2.3) are degree constraints: they specify that every vertex is entered and left exactly once. Constraints (2.4) are *subtour elimination constraints*: together with (2.2)–(2.3), they prohibit the formation of tours on subsets of V containing less than n vertices. When the degree constraints are satisfied, constraints (2.4) are equivalent to the following *connectivity constraints* which force the presence of at least one arc between every non-empty subset of V and its complement:

$$\sum_{v_i \in S} \sum_{v_j \in V \setminus S} x_{ij} \leq 1 \ (S \subset V; 2 \leq |S| \leq n-2) \,. \tag{2.6}$$

Several alternative formulations have since been proposed, but none of these seems to possess a stronger linear relaxation than TSP1 (Langevin, Soumis and Desrosiers [51]; Padberg and Sung [85]).

When constraints (2.4) are relaxed, this formulation is that of a modified *Assignment Problem* (AP), i.e., an AP in which assignments on the main diagonal are prohibited. The AP can be solved in $O(n^3)$ time (see, e.g., Carpaneto, Martello and Toth [11]). Several authors have used the AP relaxation as a basis for an enumerative algorithm (see, e.g., Eastman [22]; Little, Murty, Sweeney and Karel [63]; Shapiro [97]; Murty [74]; Bellmore and Malone [6]; Garfinkel [27]; Smith, Srinivasan and Thompson [98]; Carpaneto and Toth [12]; Balas and Christofides [4]). The last two algorithms produce very good results. In the Carpaneto and Toth [12] algorithm, the problem solved at a generic node of the search tree is a modified AP with a subset of the variables fixed at 0 or at 1, and subproblems are created by branching on the arcs of subtours. This algorithm is able to consistently solve randomly generated problems with 240 vertices or less on a CDC 6600. Memory rather than time appears to be the main limitation. Balas and Christofides [4] use a stronger relaxation. In addition to subtour elimination constraints and connectivity constraints, these authors use some linear combinations of these two types of constraints which they introduce in the objective function in a Lagrangean fashion. Using this procedure, they solve randomly generated problems containing 325 vertices or less in less than one minute on a CDC 7600.

A rather interesting development has recently been proposed by Miller and Pekny [73]. In addition to relaxing constraints, these authors initially remove from consideration a large subset of the variables, thus speeding up the subproblem resolution. As in column generation techniques, relaxed variables are gradually reintroduced into the problem until optimality is proven. To describe this algorithm, consider the dual AP:

$$\text{(DAP)} \qquad \text{Maximize} \qquad \sum_{i=1}^{N} u_i + \sum_{j=1}^{n} v_j \tag{2.7}$$

$$\text{subject to} \qquad c_{ij} - u_i - v_j \geq 0 \ \ (i,j = 1,\ldots,n;\ i\ nej) \,. \tag{2.8}$$

Denote by $z^*(\text{RSP})$ the optimal TSP solution value, by $z^*(\text{AP})$ the optimal value of the modified AP linear relaxation, and by $z^*(\text{DAP})$ the optimal value of the dual modified assignment linear relaxation. Clearly $z^*(\text{AP}) = z^*(\text{DAP})$. Moreover, note that $z^*(\text{AP}) + (c_{ij} - u_i - v_j)$ is a lower bound on the cost of an AP solution that includes arc (v_i, v_j). Miller and Pekny make use of this in an algorithm that initially removes from consideration all x_{ij} variables whose cost c_{ij} exceeds a threshold value λ. Consider a modified problem TSP′ with associated linear assignment relaxation AP′ and its dual DAP′, obtained by redefining the costs c_{ij} as follows:

$$c'_{ij} = \begin{cases} c_{ij} & \text{if } c_{ij} \leq \lambda, \\ \infty & \text{otherwise.} \end{cases}$$

The authors prove the following proposition which they use as a basis for their algorithm: an optimal solution for TSP′ is optimal for TSP if

$$z^*(\text{TSP}') - z^*(\text{AP}) \leq \lambda + 1 - u'_i - v'_{\max} \tag{2.9}$$

and

$$\lambda + 1 - u'_i - v'_{\max} \geq 0 \tag{2.10}$$

for $i = 1, \dots, n$, where $u' = (u'_j)$ and $v' = (v'_j)$ are optimal solutions to DAP′, and $v'_{\max}$ is the maximum component of v'. The quantity $\lambda + 1 - u'_i - v'_{\max}$ underestimates the smallest reduced cost of any discarded variable. The algorithm is then:

Step 1. (Initialization). Choose λ.
Step 2. (TSP solution). Construct (c'_{ij}) and solve TSP′.
Step 3. (Termination check). If (2.9) and (2.10) hold, then $z^*(\text{TSP}') = z^*(\text{TSP})$: stop. Otherwise, double λ and go to step 2.

The authors report that if λ is suitably chosen in Step 1 (e.g., the largest arc cost in a heuristic solution), there is rarely any need to perform a second iteration. To solve TSP′, the authors have developed a branch and bound algorithm based on the AP relaxation. They have applied this procedure to randomly generated problems. Instances involving up to 5000 vertices were solved within 40 seconds on a Sun 4/330 computer. The largest problem reported solved by this approach contains 500,000 vertices and requires 12,623 seconds of computing time on a Cray 2 supercomputer.

Finally, it is worth mentioning that in different papers, Miller and Pekny [72] and Pekny and Miller [86] describe parallel branch and bound algorithms based on the AP relaxation. These authors report that randomly generated asymmetrical TSPs involving up to 3000 vertices have been solved to optimality using this approach.

Another family of algorithms is based on the *Shortest Spanning r-Arborescence Problem* (r-SAP), a relaxation of TSP1. In a directed graph $G = (V, A)$, an r-arborescence is a partial graph in which the in-degree of each vertex is exactly 1 and each vertex can be reached form a given root vertex r. The shortest spanning -arborescence problem can be formulated as

$$(r\text{-SAP}) \qquad \text{Minimize} \quad \sum_{i \neq j} c_{ij} x_{ij} \tag{2.11}$$

$$\text{subject to} \quad \sum_{\substack{i=1 \\ i \neq j}}^{n} x_{ij} = 1 \qquad (j = 1, \ldots, n) \tag{2.12}$$

$$\sum_{v_i \in S} \sum_{v_j \in V \setminus S} x_{ij} \leq 1 \quad (S \subset V; r \in S) \tag{2.13}$$

$$x_{ij} \geq 0 \qquad (i, j, = 1, \ldots, n; i \neq j) . \tag{2.14}$$

The problem of determining a minimum-cost r-arborescence on G can be decomposed into two independent subproblems: determining a minimum-cost arborescence rooted at vertex r, and finding the minimum-cost arc entering vertex r. The first problem is easily solved in $O(n^2)$ time (Tarjan [102]). This relaxation can be used in conjunction with Langrangean relaxation. However, on asymmetric problems, the AP relaxation would appear empirically superior to the r-arborescence relaxation (Balas and Toth [5]).

An early reference to this lower bound is made by Held and Karp [43]. More recently, Fischetti and Toth [24] have used it within an additive bounding procedure that combines five different bounds:

- the AP bound,
- the shortest spanning 1–arborescence bound,
- the shortest spanning 1–antiarborescence bound 1–SAAP; (r-SAAP is defined in a manner similar to r-SAP but now it is required that vertex should be reached from every remaining vertex),
- for $r = 1, \ldots, n$, a bound r-SADP obtained from r-SAP by adding the constraint

$$\sum_{j \neq r} x_{rj} = 1 , \tag{2.15}$$

- for $r = 1, \ldots, n$, a bound r-SAADP obtained from r-SAAP by adding the constraint

$$\sum_{i \neq r} x_{ir} = 1 . \tag{2.16}$$

The lower bounding procedure described by Fischetti and Toth was embedded within the Carpaneto and Toth [12] branch and bound algorithm on a variety of randomly generated problems and on some problems described in the literature. The success of the algorithm depends on the type of problem considered. For the easiest problem type, the authors report having solved 2000-vertex problems in an average time of 8329 seconds on an HP 9000/840 computer.

3. Exact algorithms for the symmetrical traveling salesman problem

In symmetrical problems, the number of variables contained in TSP1 can be halved by defining x_{ij} for $i < j$ only. In other words, $x_{ij} = 1$ if and only if *edge* (v_i, v_j) appears on the optimal tour. The formulation then becomes

$$\text{(TSP2)} \quad \text{Minimize} \quad \sum_{i<j} c_{ij} x_{ij} \tag{3.1}$$

$$\text{subject to} \quad \sum_{i<k} x_{ik} + \sum_{j>k} x_{kj} = 2 \quad (k = 1, \ldots, n) \tag{3.2}$$

$$\sum_{v_i, v_j \in S} x_{ij} \le |S| - 1 \;\; (S \subset V; 3 \le |V| \le n-3) \tag{3.3}$$

$$x_{ij} \in \{0, 1\} \quad (i, j = 1, \ldots, n; i < j) \tag{3.4}$$

In this formulation, constraints (3.2) ensure that each vertex has degree 2. Subtour elimination constraints (3.3) play the same role as (2.4) and, provided constraints (3.2) are satisfied, they are equivalent to the connectivity constraints

$$\sum_{\substack{v_i \in S, v_j \in V \setminus S \\ \text{or } v_j \in S, v_i \ in V \setminus S}} x_{ij} \le 2 \quad (S \subset V; 3 \le |S| \le n-3) \tag{3.5}$$

Two classical relaxations can be extracted from TSP2. The first exploits the fact that the cost of an optimal TSP tour cannot be less than that of a shortest 1-spanning tree, i.e., a spanning tree having a vertex set $V \setminus \{v_1\}$, together with two edges incident to v_1. Several algorithms are based on this relaxation first proposed by Christofides [14]. Using a branch and bound algorithm which embeds this bound, Held and Karp [44] have solved a number of classical TSPs with very limited branching. Helbig-Hansen and Krarup [42] as well as Smith and Thompson [99] have improved upon the original algorithm by a more judicious choice of parameters in the ascent procedure. Volgenant and Jonker [104] have experimented with a new ascent procedure, upper bound computations in the branch and bound tree, and new branching schemes. Gavish and Srikanth [28] use fast sensitivity analysis techniques to increase the underlying graph sparsity and reduce the problem size. More recently, Carpaneto, Fischetti and Toth [10] have suggested a number of further improvements to the Held and Karp lower bound by making use of additive bounding procedures.

We are mostly interested here in the second relaxation which solves the 2-matching problem extracted from TSP2 by dropping constraints (3.3). Dantzig, Fulkerson and Johnson [20, 21] were the first to propose an algorithm based on this relaxation. Their method first solves a problem defined by (3.1), (3.2) and

$$0 \leq x_{ij} \leq 1 \quad (i, j = 1, \ldots, n; i < j) \tag{3.6}$$

Violated subtour elimination constraints are identified by inspection and incorporated into the general problem. Using the information provided by the cost of a feasible solution and the reduced costs of the non-basic variables, several variables can be eliminated throughout the process. With this approach, the authors solved to optimality a 42-city problem with very limited branching. Martin [66] used a similar approach except that he did not initially impose upper bounds on the variables and he generated cutting planes to eliminate fractional solutions. One drawback of his approach is that reoptimization must be executed from scratch every time a new constraint is introduced. It was Miliotis [69, 70] who, to our knowledge, developed the first completely automatic branch and cut procedure for solving symmetrical TSPs using this relaxation. He proposed several algorithms that differ in the order in which the violated constraints are introduced and in the way integer solutions are reached. Miliotis tested an algorithm in which violated subtour elimination constraints are generated even at fractional solutions. This idea is used in the seminal work of Dantzig, Fulkerson and Johnson [20, 21], but not in Martin's paper. Land [50] proposed a cutting plane algorithm incorporating some of the ideas put forward by Miliotis, as well as a number of other features:

1. the algorithm works with a subset of variables; new variables are added periodically as the need arises;
2. integrality is gradually regained using Gomory's [39] cutting planes;
3. in addition to subtour elimination constraints, 2-*matching constraints* are also used at fractional solutions. These are constraints of the form

$$\sum_{\ell=0}^{s} \sum_{v_i, v_j \in H_\ell} x_{ij} \leq |H_0| + \frac{1}{2}(s-1) \tag{3.7}$$

for all $H_0, \ldots, H_\ell \subset V$ satisfying $|H_\ell \cap H_0| = 1$ for $\ell = 1, \ldots, s$, $|H_\ell \cap H_{\ell'}| = 0$ for $1 \leq \ell \leq \ell' \leq s$, $s \geq 3$ and odd.

Using this approach, Land solved to optimality eleven 100-city problems in times ranging form 13.0 to 446.6 seconds on a CDC 7600. It was observed that problems in which the $c_{ij}s$ were randomly generated were much easier to solve than problems based on coordinates generated in a square.

This work was followed by that of Crowder and Padberg [19], Padberg and Hong [81], Padberg and Rinaldi [82, 83, 84] and Grötschel and Holland [40], among others. The basic idea behind this line of research consists of introducing several types of valid inequalities before branching on fractional variables (on this subject, see Grötschel and Padberg [41] and Padberg and Grötschel [80]). The effect of this is to increase the value of the LP relaxation and thus to limit the growth of the search tree. In addition to subtour elimination constraints, connectivity constraints and 2-matching constraints, these authors also make use of *comb inequalities* and *clique tree inequalities* which are generalizations of 2-matching inequalities. Padberg and Rinaldi [83] and Grötschel and Holland [40] propose various heuristic and exact procedures for generating violated subtour elimination constraints and 2-matching inequalities. Only heuristic approaches are known for the generation of comb and clique tree inequalities (Grötschel and Holland [40]). These authors also derive conditions under which the various inequalities are facets of the polytope of the convex hull of feasible solutions. Other types of valid inequalities have since been proposed (see, e.g., Naddef and Rinaldi [75]), but no algorithmic experience with these constraints has yet been reported. Padberg and Rinaldi [82, 83, 84] and Grötschel and Holland [40] have developed branch and cut algorithms that have been applied to a large number of instances varying from 17 to 2392 vertices. In some cases, the c_{ij}s were randomly generated. Other problems were derived from geographical or manufacturing contexts. In several cases, an optimal solution was found without branching. More recently, Applegate, Bixby, Chvátal and Cook [1] developed an alternative branch and cut method that enabled them to solve to optimality a 7397-vertex problem.

4. Approximate algorithms for the traveling salesman problem

Over the years, several heuristics having good empirical performance have been proposed for the TSP. These can be broadly classified into *tour construction procedures* which gradually build a solution by adding a vertex at each step, and *tour improvement procedures* which improve upon a feasible solution by performing various exchanges. The best methods are *composite procedures* which combine these two features. For further references and surveys on this subject, see Rosenkrantz, Stearns and Lewis [96], Golden and Stewart [38], Ong and Huang [76] and Bentley [7].

Recently, there have been some interesting developments in the areas of improvement and composite procedures. Perhaps the most famous tour improvement method is the r-opt algorithm proposed by Lin [61]. Here, r arcs on edges are removed from the tour, all possible reinsertions are attempted and the best is implemented. This operation is repeated until no further improvement is possible. Since the complexity of each step of this algorithm is $O(n^r)$, the value of r is generally taken as 2 or 3. Lin and Kernighan [62] have proposed an improvement to this method: the value of r is modified dynamically throughout the algorithm, but this procedure is generally more difficult to code than the original Lin r-opt method. Or [77] has proposed a simplified $O(n^2)$ exchange scheme that produces results nearly as good as 3-*opt* (see Golden and Stewart [38]). The *Or-opt* approach first removes strings of 3 consecutive vertices and considers all possible reinsertions. When no further improvement is possible, strings of two vertices are removed and reinserted and finally, this procedure is executed by taking only one vertex at a time. Another method in the same category is the 4-*opt** restricted exchange scheme introduced by Renaud, Boctor and Laporte [94]. This improvement scheme considers only eight of the forty-eight potential 4-*opt* moves associated with a given TSP tour. Tests show that solutions obtained with 4-*opt** are never more than 2% worse than those obtained with 4–*opt*, while the computational effort is reduced by about 75%.

Johnson [46] has developed a "randomly iterated Lin-Kernighan method". Using sophisticated data structures, restricted neighbourhoods and efficient programming techniques, this author has solved large scale problems of up to about 100,000 vertices within 2% of optimality. More recent computational results and extensive comparisons with other algorithms are reported in Johnson and McGeoch [47]. Another interesting simplification of the Lin-Kernighan heuristic was recently proposed by Mak and Morton [65]. These authors propose a streamlining of the Lin-Kernighan search procedure, resulting in a much simplified code that requires less than 100 lines in the C language. The running times of this heuristic are about 100 times faster than those of Lin and Kernighan. Using problems of the TSP library (Reinelt [92]), the method finds solution values that are in most cases within 4% of optimality.

Reinelt [93] proposes a fast composite heuristic well suited to solving in real time large scale industrial TSPs arising, for example, in the production of printed circuit boards. The construction phase uses a greedy procedure where the set of candidate neighbours is given by a Delaunay graph. The improvement phase is based on a modified version of the Lin-Kernighan heuristic which considers a restricted family of moves. In terms of solution quality, this algorithm is not the best available, but its low execution time makes it appealing. Another interesting contribution is that of Bentley [7]. Using appropriate data structures and analytical techniques, this author proposes efficient implementations of twelve construction heuristics and three improvement procedures. On uniform data, these heuristics have an empirical running time of $O(n \log n)$.

Gendreau, Hertz and Laporte [29] have developed an improved two-phase heuristic for the TSP. It uses a generalized insertion step (GENI) followed by a post-optimization procedure (US). Combining GENI and US yields GENIUS, a powerful algorithm that outperforms some of the best known TSP heuristics. At a general step GENI, some vertices already belong to a partial tour while others are free. To perform a generalized insertion, consider the partially constructed tour $(v_1, v_2, \ldots, v_{t-1}, v_t, v_{t+1}, \ldots, v_h, v_1)$ with a given orientation. For any vertex v, define $N_p(v)$ as the set of the p vertices closest to v already on the tour if $p \leq h$, or as the set of all vertices on the tour if $p > h$. Let P_{rs} be the set of vertices on the path from v_r to v_s for a given orientation of the tour. For a vertex v not yet on the tour, GENI considers two types of insertion:

Type I: Select vertices $v_i, v_j \in N_p(v)$ and $v_k \in N_p(v_{j+1}) \cap P_{ji}$. Delete arcs (v_i, v_{i+1}), (v_j, v_{j+1}) and (v_k, v_{k+1}); insert arcs (v_i, v), (v, v_j), (v_{i+1}, v_k) and (v_{j+1}, v_{k+1}).

Type II: Select vertices $v_i, v_j \in N_p(v)$, $v_k \in (v_{i+1}) \cap P_{ji} \setminus \{v_j, v_{j+1}\}$ and $v_\ell \in N_p(v_{j+1}) \cap P_{ij} \setminus \{v_i, v_{i+1}\}$. Delete arcs (v_i, v_{i+1}), $(v_{\ell-1}, v_\ell)$, (v_j, v_{j+1}) and (v_{k_1}, v_k); insert arcs (v_i, v), (v, v_j), (v_ℓ, v_{j+1}), $(v_{k-1}, v_{\ell-1})$ and (v_{i+1}, v_k).

To determine the best move, it is necessary to compute the cost of the tour corresponding to each insertion, to each orientation of the tour, and to each possible choice of v_i, v_j, v_k, v_ℓ. GENI can be executed in $O(np^4 + n^2)$ operations. In the post-optimization phase US, each vertex is in turn removed from the tour which is then reoptimized locally, using the reverse GENI operation, and the vertex is then reinserted in the tour using GENI. The procedure ends when it yields no further improvement. On randomly generated problems ranging from 100 to 500 vertices GENIUS procedures solutions believed to be close to optimality. It also yields optimal or near-optimal solutions for some of the TSPLIB problems (Reinelt [92]).

Another interesting composite algorithm is I^3, developed by Renaud, Boctor and Laporte [94]. It constructs an initial tour using a simple geometrical procedure, inserts the remaining vertices and applies 4-*opt** to the tour. I^3 dominates GENIUS for $p = 2$ and 3. When $p = 5$, it produces solutions of quality comparable to those of GENIUS for about one quarter of the computing time. When $p > 5$ GENIUS seems to be better than I^3.

Recently, Codenotti et al. [17] have proposed a perturbation heurist ic for the TSP, rooted in the work of Charon and Hudry [13] and of Storer, Wu and Vaccari [100]. The main idea is to escape from local optima by acting on the instance itself rather than on the solution. More precisely, given a local optimum s to an instance P, the method creates a perturbed instance P' and a one-to-one correspondance rule f between the vertices of P and those of P'. The solution $s' = f(s)$ is constructed, local search is applied to s' to yield a new local optimum s'' for P', and a new solution $t = f^{-1}(s'')$ is then derived for P. The authors consider two types of move to introduce perturbation: moving the vertices of P by a small amount in a random direction; removing vertices from P and reintroducing them near one of their closest neighbors. Tests performed on instances containing up to 100,000 vertices indicate that this method outperforms the best known heuristics, including the Lin and Kerninghan method.

5. Exact algorithms for the symmetric vehicle routing problem

Like the symmetric TSP, the symmetric VRP can be solved by branch and cut, using the two-index formulation proposed by Laporte, Nobert and Desrochers [57]. Here, x_{ij} represents the number of times a vehicle travels between v_i and v_j. Thus, $x_{1j} = 0$, 1 or 2, $x_{ij} = 0$ or 1, if $i > 1$. The formulation is

$$\text{(VRP)} \quad \text{Minimize} \sum_{i<j} c_{ij} x_{ij} \tag{5.1}$$

$$\text{subject to} \quad \sum_{j=2}^{n} x_{1j} = 2m \tag{5.2}$$

$$\sum_{i<k} x_{ik} + \sum_{j>k} x_{kj} = 2 \quad (v_k \in V \setminus \{v_1\}) \tag{5.3}$$

$$\sum_{v_i, v_j \in S} x_{ij} \leq |S| - V(S) \quad (S \subset V \setminus \{1\};\ 2 \leq |S| \leq n-2) \tag{5.4}$$

$$x_{1j} = 0, 1 \text{ or } 2 \quad (v_j \in V \setminus \{v_1\}) \tag{5.5}$$

$$x_{ij} = 0 \text{ or } 1 \quad (v_i, v_j \in V \setminus \{v_1\}) . \tag{5.6}$$

In this formulation, constraints (5.2) and (5.3) are degree constraints. Constraints (5.5) and (5.6) are integrality constraints. Constraints (5.4) are subtour elimination constraints preventing the creation of subtours disconnected from v_1. They also eliminate subtours containing v_1, but whose total demand exceeds Q or whose duration exceeds L. They can be explained as follows: $V(S)$ is a lower bound on the number of vehicles required to serve all vertices of S. Each vehicle visiting S generates at least two edges between S and $\overline{S}$, hence the constraint. Equivalent constraints are the connectivity constraints

$$\sum_{\substack{v_i \in S, v_j \in V \setminus S \\ \text{or } v_j \in S, v_i \in V \setminus S}} x_{ij} \geq 2V(S) \quad (S \subset V \setminus \{v_1\}; 2 \leq |S| \leq n-2) . \tag{5.7}$$

In the remainder of this section, we will concentrate on the capacitated VRP. In this case, the value of $V(S)$ in (5.4) and (5.7) is obtained by solving a bin packing with the demands of S and bins of size Q. Alternatively, $V(S)$ can be computed as

$$V(S) = \left\lceil \sum_{v_i \in S} q_i / Q \right\rceil . \tag{5.8}$$

This model can again be solved by branch and cut by first relaxing subtour elimination and integrality constraints. However, the quality of this relaxation is not as good as for the TSP and fewer valid inequalities are known for the

VRP. Using this approach, Laporte, Nobert and Desrochers [57] have solved to optimality several loosely constrained instances containing up to 60 vertices. In these instances, vehicles are filled at about 70% of their capacity. Laporte and Nobert [56] have introduced comb inequalities for the VRP. Let $H_0 \subseteq V \setminus \{v_1\}$ and $H_1 \subseteq V \setminus \{v_1\}$ be the handle and teeth of a comb, respectively, and let these sets satisfy $|H_\ell \setminus H_0| \geq 1$, $|H_\ell \cap H_0| \geq 1$ for $\ell = 1, \ldots, s$, and $|H_\ell \cap H_{\ell'}| = 0$ for $1 \leq \ell \leq \ell' \leq s$. Let s' be the number of teeth H_ℓ for which the relation

$$V(H_\ell) < V(H_\ell \setminus H_0) + V(H_\ell \cap H_0) \tag{5.9}$$

holds. Then the following comb inequalities are valid for the VRP:

$$\sum_{\ell=0}^{s} \sum_{v_i, v_j \in H_\ell} x_{ij} \leq |H_0| + \sum_{\ell=1}^{s} \Big(|H_\ell| - V(H_\ell)\Big) - \left\lceil \frac{1}{2} s' \right\rceil . \tag{5.10}$$

Laporte and Nobert exhibit a 12-vertex example for which none of the constraints (3.7) and (5.4) are violated, but for which (5.10) does not hold. Cornuéjols and Harche [18] have generalized constraints (5.10) to the case where $H_\ell \subseteq V$ for $\ell = 1, \ldots, s$ (i.e., the depot may belong to the comb) and have embedded these comb inequalities within a branch and cut algorithm. They report exact solutions for two examples of sizes 18 and 50.

Araque *et al.* [2] describe "multistar inequalities" for the case where $q_i = 1$ for $i = 2, \ldots, n$. Embedding these constraints within branch and cut yields a procedure that can solve 60-vertex instances to optimality consistently.

Augerat *et al.* [3] have introduced a new class of valid inequalities, called "hypotour inequalities", extending those of Grötschel and Padberg [41] for the TSP. Branch and cut was applied to thirteen instances $22 \leq n \leq 135$. Six of these ($n = 22, 23, 30, 33, 45, 101$) were solved to optimality at the root of the search tree. Three more ($n = 51, 72, 135$) were solved by branching.

More constraints were introduced by Hill [45] in his Ph.D. thesis. Using these, he solved to optimality two instances involving 44 and 71 vertices. Another instance ($n = 100$) was solved for the case where return routes to and from the depot are prohibited.

Fisher [25] used a formulation based on that proposed by Laporte, Nobert and Desrochers [57], except that he prohibits return trips and uses slightly stronger subtour elimination constraints. Two interesting features of his branch and cut algorithm are the use of shortest spanning trees for the detection of violated subtour elimination constraints, and the Lagrangean dualization of side constraints. The largest instance solved with this approach contains 100 vertices.

Finally, Miller [71] dualizes constraints (5.4) in a Lagrang ean fashion and works on the resulting "b-matching" relaxation which can be solved relatively efficiently. Using this approach, he solves several problems containing 51 vertices or less.

6. Heuristic algorithms for the vehicle routing problem

Heuristic algorithms for the VRP can broadly be classified into *constructive heuristics, two-phase algorithms, incomplete optimization algorithms* and *improvement methods* (see Christofides [15], and Gendreau, Hertz and Laporte [30]). A number of interesting developments have recently taken place in the area of improvement metaheuristics such as *simulated annealing, tabu search, genetic algorithms* and *neural networks* (see Pirlot [87], Reeves [90], Laporte and Osman [58]). Simulated annealing and tabu search are search schemes in which successive neighbours of a solution are examined and the objective is allowed to deteriorate in order to avoid local optima. Simulated annealing is based on an analogy with a material annealing process used in mechanics (Metropolis *et al.* [68]; Kirkpatrick, Gelatt and Vecchi [49]). It ensures that the probability of attaining a worse solution tends to zero as the number of iterations grows. Such a method was applied to the VRP by Osman [78, 79]. Tabu search was proposed by Glover (see Glov er [32, 33, 34]; Glover, Taillard and de Werra [35]). Here, successive neighbours of a solution are examined and the best is selected . To prevent cycling, solutions that were recently examined are forbidden and inserted in a constantly updated *tabu list*. Various implementations are possible depending on how neighbourhoods are defined, on various parameters used to govern the search, and on a host of other rules. In recent years, various tabu search algorithms for the VRP have been proposed by Willard [105], Pureza and Franca [89], Osman [78, 79], Taillard [101], Gendreau, Hertz and Laporte [30], Rego and Roucairol [91], Xu and Kelly [106]. Genetic algorithms are rooted in the work of Fogel, Owens and Walsh (1966) and of Holland (1975). They work on a population of solutions. At each step, a new population is derived from the preceeding one by combining some of its best elements, and discarding the worst. Two representative implementations of genetic algorithms for the VRP with time windows are GIDEON (Thangiah [103]) and GENEROUS (Potvin and Bengio [88]). Neural networks are a mechanism in which weights are gradually adjusted, according to a learning process, until a satisfactory solution is reached. They have been applied to the VRP with time windows by El Ghaziri [23].

By and large, tabu search heuristics are the most successful of these techniques as applied to the VRP (Gendreau, Laporte and Potvin [31]). We now summarize TABUROUTE, developed by Gendreau, Hertz and Laporte [30], and the tabu search algorithm of Taillard [101].

At a general step of TABUROUTE, consider the current solution and randomly select q vertices among a subset of $V \setminus \{v_1\}$. For each selected vertex, compute the cost of the solution obtained by removing it from its current route, and inserting it using GENI into another route containing one of its p closest neighbours. Perform the best non-tabu insertion. Whenever a vertex v is moved from route r to route s, its reinsertion into route r is tabu for the next θ iterations, where θ is randomly selected in some interval $[\theta_{\min}, \theta_{\max}]$.

TABUROUTE contains several other features.

1. Initially, $\sqrt{n}/2$ trial initial solutions are created and a limited se arch is conducted for each of them; the most promising solution is then selected as a starting point for the algorithm.
2. Route infeasibilities due to excess weight or excess length are allowed during the course of the search; the excess capacity and length are multiplied by two positive penalty factors α and β which are self-adjusted during the course of the algorithm.
3. Whenever the best potential move produces a better incumbent, it is implemented even if it is tabu.
4. *Diversification* is used: as the search progresses, vertices which have been moved often are given a lesser likelihood of being selected for reinsert ion into a different route.
5. *Intensification* is also used: when the main search is completed, an intensive search of short duration is carried out in order to improve upon the current best solution.
6. When the search is completed, each individual route is reoptimized using the US procedure.

TABUROUTE was compared with twelve alternative heuristics on the fourteen test problems described in Christofides, Mingozzi and Toth [16]. Results show that simulated annealing and tabu search perform much better than previous heuristics. Also, TABUROUTE produces a best known solution for 11 of the 14 problems.

Taillard's [101] tabu search implementation also uses random tabu durations and diversification. It defines neighbourhood solutions by allowing a vertex to be inserted into a different route, or by swapping the routes associated with two vertices. Insertions are performed in the standard way and feasibility is always maintained. Periodic route reoptimizations also take place. Taillard applies a decomposition process to the problem which allows parallel application of tabu search to the various subproblems. Periodically, vertices are moved an adjacent region during the search process. This algorithms has yielded best known solutions for all of the fourteen Christofides, Mingozzi and Toth [16]test problems.

Finally, Rochat and Taillard [95] report on a generic technique aime d at generating sets of good solutions in local search. It can be used in conjunction with most search techniques like simulated annealing and tabu search, for example. The basic idea is to produce several good solutions using local search, keep a list of the best solutions, and combine these to produce new good solutions. The combination process uses some random mechanisms. As new solutions are produced, old ones are discarded. In a sense, this technique is akin to genetic algorithms insofar as it acts on a population of solutions by combining several parent solutions to produce several offspring. The authors show that their method help enhance classical local search processes.

Acknowledgement. This research was supported in part by the Canadian Natural Sciences and Engineering Research Council under grant OGP0039682. This support is gratefully acknowledged.

References

1. Applegate, D., Bixby, R., Chvátal, Cook, W. (1994): Solving Traveling Salesman Problems. Presented at Fifth International Symposium on Mathematical Programming, Ann Arbor, MI.
2. Araque, J.R., Kudva, G., Morin, T.L., Pekny, J.F. (1994): A Branch-and-Cut Algorithm for Vehicle Routing Problems. Annals of Operations Research **50**, 37–59.
3. Augerat, P., Belenguer, J.M., Benavent, E., Corberán, A., Naddef, D., Rinaldi, G. (1995): Computational Results with a Branch and Cut Code for the Capacitated Vehicle Routing Problem. Working Paper.
4. Balas, E., Christofides, N. (1981): A Restricted Langrangean Approach to the Traveling Salesman Problem. Mathematical Programming **21**, 19–46.
5. Balas, E., Toth, P. (1985): Branch and Bound Methods, In: Lawler, E.L., Lenstra, J.K., Rinnooy Kan, A.H.G., Shmoys, D.B. (eds). The Traveling Salesman Problem. A Guided Tour of Combinatorial Optimization. Wiley, Chicester, pp. 361–401.
6. Bellmore, M., Malone, J.C. (1971): Pathology of Traveling-Salesman Subtour-Elimination Algorithnms. Operations Research **19**, 278–307.
7. Bentley, J.J. (1992): Fast Algorithms for Geometric Traveling Salesman Problems. ORSA Journal on Computing **4**, 387–411.
8. Bodin, L.D. (1990): Twenty Years of Routing and Scheduling. Operations Research **38**, 571–579.
9. Bodin, L.D., Golden, B.L., Assad, A.A., Ball, M.O. (1983): Routing and Scheduling of Vehicles and Crews. The State of the Art. Computers & Operations Research **10**, 69–211.
10. Carpaneto, G., Fischetti, M., Toth, P. (1989): New Lower Bounds for the Symmetric Travelling Salesman Problem. Mathematical Programming **45**, 233–254.
11. Carpaneto, G., Martello, S., Toth, P. (1988): Algorithms and Codes for the Assignment Problem. In: Simeone, B., Toth, P., Gallo, G., Maffioli, F., Pallottino, S. (eds). FORTRAN Codes for Network Optimization. Annals of Operations Research **13**, 193–223.
12. Carpaneto, G., Toth, P. (1980): Some New Branching and Bounding Criteria for the Asymmetric Travelling Salesman Problem. Management Science **26**, 736–743.
13. Charon, L., Hudry, O. (1993): The Noising Method: A New Method for Combinatorial Optimization. Operations Research Letters **14**, 133–137.
14. Christofides, N. (1970): The Shortest Hamiltonian Chain of a graph. SIAM Journal on Applied Mathematics **19**, 689–696.
15. Christofides, N. (1985): Vehicle Routing. In Lawler, E.L., Lenstra, J.K., Rinnooy Kan, A.H.G., Schmoys, D.B. (eds). The Traveling Salesman Problem. A Guided Tour of Combinatorial Optimization. Wiley, Chichester, pp. 431–448.
16. Christofides, N., Mingozzi, A., Toth, P., (1979): The Vehicle Routing Problem. In: Christofides, N., Mingozzi, A., Toth, P., Sandi, C. (eds). Combinatorial Optimization. Wiley, Chichester, pp. 315–338.

17. Codenotti, B., Manzini, G., Margara, L., Resta, G. (1996): Perturbation: An Efficient Technique for the Solution of Very Large Instances of the Euclidean TSP. INFORMS Journal on Computing **8**, 125–133.
18. Cornuéjols, G., Harche, F. (1993): Polyhedral Study of the Capacitated Vehicle Routing Problem. Mathematical Programming **60**, 21–52.
19. Crowder, H., Padberg, M.W. (1980): Solving Large-Scale Symmetric Travelling Salesman Problems to Optimality. Management Science **26**, 495–509.
20. Dantzig, G.B., Fulkerson, D.R., Johnson, S.M. (1954): Solution of a Large-Scale Traveling-Salesman Problem. Operations Research **2**, 393–410.
21. Dantzig, G.B., Fulkerson, D.R., Johnson, S.M. (1959): On a Linear-Programming Combinatorial Approach to the Traveling-Salesman Problem. Operations Research **7**, 58–66.
22. Eastman, W.L. (1958): Linear Programming with Pattern Constraints. Ph.D. Thesis. Harvard University, Cambridge, MA.
23. El Ghaziri, H. (1993): Algorithmes connexionistes pour l'optimisation combinatoire. Ph.D. Dissertation 1167, École Polytechnique Fédérale de Lausanne, Switzerland.
24. Fischetti, M., Toth, P. (1992): An Additive Bounding Procedure for the Asymmetric Travelling Salesman Problem. Mathematical Programming **53**, 173–197.
25. Fisher, M.L. (1994): Optimal Solution of Vehicle Routing Problems Using Minimum K-Trees. Operations Research **42**, 626–642.
26. Fisher, M.L. (1995): Vehicle Routing, In: Ball, M.O., Magnanti, T.L., Monma, C.L., Nemhauser, G.L. (eds). Network Routing, Handbooks in Operations Research and Management Science 8. North-Holland, Amsterdam, pp. 1–33.
27. Garfinkel, R.S. (1973): On Partitioning the Feasible Set in a Branch-and-Bound Algorithm for the Asymmetric Traveling-Salesman Problem. Operations Research **21**, 340–343.
28. Gavish, B., Srikanth, K.N. (1986): An Optimal Solution Method for Large-Scale Multiple Traveling Salesman Problems. Operations Research **34**, 698–717.
29. Gendreau, M., Hertz, A., Laporte, G. (1992): New Insertion and Post-Optimization Procedures for the Traveling Salesman Problem. Operations Research **40**, 1086–1094.
30. Gendreau, M., Hertz, A., Laporte, G. (1994): A Tabu Search Heuristic for the Vehicle Routing Problem. Management Science **40**, 1276–1290.
31. Gendreau, M., Laporte, G., Potvin, J.-Y. (1997): Vehicle Routing: Modern Heuristics, In: Aarts, E.H.L., Lenstra, J.K. (eds). Local Search in Combinatorial Optimization. Wiley, Chichester, pp. 311–336.
32. Glover, F. (1977): Heuristic for Integer Programming Using Surrogate Constraints. Decision Sciences **8**, 156–166.
33. Glover, F. (1989): Tabu Search, Part I. ORSA Journal of Computing **1**, 190–206.
34. Glover, F. (1990): Tabu Search, Part II. ORSA Journal of Computing **2**, 4–32.
35. Glover, F., Taillard, E., deWerra, D. (1993): A User's Guide to Tabu Search. Annals of Operations Research **41**, 3–28.
36. Golden, B.L., Assad, A.A. (1986): Perspectives on Vehicle Routing: Exciting New Developments. Operations Research **34**, 803–810.
37. Golden, B.L., Assad, A.A. (1988): Vehicle Routing: Methods and Studies. North-Holland, Amsterdam.
38. Golden, B.L., Stewart, W.R. Jr. (1985): Empirical Analysis of Heuristics. In: Lawler, E.L., Lenstra, J.K., Rinnooy Kan, A.H.G., Shmoys, D.B. (eds). The Traveling Salesman Problem. A Guided Tour of Combinatorial Optimization. Wiley, Chichester, pp. 207–249.

39. Gomory, R.E. (1963): An Algorithm for Integer Solutions to Linear Program. In: Graves, R.L., Wolfe, P. (eds). Recent Advances in Mathematical Programming. McGraw-Hill, New York, pp. 269–302.
40. Grötschel, M., Holland, O. (1991): Solution of Large-Scale Symmetric Travelling Salesman Problems. Mathematical Programming **51**, 141–202.
41. Grötschel, M., Padberg, M.W. (1985): Polyhedral Theory, In: Lawler, E.L., Lenstra, J.K., Rinnooy Kan, A.H.G., Shmoys, D.B. (eds). The Traveling Salesman Problem. A guided Tour of Combinatorial Optimization. Wiley, Chichester, pp. 251–305.
42. Helbig-Hansen, K., Krarup, J. (1974): Improvements of the Held-Karp Algorithm for the Symmetric Traveling-Salesman Problem. Mathematical Programming **7**, 87–96.
43. Held, M., Karp, R.M. (1970): The Traveling Salesman Problem and Minimum Spanning Trees. Operations Research **18**, 1138–1162.
44. Held, M., Karp, R.M. (1971): The Traveling-Salesman Problem and Minimum Spanning Trees. Part II. Mathematical Programming **1**, 6–25.
45. Hill, S.P. (1995): Branch-and-Cut Method for the Symmetric Capacitated Vehicle Routing Problem. Ph.D. Thesis, Curtin University of Technology, Australia.
46. Johnson, D.L. (1990): Local Optimization and the Traveling Salesman Problem, In: Paterson, E.M. (ed). Proceedings of the 17th International Colloquium on Automata, Languages and Programming. Lecture Notes in Computer Science. Springer-Verlag, Berlin, pp. 446–461.
47. Johnson, D.L., McGeoch, L.A. (1997): The Traveling Salesman Problem: A Case Study in Local Optimization. In: Aarts, E.H.L., Lenstra, J.K. (eds). Local Search in Combinatorial Optimization. Wiley, Chichester. Forthcoming.
48. Jünger, M., Reinelt, G., Rinaldi, G. (1995): The Traveling Salesman Problem, In: Ball,M.O., Magnanti, T.L., Monma, C.L., Nemhauser, G.L. (eds). Network Models, Handbooks in Operations Research and Management Science 7. North-Holland, Amsterdam, pp. 225–330.
49. Kirkpatrick, S., Gelatt, C.D. Jr., Vecchi, M.P. (1983): Optimization by Simulated Annealing. Science **220**, 671–680.
50. Land, A.H. (1979): The Solution of Some 100-City Travelling Salesman Problems. Working Paper. London School of Economics.
51. Langevin, A., Soumis, F., Desrosiers, J. (1990): Classification of Travelling Salesman Problem Formulations. Operations Research Letters **9**, 127–132.
52. Laporte, G. (1992a): The Traveling Salesman Problem: An Overview of Exact and Approximate Algorithms. European Journal of Operational Research **59**, 231–247.
53. Laporte, G. (1992b): The Vehicle Routing Problem: An Overview of Exact and Approximate Algorithms. European Journal of Operational Research **59**, 345–358.
54. Laporte, G. (1993): Recent Algorithmic Developments for the Traveling Salesman Problem and the Vehicle Routing Problem. Ricerca Operativa **25**, 5–27.
55. Laporte, G. (1997): Vehicle Routing, In: Dell'Amico, M., Maffioli, F., Martello, S. (eds). Annotated Bibliographies in Combinatorial Optimization. Wiley, Chichester. Forthcoming.
56. Laporte, G., Nobert, Y. (1984): Comb Inequalities for the Vehicle Routing Problem. Methods of Operations Research **51**, 271–276.
57. Laporte, G., Nobert, Y., Desrochers, M. (1985): Optimal Routing under Capacity and Distance Restrictions. Operations Research **33**, 1050–1073.
58. Laporte, G., Osman, I.H. (1996): Metaheuristics in Combinatorial Optimization. Baltzer, Amsterdam.

59. Laporte, G., Osman, I.H. (1995): Routing Problems: A Bibliography. Annals of Operations Research **61**, 227–262.
60. Lawler, E.L., Lenster, J.K., Rinnooy Kan, A.H.G., Shmoys, D.B. (1985): The Traveling Salesman Problem. A Guided Tour of Combinatorial Optimization. Wiley, Chichester.
61. Lin, S. (1965): Computer Solutions of the Traveling Salesman Problem. Bell System Computer Journal **44**, 2245–2269.
62. Lin, S., Kernighan, B.W. (1973): An Effective Heuristic Algorithm for the Traveling-Salesman Problem. Operations Research **21**, 498–516.
63. Little, J.D.C., Murty, K.G., Sweeney, D.W., Karel, C. (1963): An Algorithm for the Traveling Salesman Problem. Operations Research **11**, 972–989.
64. Magnanti, T.L. (1981): Combinatorial Optimization and Vehicle Fleet Planning: Perspectives and Prospects. Networks **11**, 179–214.
65. Mak, K.-T., Morton, A.J. (1993): A Modified Lin-Kernighan Traveling-Salesman Heuristic. Operations Research Letters **13**, 127–132.
66. Martin, G.T. (1966): Solving the Travelling Salesman Problem by Integer Programming. Working Paper. CEIR, New York.
67. Menger, K. (1932): Das Botenproblem, In: K. Menger (ed). Ergebnisse eines Mathematischen Kolloquiums 2, Teunber, Leipzig, 12.
68. Metropolis, N., Rosenbluth, A.W., Rosenbluth, M.N., Teller, A.H., Teller, E. (1953): Equation of State Calculations by Fast Computing Machines. Journal of Chemical Physics **21**, 1087–1091.
69. Miliotis, P. (1976): Integer Programming Approaches to the Travelling Salesman Problem. Mathematical Programming **10**, 367–378.
70. Miliotis, P. (1978): Using Cutting Planes to Solve the Symmetric Travelling Salesman Problem. Mathematical Programming **15**, 177–188.
71. Miller, D.L. (1995): A Matching Based Exact Algorithm for Capacitated Vehicle Routing Problems. ORSA Journal on Computing **7**, 1–9.
72. Miller, D.L., Pekny, J.F. (1989): Results from a Parallel Branch and Bound Algorithm for Solving Large Asymmetrical Traveling Salesman Problems. Operations Research Letters **8**, 129–135.
73. Miller, D.L. Pekny, J.F. (1991): Exact Solution of Large Asymmetric Traveling Salesman Problems. Science **251**, 754–761.
74. Murty, K.G. (1968): An Algorithm for Ranking all the Assignments in Order of Increasing Cost. Operations Research **16**, 682–687.
75. Naddef, D., Rinaldi, G. (1991): The Symmetric Traveling Salesman Polytope and its Graphical Relaxation: Composition of Valid Inequalities. Mathematical Programming **51**, 359–400.
76. Ong, H.L., Huang, H.C. (1989): Asymptotic Expected Performance of Some TSP Heuristics. European Journal of Operational Research **43**, 231–238.
77. Or, I. (1976): Traveling Salesman-Type Combinatorial Problems and their Relation to the Logistics of Regional Blood Banking. Ph.D. Dissertation, Northwestern University, Evanston, IL.
78. Osman, I.H. (1991): Metastrategy Simulated Annealing and Tabu Search Algorithms for Combinatorial Optimization Problems. Ph.D. Thesis, Imperial College, London.
79. Osman, I.H. (1993): Metastrategy Simulated Annealing and Tabu Search Algorithms for the Vehicle Routing Problem. Annals of Operations Research **41**, 421–451.
80. Padberg, M.W., Grötschel, M. (1985): Polyhedral Computations, In: Lawler, E.L., Lenstra, J.K., Rinnooy Kan, A.H.G., Shmoys, D.B. (eds). The Traveling Salesman Problem. A Guided Tour of Combinatorial Optimization. Wiley, Chichester, pp. 307–360.

81. Padberg, M.W., Hong, S. (1980): On the Symmetric Travelling Salesman Problem: A Computational Study. Mathematical Programming Study **12**, 78–107.
82. Padberg, M.W., Rinaldi, G. (1987): Optimization of a 532–City Symmetric Traveling Salesman Problem by Branch and Cut. Operations Research Letters **6**, 1–7.
83. Padberg, M.W., Rinaldi, G. (1990): Facet Identification for the Symmetric Traveling Salesman Problem. Mathematical Programming **47**, 219–257.
84. Padberg, M.W., Rinaldi, G. (1991): A Branch-and-Cut Algorithm for the Resolution of Large-Scale Symmetric Traveling Salesman Problems. SIAM Review **33**, 60–100.
85. Padberg, M.W., Sung, T.-Y. (1991): An Analytical Comparison of Different Formulations of the Travelling Salesman Problem. Mathematical Programming **52**, 315–357.
86. Pekny, J.F., Miller, D.L. (1992): A Parallel Branch and Bound Algorithm for Solving Large Asymmetric Traveling Salesman Problems. Mathematical Programming **55**, 17–33.
87. Pirlot, M. (1992): General Local Search Heuristics in Combinatorial Optimization: A Tutorial. Belgian Journal of Operations Research, Statistics and Computer Science **32**, 8–67.
88. Potvin, J.-Y., Bengio, S. (1996): The Vehicle Routing with Time Windows—Part II: Genetic Search. INFORMS Journal on Computing **8**, 165–172.
89. Pureza, V.M., Franca, P.M. (1991): Vehicle Routing Problems via Tabu Search Metaheuristic. Publication CRT–747, Centre de Recherche sur les Transports, Montreal.
90. Reeves, C.R. (1993): Modern Heuristic Techniques for Combinatorial Optimization. Blackwell, Oxford.
91. Rego, C., Roucairol, C. (1996): A Parallel Tabu Search Algorithm Using Ejection Chains for the Vehicle Routing Problem, In: Osman, I.H., Kelly, J.P. (eds). Metaheuristics: Theory & Applications. Kluwer, Norwell, MA, pp. 661–675.
92. Reinelt, G. (1991): TSPLIB — A Traveling Salesman Problem Library. ORSA Journal on Computing **3**, 376–384.
93. Reinelt, G. (1992): Fast Heuristics for Large Scale Geometric Traveling Salesman Problems. ORSA Journal on Computing **4**, 206–217.
94. Renaud, J., Boctor, F.F., Laporte, G. (1996): A Fast Composite Heuristic for the Symmetric Traveling Salesman Problem. INFORMS Journal on Computing **8**, 134–143.
95. Rochat, Y., Taillard, E.D. (1995): Probabilistic Diversification and Intensification in Local Search for Vehicle Routing. Journal of Heuristics **1**, 147–167.
96. Rosenkrantz, D.J., Stearns, R.E., Lewis, P.M. II (1977): An Analysis of Several Heuristics for the Traveling Salesman Problem. SIAM Journal on Computing **6**, 563–581.
97. Shapiro, D.M. (1966): Algorithms for the Solution of the Optimal Cost and Bottleneck Traveling Salesman Problems. Sc.D. Thesis. Washington University, St.Louis, MO.
98. Smith, T.H.C., Srinivasan, V., Thompson, G.L. (1977): Computational Performance of Three Subtour Elimination Algorithms for Solving Asymmetric Traveling Salesman Problems. Annals of Discrete Mathematics **1**, 495–506.
99. Smith, T.H.C., Thompson, G.L. (1977): A LIFO Implicit Enumeration Search Algorithm for the Symmetric Traveling Salesman Problem Using Held and Karp's 1–Tree Relaxation. Annals of Discrete Mathematics **1**, 479–493.
100. Storer, R.H., Wu, S.D., Vaccari, R. (1992): New Search Spaces for Sequencing Problems with Application to Job Chop Scheduli ng. Management Science **38**, 1495–1509.

101. Taillard, E. (1993): Parallel Iterative Search Methods for Vehicle Routing Problems. Networks **23**, 661–676.
102. Tarjan, R.E., (1977): Finding Optimum Branchings. Networks **7**, 25–35.
103. Thangiah, S.R. (1993): Vehicle Routing with Time Windows Using Genetic Algorithms. Technical Report SRU-SpSc-TR-93–23, Slippery Rock University, Slippery Rock, PA.
104. Volgenant, T., Jonker, R. (1982): A Branch and Bound Algorithm for the Symmetric Traveling Salesman Problem Based on the 1–Tree Relaxation. European Journal of Operational Research **9**, 83–89.
105. Willard, J.A.G. (1989): Vehicle Routing Using r-Optimal Tabu Search. M.Sc. Dissertation, The Management School, Imperial College, London.
106. Xu, J., Kelly, J.P. (1996): A Network Flow-Based Tabu Search Heuristic for the Vehicle Routing Problem. Transportation Science **30**, 379–393.

Discrete Choice Models

Michel Bierlaire

Intelligent Transportation Systems Program
Massachusetts Institute of Technology
E-mail: mbi@mit.edu
URL: http://web.mit.edu/mbi/www/michel.html

Summary. Discrete choice models have played an important role in transportation modeling for the last 25 years. They are namely used to provide a detailed representation of the complex aspects of transportation demand, based on strong theoretical justifications. Moreover, several packages and tools are available to help practionners using these models for real applications, making discrete choice models more and more popular.

Discrete choice models are powerful but complex. The art of finding the appropriate model for a particular application requires from the analyst both a close familiarity with the reality under interest and a strong understanding of the methodological and theoretical background of the model.

The main theoretical aspects of discrete choice models are reviewed in this paper. The main assumptions used to derive discrete choice models in general, and random utility models in particular, are covered in detail. The Multinomial Logit Model, the Nested Logit Model and the Generalized Extreme Value model are also discussed.

1. Introduction

In the context of transportation demand analysis, disaggregate models have played an important role these last 25 years. These models consider that the demand is the result of several decisions of each individual in the population under consideration. These decisions usually consist of a choice made among a finite set of alternatives. An example of sequence of choices in the context of transportation demand is described in Figure 1.1: choice of an activity (playyard), choice of destination (6th street), choice of departure time (early), choice of transportation mode (bike) and choice of itinerary (local streets). For this reason, discrete choice models have been extensively used in this context.

A model, as a simplified description of the reality, provides a better *understanding* of complex systems. Moreover, it allows for obtaining *prediction* of future states of the considered system, *controlling* or *influencing* its behavior and *optimizing* its performances.

The complex system under consideration here is a specific aspect of human behavior dedicated to choice decisions. The complexity of this "system" clearly requires many simplifying assumptions in order to obtain operational models. A specific model will correspond to a specific set of assumptions, and it is important from a practical point of view to be aware of these assumptions when prediction, control or optimization is performed.

> This morning, we have decided to go to the play-yard with the children. There are several of them in the city, but the one on 6th street has a very nice climbing structure that the children love. This place is very popular and is usually crowded at late morning. Therefore, we decide to depart early. The day is gorgeous and we prefer to bike to the play-yard. For safety reasons, we decide to avoid the main avenue, and follow a longer itinerary using local streets.

Fig. 1.1. A sequence of choices

The assumptions associated with discrete choice models in general are detailed in Section 2. Section 3 focuses specifically on assumptions related to random utility models. Some of the most used models, the Multinomial Logit Model (Section 4), the Nested Logit Model (Section 5) and the Generalized Extreme Value Model (Section 6), are then introduced, with special emphasis on the Nested Logit model.

Among the many publications that can be found in the literature, we refer the reader to Ben-Akiva and Lerman [1], Anderson, De Palma and Thisse [1], Hensher and Johnson [13] and Horowitz, Koppelman and Lerman [14] for more comprehensive developments.

2. Modeling assumptions

In order to develop models capturing how individuals are making choices, we have to make specific assumptions. We will distinguish here among assumptions about

1. the decision-maker: these assumptions define who is the decision-maker, and what are his/her characteristics;
2. the alternatives: these assumptions determine what are the possible options of the decision-maker;
3. the attributes: these assumptions identify the attributes of each potential alternative that the decision-maker is taking into account to make his/her decision;
4. the decision rules: they describe the process used by the decision-maker to reach his/her choice.

In order to narrow down the huge number of potential models, we will consider some of these assumptions as fixed throughout the paper. It does not mean that there is no other valid assumption, but we cannot cover everything in this context. For example, even if continuous models will be briefly described, discrete models will be the primary focus of this paper.

2.1 Decision-maker

As mentioned in the introduction, choice models are referred to as *disaggregate* models. It means that the decision-maker is assumed to be an *individual*. In general, for most practical applications, this assumption is not restrictive. The concept of "individual" may easily been extended, depending on the particular application. We may consider that a group of persons (a household or a government, for example) is the decision-maker. In doing so, we decide to ignore all internal decisions within the group, and to consider only the decision of the group as a whole. The example described in Figure 1.1 reflects the decisions of a household, without accounting for all potential negotiations among the parents and the children. We will refer to "decision-maker" and "individual" interchangeably throughout the rest of the paper.

Because of its disaggregate nature, the model has to include the characteristics, or attributes, of the individual. Many attributes, like age, gender, income, eyes color or social security number may be considered in the model[1].

The analyst has to identify those that are likely to explain the choice of the individual. There is no automatic process to perform this identification. The knowledge of the actual application and the data availability play an important role in this process.

2.2 Alternatives

Analyzing the choice of an individual requires the knowledge of what has been chosen, but also of what has *not* been chosen. Therefore, assumptions must be made about options, or *alternatives*, that were considered by the individual to perform the choice. The set containing these alternatives, called the *choice set*, must be characterized.

The characterization of the choice set depends on the context of the application. If we consider the example described in Figure 2.1, the time spent on each Internet site may be anything, as far as the total time is not more than two hours. The resulting choice set $\mathcal{C}$ is represented in Figure 2.2, and is defined by

$$\mathcal{C} = \{(t_{\text{Museum}}, t_{\text{News}}) \mid t_{\text{Museum}} + t_{\text{News}} \leq 2, t_{\text{Museum}} \geq 0, t_{\text{News}} \geq 0\} \qquad (2.1)$$

It is a typical example of a continuous choice set, where the alternatives are defined by some constraints and cannot be enumerated.

In this paper, we focus on discrete choice sets. A discrete choice set contains a finite number of alternatives that can be explicitly listed. The corresponding choice models are called *discrete choice models*. The choice of a transportation mode is a typical application leading to a discrete choice set.

[1] However, if you develop a choice model including eyes color or social security number, please let me know.

> **I have decided to spend a maximum of two hours on Internet, today. There are many interesting services available, but I really like to access the site of the Museum of Fine Arts, Boston, and to read the electronic version of a Belgian newspaper. How much time am I going to spend on each of these sites?**

Fig. 2.1. Choice on Internet

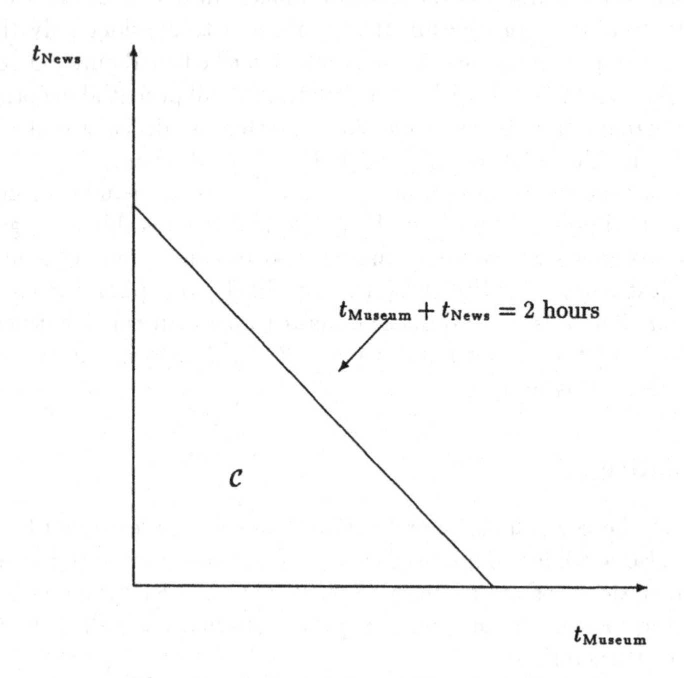

Fig. 2.2. Example of a continuous choice set

In this context, the characterization of the choice set consists in the identification of the list of alternatives. To perform this task, two concepts of choice set are considered: the *universal* choice set and the *reduced* choice set.

The universal choice set contains all potential alternatives in the context of the application. Considering the mode choice in the example of Figure 1.1, the universal choice set may contain all potential transportation modes, like `walk`, `bike`, `bus`, `car`, etc. The alternative `plane`, which is also a transportation mode, is clearly not an option in this context and, therefore, is not included in the universal choice set.

The reduced choice set is the subset of the universal choice set considered by a particular individual. Alternatives in the universal choice set that are not available to the individual under consideration are excluded (for example, the alternative `car` may not be an option for individuals without a driver license). The awareness of the availability of the alternative by the decision-maker should be considered as well. The reader is referred to Swait [21] for more details on choice set generation. In the following, "choice set" will refer to the reduced choice set, except when explicitly mentioned.

2.3 Attributes

Each alternative in the choice set must be characterized by a set of attributes. Similarly to the characterization of the decision-maker described in Section 2.1, the analyst has to identify the attributes of each alternative that are likely to affect the choice of the individual. In the context of a transportation mode choice, the list of attributes for the mode **car** could include the travel time, the out-of-pocket cost and the comfort. The list for **bus** could include the travel time, the out-of-pocket cost, the comfort and the bus frequency. Note that some attributes may be generic to all alternatives, and some may be specific to an alternative (bus frequency is specific to **bus**). Also, qualitative attributes, like comfort, may be considered.

An attribute is not necessarily a directly observed quantity. It can be any function of available data. For example, instead of considering travel time as an attribute, the logarithm of the travel time may be considered. The out-of-pocket cost may be replaced by the ratio between the out-of-pocket cost and the income of the individual. The definition of attributes as a function of available data depends on the problem. Several definitions must usually be tested to identify the most appropriate.

2.4 Decision rules

At this point, we have identified and characterized both the decision-maker and all available alternatives. We will now focus on the assumptions about the rules used by the decision-maker to come up with the actual choice. Different sets of assumptions can be considered, that leads to different family of models. We will describe here three theories on decision rules, and the corresponding models. The neoclassical economic theory, described in Section 2.4.1, introduces the concept of *utility*. The Luce model (Section 2.4.2) and the random utility models (introduced in Section 2.4.3 and developed in Section 3) are designed to capture uncertainty.

2.4.1 Neoclassical Economic Theory. The neoclassical economic theory assumes that each decision-maker is able to compare two alternatives a and b in the choice set C using a preference-indifference operator $\succeq$. If $a \succeq b$, the decision-maker either prefers a to b, or is indifferent. The preference-indifference operator is supposed to have the following properties:

1. Reflexivity:

$$a \succeq a, \quad \forall a \in C.$$

2. Transitivity:

$$a \succeq b \text{ and } b \succeq c \Rightarrow a \succeq c, \quad \forall a, b, c \in \mathcal{C}.$$

3. Comparability:

$$a \succeq b \text{ or } b \succeq a, \quad \forall a, b \in \mathcal{C}.$$

Because the choice set $\mathcal{C}$ is finite, the existence of an alternative which is preferred to all of them is guaranteed, that is

$$\exists a^* \text{ s.t. } a^* \succeq a \quad \forall a \in \mathcal{C}. \tag{2.2}$$

More interestingly, and because of the three properties listed above, it can be shown that the existence of a function

$$U : \mathcal{C} \longrightarrow \mathbf{R} : a \rightsquigarrow U(a) \tag{2.3}$$

such that

$$a \succeq b \Leftrightarrow U(a) \geq U(b) \quad \forall a, b \in \mathcal{C} \tag{2.4}$$

is guaranteed. Therefore, the alternative a^* defined in (2.2) may be identified as

$$a^* = \arg\max_{a \in \mathcal{C}} U(a). \tag{2.5}$$

It results that using the preference-indifference operator $\succeq$ to make a choice is equivalent to assigning a value, called *utility*, to each alternative, and selecting the alternative a^* associated with the highest utility.

The concept of utility associated with the alternatives plays an important role in the context of discrete choice models. However, the assumptions of neoclassical economic theory presents strong limitations for practical applications. Indeed, the complexity of human behavior suggests that a choice model should explicitly capture some level of uncertainty. The neoclassical economic theory fails to do so.

The exact source of uncertainty is an open question. Some models assume that the decision rules are intrinsically stochastic, and even a complete knowledge of the problem would not overcome the uncertainty. Others consider that the decision rules are deterministic, and motivate the uncertainty from the impossibility of the analyst to observe and capture all dimensions of the problem, due to its high complexity. Anderson et al. [1] compare this debate with the one between Einstein and Bohr, about the uncertainty principle in theoretical physics. Bohr argued for the intrinsic stochasticity of nature and Einstein claimed that "Nature does not play dice".

Two families of models can be derived, depending on the assumptions about the source of uncertainty. Models with stochastic decision rules, like the model proposed by Luce [16], described in Section 2.4.2, or the "elimination by aspects" approach, proposed by Tverski [22], assumes a deterministic utility and a probabilistic decision process. Random Utility Models, introduced in Section 2.4.3 and developed in Section 3, are based on the deterministic decision rules from the neoclassical economic theory, where uncertainty is captured by random variables representing utilities.

2.4.2 The Luce model. An important characteristic of models dealing with uncertainty is that, instead of identifying one alternative as the chosen option, they assign to each alternative a *probability* to be chosen.

Luce [16] proposed the *choice axiom* to characterize a choice probability law. The choice axiom can be stated as follow.

Denoting $\mathrm{P}_{\mathcal{C}}(a)$ the probability of choosing a in the choice set $\mathcal{C}$, and $\mathrm{P}_{\mathcal{C}}(\mathcal{S})$ the probability of choosing one element of the subset $\mathcal{S}$ within $\mathcal{C}$, the two following properties hold for any choice set $\mathcal{U}$, $\mathcal{C}$ and $\mathcal{S}$, such that $\mathcal{S} \subseteq \mathcal{C} \subseteq \mathcal{U}$.

1. If an alternative $a \in \mathcal{C}$ is dominated, that is if there exists $b \in \mathcal{C}$ such that b is always preferred to a or, equivalently, $\mathrm{P}_{\{a,b\}}(a) = 0$, then removing a from $\mathcal{C}$ does not modify the probability of any other alternative to be chosen, that is
$$\mathrm{P}_{\mathcal{C}}(\mathcal{S}) = \mathrm{P}_{\mathcal{C}\setminus\{a\}}(\mathcal{S} \setminus \{a\}). \tag{2.6}$$
2. If no alternative is dominated, that is if $0 < \mathrm{P}_{\{a,b\}}(a) < 1$ for all $a, b \in \mathcal{C}$, then the choice probability is independent from the sequence of decisions, that is
$$\mathrm{P}_{\mathcal{C}}(a) = \mathrm{P}_{\mathcal{C}}(\mathcal{S})\mathrm{P}_{\mathcal{S}}(a). \tag{2.7}$$

The independence described by (2.7) can be illustrated using a example of transportation mode choice, where we consider $\mathcal{C} = \{\text{Car, Bike, Bus}\}$. We apply two different assumptions to compute the probability of choosing "car" as a transportation mode.

1. The decision-maker may decide first to use a motorized mode (car or bus, in this case). The probability of choosing "car" is then given by
$$\mathrm{P}_{\{\text{car,bus,bike}\}}(\text{car}) = \mathrm{P}_{\{\text{car,bus,bike}\}}(\{\text{car,bus}\})\mathrm{P}_{\{\text{car,bus}\}}(\text{car}). \tag{2.8}$$
2. Alternatively, the decision-maker may decide first to use a private transportation mode (car or bike, in this case). The probability of choosing "car" is then given by
$$\mathrm{P}_{\{\text{car,bus,bike}\}}(\text{car}) = \mathrm{P}_{\{\text{car,bus,bike}\}}(\{\text{car,bike}\})\mathrm{P}_{\{\text{car,bike}\}}(\text{car}). \tag{2.9}$$

Equation (2.7) of the choice axiom imposes that both assumptions produce the same probability, that is

$$\begin{aligned} \mathrm{P}_{\{\text{car,bus,bike}\}}(\text{car}) &= \mathrm{P}_{\{\text{car,bus,bike}\}}(\{\text{car,bus}\})\mathrm{P}_{\{\text{car,bus}\}}(\text{car}) \\ &= \mathrm{P}_{\{\text{car,bus,bike}\}}(\{\text{car,bike}\})\mathrm{P}_{\{\text{car,bike}\}}(\text{car}). \end{aligned} \tag{2.10}$$

The second part of the choice axiom can be interpreted in a different way. Luce [16] has shown that (2.7) is a sufficient and necessary condition for the existence of a function $v : \mathcal{C} \longrightarrow \boldsymbol{R}$, such that, for all $\mathcal{S} \subseteq \mathcal{C}$, we have

$$\mathrm{P}_{\mathcal{S}}(a) = \frac{v(a)}{\sum_{b\in\mathcal{S}} v(b)}. \tag{2.11}$$

Also, function v is unique up to a proportionality factor. If there exists $v' : \mathcal{C} \longrightarrow \mathbf{R}$ verifying (2.11), then

$$v(a) = kv'(a), \forall a \in \mathcal{C}, \tag{2.12}$$

where $k \in \mathbf{R}$. Similarly to (2.3), $v(\cdot)$ may be interpreted as a utility function. We will elaborate more on this result in Section 4.

2.4.3 Random Utility Models. Random utility models assume, as neoclassical economic theory, that the decision-maker has a perfect discrimination capability. In this context, however, the analyst is supposed to have incomplete information and, therefore, uncertainty must be taken into account. Manski [18] identifies four different sources of uncertainty: unobserved alternative attributes, unobserved individual attributes (called "unobserved taste variations" by Manski [18]), measurement errors and proxy, or instrumental, variables.

The utility is modeled as a random variable in order to reflect this uncertainty. More specifically, the utility that individual i is associating with alternative a is given by

$$U_a^i = V_a^i + \varepsilon_a^i, \tag{2.13}$$

where V_a^i is the deterministic part of the utility, and ε_a^i is the stochastic part, capturing the uncertainty. Similarly to the neoclassical economic theory, the alternative with the highest utility is supposed to be chosen. Therefore, the probability that alternative a is chosen by decision-maker i within choice set $\mathcal{C}$ is

$$\mathrm{P}_{\mathcal{C}}^i(a) = \mathrm{P}\left[U_a^i = \max_{b \in \mathcal{C}} U_b^i\right]. \tag{2.14}$$

Random utility models are the most used discrete choice models for transportation applications. Therefore, the rest of the paper is devoted to them.

3. Random utility models

The derivation of random utility models is based on a specification of the utility as defined by (2.13). Different assumptions about the random term ε_a^i and the deterministic term V_a^i will produce specific models. We present here the most usual assumptions that are used in practice. In Section 3.1, common assumptions about the random part of the utility are discussed. The deterministic part is treated in Section 3.2.

We consider a choice between two alternatives, that is $\mathcal{C} = \{1, 2\}$. The probability for a given decision-maker to choose alternative 1 is

$$\begin{aligned} P_{\{1,2\}}(1) &= P[U_1 \geq U_2] \\ &= P[V_1 + \varepsilon_1 \geq V_2 + \varepsilon_2] \\ &= P[V_1 - V_2 \geq \varepsilon_2 - \varepsilon_1]. \end{aligned}$$

Fig. 3.1. A binary model

3.1 Assumptions on the random term

We will focus here on assumptions about the mean, the variance and the functional form of the random term.

For all practical purposes, the mean of the random term is usually supposed to be zero. It can be shown that this assumption is not restrictive. We do it here on a simple example. Considering the example described in Figure 3.1, we denote the mean of the error term of each alternative by $m_1 = E[\varepsilon_1]$ and $m_2 = E[\varepsilon_2]$, respectively. Then, the error terms can be specified as

$$\varepsilon_1 = m_1 + \varepsilon_1' \tag{3.1}$$

and

$$\varepsilon_2 = m_2 + \varepsilon_2', \tag{3.2}$$

where ε_1' and ε_2' are random variables with zero mean. Therefore,

$$\begin{aligned} P_{\{1,2\}}(1) &= P[V_1 - V_2 \geq \varepsilon_2 - \varepsilon_1] \\ &= P[V_1 - V_2 \geq (m_2 + \varepsilon_2') - (m_1 + \varepsilon_1')] \\ &= P[(V_1 + m_1) - (V_2 + m_2) \geq \varepsilon_2' - \varepsilon_1'] \end{aligned} \tag{3.3}$$

The terms m_1 and m_2, called Alternative Specific Constants (ASC), are capturing the mean of the error term. Therefore, it can be assumed without loss of generality, that the error terms have zero mean if the model specification includes these ASCs.

The zero mean assumption is valid if the deterministic part of the utility function of each alternative includes an Alternative Specific Constant.

In practice, it is impossible to estimate the value of all ASCs from observed data. Considering again the example of Figure 3.1, the probability of choosing alternative 1, say, is not modified if an arbitrary constant K is added to both utilities. Therefore, only the *difference* between the two ASCs can be identified. Indeed, from (3.3), we have

$$\begin{aligned} P_{\{1,2\}}(1) &= P[(V_1 + m_1) - (V_2 + m_2) \geq \varepsilon_2' - \varepsilon_1'] \\ &= P[V_1 + m_1 + \varepsilon_1' \geq V_2 + m_2 + \varepsilon_2'] \\ &= P[V_1 + m_1 + \varepsilon_1' + K \geq V_2 + m_2 + \varepsilon_2' + K], \end{aligned} \tag{3.4}$$

for any $K \in \boldsymbol{R}$. If $K = -m_1$, we obtain

$$\mathrm{P}_{\{1,2\}}(1) = \mathrm{P}[V_1 + \varepsilon_1' \geq V_2 + (m_2 - m_1) + \varepsilon_2'],$$

or, equivalently, defining $M = m_2 - m_1$,

$$\mathrm{P}_{\{1,2\}}(1) = \mathrm{P}[V_1 + \varepsilon_1' \geq V_2 + M + \varepsilon_2'].$$

Defining $K = -m_2$ produces the same result. This property can be generalized easily to models with more than two alternatives, where only differences between ASCs can be identified.

It is common practice to constrain one ASC in the model to zero. From a modeling viewpoint, the choice of the particular alternative whose ASC is constrained is purely arbitrary. However, Bierlaire, Lotan and Toint [6] have shown that the estimation process is influenced by this choice. They propose a different technique of ASC specification which is optimal from an estimation perspective.

To derive assumptions about the variance of the random term, we observe that the scale of the utility may be arbitrarily specified. Indeed, for any $\alpha \in \boldsymbol{R}, \alpha > 0$, we have

$$\begin{aligned} \mathrm{P}_{\{1,2\}}(1) &= \mathrm{P}[U_1 \geq U_2] \\ &= \mathrm{P}[\alpha U_1 \geq \alpha U_2] \\ &= \mathrm{P}[\alpha V_1 - \alpha V_2 \geq \alpha(\varepsilon_2 - \varepsilon_1)]. \end{aligned} \tag{3.5}$$

The arbitrary decision about α is equivalent to assuming a particular variance v of the distribution of the error term. Indeed, if

$$\mathrm{Var}[\alpha(\varepsilon_2 - \varepsilon_1)] = v, \tag{3.6}$$

we have also

$$\alpha = \frac{v}{\sqrt{\mathrm{Var}[(\varepsilon_2 - \varepsilon_1)]}}. \tag{3.7}$$

We will illustrate this relationship with several examples in the remaining of this section.

Once assumptions about the mean and the variance of the error term distribution have been defined, the focus is now on the actual functional form of this distribution. We will consider here three different distributions yielding to three different families of models: linear, probit and logit models.

The linear model is obtained from the assumption that the density function of the error term is given by

$$f(x) = \begin{cases} \frac{1}{2L} & \text{if } x \in [-L, L] \\ 0 & \text{elsewhere} \end{cases} \tag{3.8}$$

where $L \in \boldsymbol{R}, L \geq 0$, is an arbitrary constant. This density function is used to derive the probability of choosing one particular alternative. Considering the example presented in Figure 3.1, the probability is given by (3.9) (see Figure 3.2).

$$P_{\{1,2\}}(1) = \begin{cases} 0 & \text{if } V_1 - V_2 < -L \\ \dfrac{V_1 - V_2 + L}{2L} & \text{if } -L \leq V_1 - V_2 \leq L \\ 1 & \text{if } V_1 - V_2 > L \end{cases} \tag{3.9}$$

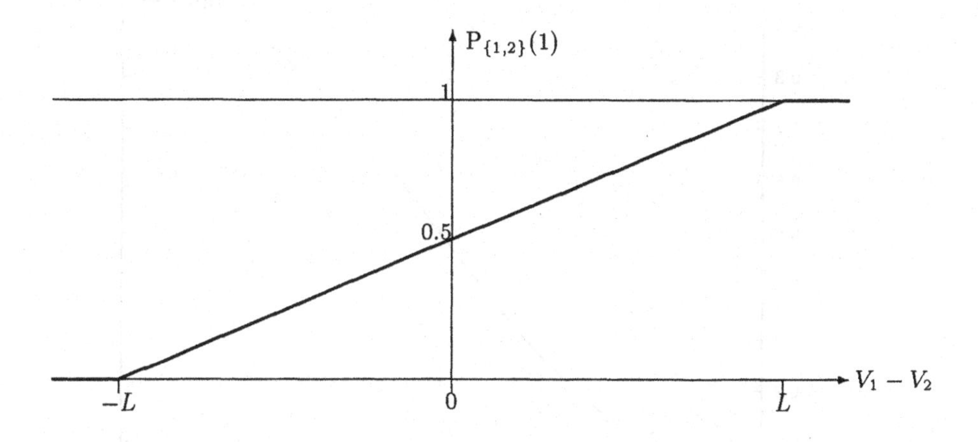

Fig. 3.2. Linear model

The linear model presents some problem for real applications. First, the probability associated with extreme values ($|V_1 - V_2| \geq L$ in the example) is exactly zero. Therefore, if any extreme event happens in the reality, the model will never capture it. Second, the discontinuity of the derivatives at $-L$ and L causes problems to most of the estimation procedures. We conclude the presentation of the linear model by emphasizing that the constant L determines the scale of the distribution. For the binary example, $\text{Var}(\varepsilon_2 - \varepsilon_1) = L^2/3$. Using (3.7), we have that assuming $\text{Var}(\alpha(\varepsilon_2 - \varepsilon_1)) = 1$ is equivalent to assuming $\alpha = \sqrt{3}/L$. A common value for L is 1/2, that is $\alpha = 2\sqrt{3}$.

The Normal Probability Unit, or Probit, model is derived from the assumption that the error terms are normally distributed, that is

$$f(x) = \frac{1}{\sigma\sqrt{2\pi}} e^{-\frac{1}{2}\left(\frac{x}{\sigma}\right)^2}, \tag{3.10}$$

where $\sigma \in \boldsymbol{R}, \sigma > 0$ is an arbitrary constant. This density function is used to derive the probability of choosing one particular alternative. Considering the example presented in Figure 3.1, and assuming that ε_1 and ε_2 are normally distributed with zero mean, variances σ_1^2 and σ_2^2 respectively, and covariance σ_{12}, the probability is given by (3.11) (see Figure 3.3).

$$P_{\{1,2\}}(1) = \int_{x=-\infty}^{V_1-V_2} \frac{1}{\sigma\sqrt{2\pi}} e^{-\frac{1}{2}\left(\frac{x}{\sigma}\right)^2} dx, \tag{3.11}$$

where $\sigma^2 = \sigma_1^2 + \sigma_2^2 - 2\sigma_{12}$ is the variance of $(\varepsilon_2 - \varepsilon_1)$

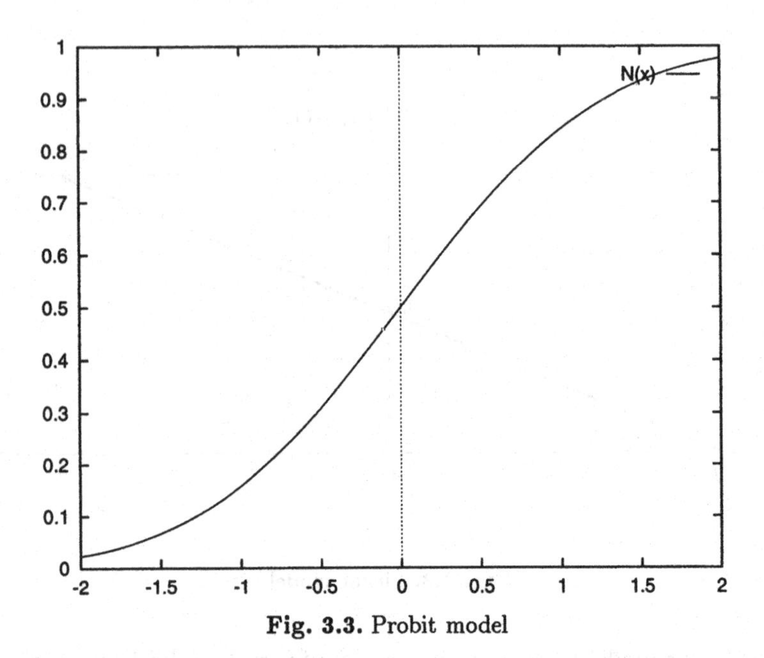

Fig. 3.3. Probit model

The probit model is motivated by the Central Limit Theorem[2], assuming that the error terms are the sum of independent unobserved quantities. Unfortunately, the probability function (3.11) has no closed analytical form, which limits practical use of this model. We refer the reader to Daganzo [11] for a comprehensive development of probit models. We conclude this short introduction of the probit model by looking at the scale parameter. Considering again the binary example presented in Figure 3.1 in the probit context, we have $\text{Var}(\varepsilon_2 - \varepsilon_1) = \sigma^2$. Using (3.7), we have that assuming $\text{Var}(\alpha(\varepsilon_2 - \varepsilon_1)) = 1$ is equivalent to assuming $\alpha = 1/\sigma$. It is common practice to arbitrary define $\sigma = 1$, that is $\alpha = 1$.

Despite its complexity, the probit model has been applied to many practical problems (see Whynes, Reedand and Newbold [24], Bolduc, Fortin and

[2] The Central Limit Theorem states that the sum of a large number of independent random variables approximates the asymptotically normal distribution. It has been proved by Markov [19].

Fournier [8], Yai, Iwakura and Morichi [25] among recent publications). However, the most widely used model in practical applications is probably the Logistic Probability Unit, or Logit, model. The error terms are now assumed to be independent and identically Gumbel distributed. The density function of the Gumbel distribution is given by (3.12) (see Figure 3.4).

$$f(x) = \mu e^{-\mu(x-\eta)} e^{-e^{\mu(x-\eta)}}, \tag{3.12}$$

where $\eta \in \boldsymbol{R}$ is the location parameter, and $\sigma \in \boldsymbol{R}, \sigma > 0$ is the scale parameter.

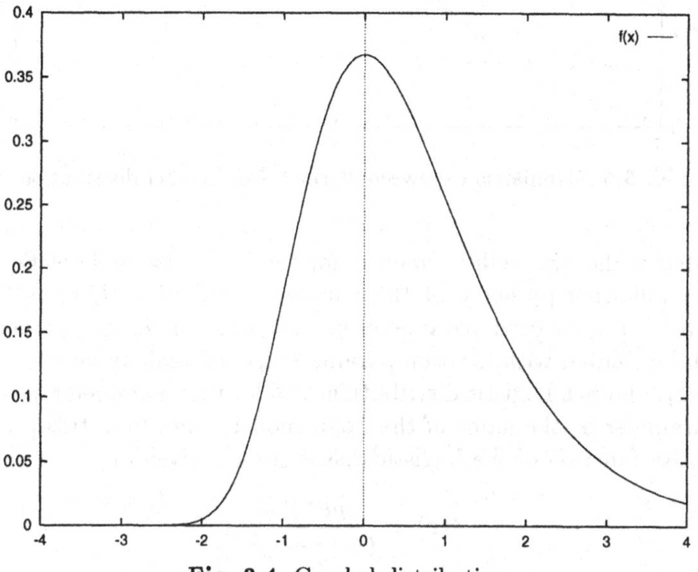

Fig. 3.4. Gumbel distribution

The mean of the Gumbel distribution is

$$\eta + \frac{\gamma}{\mu}, \tag{3.13}$$

where

$$\gamma = \lim_{n \to \infty} \sum_{i=1}^{n} \frac{1}{i} - \ln(n) \approx 0.5772 \tag{3.14}$$

is the Euler constant. The variance is

$$\frac{\pi^2}{6\mu^2}. \tag{3.15}$$

The Gumbel distribution is an approximation of the Normal law, as shown in Figure 3.5, where the plain line represents the Normal distribution, and the dotted line the Gumbel distribution.

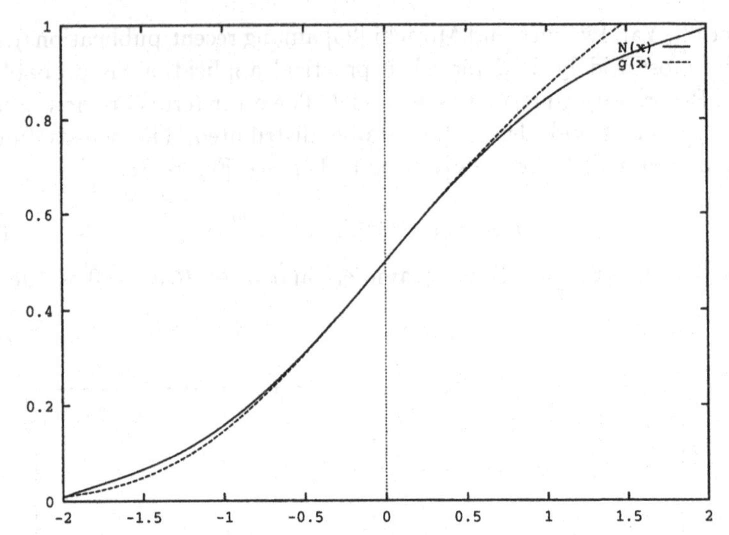

Fig. 3.5. Comparison between Normal and Gumbel distribution

We derive the probability function for the binary example of Figure 3.1 from the following property of the Gumbel distribution. If ε_1 is Gumbel distributed with location parameter η_1 and scale parameter μ, and ε_2 is Gumbel distributed with location parameter η_2 and scale parameter μ, then $\varepsilon = \varepsilon_2 - \varepsilon_1$ follows a Logistic distribution with location parameter $\eta_2 - \eta_1$ and scale parameter μ (the name of the Logit model comes from this property). The density function of the Logistic distribution is given by

$$f(x) = \frac{\mu e^{-\mu x}}{(e^{-\mu x} + 1)^2}, \tag{3.16}$$

where $\mu \in \boldsymbol{R}, \mu > 0$ is the scale parameter. As a consequence, we have,

$$\mathrm{P}_{\{1,2\}}(1) = \frac{1}{e^{-\mu(V_1 - V_2)} + 1} \tag{3.17}$$

or, equivalently,

$$\mathrm{P}_{\{1,2\}}(1) = \frac{e^{\mu V_1}}{e^{\mu V_1} + e^{\mu V_2}}. \tag{3.18}$$

In order to determine the relationship between the scale parameter and the variance of the distribution, we compute $\mathrm{Var}(\varepsilon_2 - \varepsilon_1) = \mathrm{Var}(\varepsilon_2) + \mathrm{Var}(\varepsilon_1) = 2\pi^2/6\mu^2$. Using (3.7), we have that assuming $\mathrm{Var}(\alpha(\varepsilon_2 - \varepsilon_1)) = 1$ is equivalent to assuming $\alpha = \mu\sqrt{3}/\pi$. It is common practice to arbitrary define $\mu = 1$, that is $\alpha = \sqrt{3}/\pi$.

In most cases, the arbitrary decision about the scale parameter does not matter and can be safely ignored. But it is important not to completely forget its existence. Indeed, it may sometimes play an important role. For example, utilities derived from different models can be compared only if the value of α

is the same for all of them. It is usually not the case with the scale parameters commonly used in practice, as shown in Table 3.1. Namely, a utility estimated with a logit model has to be divided by $\sqrt{3}/\pi$ before being compared with a utility estimated with a probit model.

Model	Arbitrary value	Estimated utility
Linear	$L = \frac{1}{2}$	$2\sqrt{3}$ V
Probit	$\sigma = 1$	1 V
Logit	$\mu = 1$	$\frac{\sqrt{3}}{\pi}$ V

Table 3.1. Model comparison

The list of models presented here above is not exhaustive. Other assumptions about the distribution of the error term will lead to other families of models. For instance, Ben-Akiva and Lerman [1] cite the arctan and the truncated exponential models. These models are not often used in practice and we will not consider them here.

3.2 Assumptions on the deterministic term

The utility of each alternative must be a function of the attributes of the alternative itself and of the decision-maker identified in Sections 2.1 and 2.3. We can write the deterministic part of the utility that individual i is associating with alternative a as

$$V_a^i = V_a^i(x_a^i), \tag{3.19}$$

where x_a^i is a vector containing all attributes, both of individual i and alternative a. The function defined in (3.19) is commonly assumed to be linear in the parameters, that is, if n attributes are considered,

$$V_a^i(x_a^i) = \beta_1 x_a^i(1) + \beta_2 x_a^i(2) + \cdots + \beta_n x_a^i(n) = \sum_{k=1}^{n} \beta_k x_a^i(k), \tag{3.20}$$

where $\beta_1, \ldots, \beta_n$ are parameters to be estimated. This assumption simplifies the formulation and the estimation of the model, and is not as restrictive as it may seem. Indeed, nonlinear effects can still be captured in the attributes definition, as mentioned in Section 2.3.

Nonlinear effects can be captured with a linear-in-parameters utility function, using an appropriate definition of attributes.

4. Multinomial logit model

As introduced in the previous section, the logit model is derived from the assumption that the error terms of the utility functions are independent and identically Gumbel distributed. These models were first introduced in the context of binary choice models, where the logistic distribution is used to derive the probability. Their generalization to more than two alternatives is referred to as *multinomial* logit models.

If the error terms are independent and identically Gumbel distributed, with location parameter 0 and scale parameter μ, the probability that a given individual choose alternative i within $\mathcal{C}$ is given by

$$\mathrm{P}_{\mathcal{C}}(i) = \frac{e^{\mu V_i}}{\sum_{k \in \mathcal{C}} e^{\mu V_k}}. \tag{4.1}$$

The derivation of this result is attributed to Holman and Marley by Luce and Suppes [17]. We refer the reader to Ben-Akiva and Lerman [1] and Anderson et al. [1] for additional details.

It is interesting to note that the multinomial logit model can also be derived from the choice axiom defined by (2.6) and (2.7). Indeed, defining $\mathcal{S} = \mathcal{C}$ and $v(a) = e^{\mu V_a}$, we have that (2.11) is equivalent to (4.1).

The multinomial logit model can be derived both from random utility theory and from the choice axiom.

An important property of the multinomial logit model is the Independence from Irrelevant Alternatives (IIA). This property can be stated as follows. *The ratio of the probabilities of any two alternatives is independent from the choice set.* That is, for any choice sets $\mathcal{S}$ and $\mathcal{T}$ such that $\mathcal{S} \subseteq \mathcal{T} \subseteq \mathcal{C}$, for any alternative a_1 and a_2 in $\mathcal{S}$, we have

$$\frac{\mathrm{P}_{\mathcal{S}}(a_1)}{\mathrm{P}_{\mathcal{S}}(a_2)} = \frac{\mathrm{P}_{\mathcal{T}}(a_1)}{\mathrm{P}_{\mathcal{T}}(a_2)}. \tag{4.2}$$

This result can be proven easily using (4.1). Ben-Akiva and Lerman [1] propose an equivalent definition: *The ratio of the choice probabilities of any two alternatives is entirely unaffected by the systematic utilities of any other alternatives.*

The IIA property of multinomial logit models is a limitation for some practical applications. This limitation is often illustrated by the red bus/blue bus paradox (see, for example, Ben-Akiva and Lerman [1]) in the modal choice context. We prefer here the path choice example presented in Figure 4.1.

The probability provided by the multinomial logit model (4.1) for this example are

We consider a commuter traveling from an origin O to a destination D. He/she is confronted with the path choice problem described below, where the choice set is $\{1, 2a, 2b\}$ and the only attribute considered for the choice is travel time. We assume furthermore that the travel time for any alternative is the same, that is $V_1 = V_{2a} = V_{2b} = T$. Finally, the travel time δ on the small sections a and b is supposed to be significantly smaller than the total travel time T. As a result, we expect the probability of choosing path 1 or path 2 to be almost 50%, irrespectively of the choice between a and b.

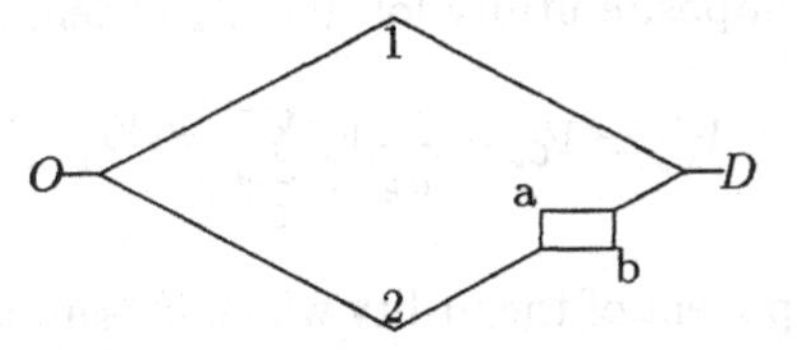

Fig. 4.1. A path choice example

$$\mathrm{P}_{\{1,2a,2b\}}(1) = \mathrm{P}_{\{1,2a,2b\}}(2a) = \mathrm{P}_{\{1,2a,2b\}}(2b) = \frac{e^{\mu T}}{\sum_{k\in\{1,2a,2b\}} e^{\mu T}} = \frac{1}{3}, \tag{4.3}$$

which is not consistent with the intuitive result. This situation appears in choice problems with significantly correlated alternatives, as it is clearly the case in the example. Indeed, alternatives $2a$ and $2b$ are so similar that their utilities share many unobserved attributes of the path and, therefore, the assumption of independence of the random part of these utilities is not valid in this context.

The Nested Logit Model, presented in the next section, partly overcomes this limitation of the multinomial logit model.

5. Nested logit model

The nested logit model, first derived by Ben-Akiva [2], is an extension of the multinomial logit model designed to capture correlations among alternatives. It is based on the partitioning of the choice set $\mathcal{C}$ into several nests $\mathcal{C}_k$ such that

$$\mathcal{C} = \bigcup_k \mathcal{C}_k, \tag{5.1}$$

and

$$\mathcal{C}_k \cap \mathcal{C}_l = \emptyset, \ \forall k \neq l. \tag{5.2}$$

The utility function of each alternative is composed of a term specific to the alternative, and a term associated with the nest. If $i \in \mathcal{C}_k$, we have

$$U_i = V_i + \varepsilon_i + V_{\mathcal{C}_k} + \varepsilon_{\mathcal{C}_k}. \tag{5.3}$$

The error terms ε_i and ε_{C_k} are supposed to be independent. As for the multinomial logit model, error terms ε_i are supposed to be independent and identically Gumbel distributed, with scale parameter σ_k. The distribution of ε_{C_k} is such that the random variable $\max_{j \in C_k} U_j$ is Gumbel distributed with scale parameter μ.

Each nest within the choice set is associated with a pseudo-utility, called *composite utility, expected maximum utility, inclusive value* or *accessibility* in the literature. The composite utility for nest C_k is defined as

$$V'_{C_k} = V_{C_k} + \frac{1}{\sigma_k} \ln \sum_{j \in C_k} e^{\sigma_k V_j}, \tag{5.4}$$

where V_{C_k} is the component of the utility which is common to all alternatives in the nest C_k.

The probability model is then given by

$$\mathrm{P}_C(i) = \mathrm{P}_C(C_k)\mathrm{P}_{C_k}(i), \tag{5.5}$$

where

$$\mathrm{P}_C(C_k) = \frac{e^{\mu V'_{C_k}}}{\sum_{l=1}^{n} e^{\mu V'_{C_l}}}, \tag{5.6}$$

and

$$\mathrm{P}_{C_k}(i) = \frac{e^{\sigma_k V_i}}{\sum_{j \in C_k} e^{\sigma_k V_j}}. \tag{5.7}$$

The parameters μ and σ_k reflect the correlation among alternatives in the nest C_k. Indeed, if $i, j \in C_k$, we have

$$\frac{\mu}{\sigma_k} = \sqrt{1 - \operatorname{corr}(U_i, U_j)}. \tag{5.8}$$

Clearly, we have

$$0 \leq \frac{\mu}{\sigma_k} \leq 1. \tag{5.9}$$

Ben-Akiva and Lermand [1] derive condition (5.9) directly from utility theory. Note also that if $\frac{\mu}{\sigma_k} = 1$, we have $\operatorname{corr}(U_i, U_j) = 0$.

When $\frac{\mu}{\sigma_k} = 1$ for all k, the nested logit model is equivalent to the multinomial logit model.

The parameters μ and σ_k are closely related in the model. Actually, only their ratio is meaningful. It is not possible to identify them separately. A common practice is to arbitrarily constrain one of them to a value (usually 1). The impacts of this arbitrary decision on the model are briefly discussed in Section 5.1. We illustrate here the Nested Logit Model with the path choice example described in Figure 4.1. First, the choice set $C = \{1, 2a, 2b\}$

is divided into $\mathcal{C}_1 = \{1\}$ and $\mathcal{C}_2 = \{2a, 2b\}$. The deterministic components of the utilities are $V_{\mathcal{C}_1} = T$, $V_1 = 0$, $V_{\mathcal{C}_2} = T - \delta$ and $V_{2a} = V_{2b} = \delta$. The composite utilities of each nest are

$$V'_{\mathcal{C}_1} = V_{\mathcal{C}_1} = T, \tag{5.10}$$

and

$$V'_{\mathcal{C}_2} = T - \delta + \frac{1}{\sigma_2} \ln(e^{\sigma_2 \delta} + e^{\sigma_2 \delta}) = T + \frac{1}{\sigma_2} \ln 2. \tag{5.11}$$

The probability of choosing each nest is then

$$\mathrm{P}_{\mathcal{C}}(\mathcal{C}_1) = \frac{e^{V'_{\mathcal{C}_1}}}{e^{V'_{\mathcal{C}_1}} + e^{V'_{\mathcal{C}_2}}} = \frac{e^T}{e^T + e^{(T + \frac{\ln 2}{\sigma_2})}} = \frac{1}{1 + e^{\frac{\ln 2}{\sigma_2}}} = \frac{1}{1 + 2^{\frac{1}{\sigma_2}}}, \tag{5.12}$$

and

$$\mathrm{P}_{\mathcal{C}}(\mathcal{C}_2) = 1 - \mathrm{P}_{\mathcal{C}}(\mathcal{C}_1) = \frac{2^{\frac{1}{\sigma_2}}}{1 + 2^{\frac{1}{\sigma_2}}}, \tag{5.13}$$

where the value of μ has been assumed to be 1, without loss of generality. The probability of each alternative is then computed. We obtain

$$\mathrm{P}_{\mathcal{C}}(1) = \mathrm{P}_{\mathcal{C}}(\mathcal{C}_1) = \frac{1}{1 + 2^{\frac{1}{\sigma_2}}}, \tag{5.14}$$

and

$$\mathrm{P}_{\mathcal{C}}(2a) = \mathrm{P}_{\mathcal{C}}(2b) = \frac{e^{\sigma_2 \delta}}{e^{\sigma_2 \delta} + e^{\sigma_2 \delta}} \mathrm{P}_{\mathcal{C}}(\mathcal{C}_2) = \frac{1}{2} \frac{2^{\frac{1}{\sigma_2}}}{1 + 2^{\frac{1}{\sigma_2}}}. \tag{5.15}$$

The values of $\mathrm{P}_{\mathcal{C}}(1)$, $\mathrm{P}_{\mathcal{C}}(2a)$ and $\mathrm{P}_{\mathcal{C}}(2b)$ as a function of $\frac{1}{\sigma_2}$ are plotted on Figure 5.1. From (5.9), we have that $0 \leq \frac{1}{\sigma_2} \leq 1$ because μ has been arbitrarily defined as 1. We observe that, when $\frac{1}{\sigma_2} = 1$, the nested logit model produces the same results as the multinomial logit model (4.3), and all probabilities are $\frac{1}{3}$. On the other hand, when σ_2 goes to infinity, and $1/\sigma_2$ goes to 0, the probability of each nest is closer and closer to 1/2. At the limit, the model is becoming a binary choice model, where the small detours a and b are ignored in the choice process.

5.1 Normalization of nested logit models

In order to compute the probabilities in the previous example, we have arbitrarily decided to constraint μ to 1. Alternatively, we could have decided to constraint σ_2 to 1. It is easy to show that, in this case, we have

$$\mathrm{P}_{\mathcal{C}}(1) = \frac{1}{1 + 2^{\mu}} \tag{5.16}$$

and

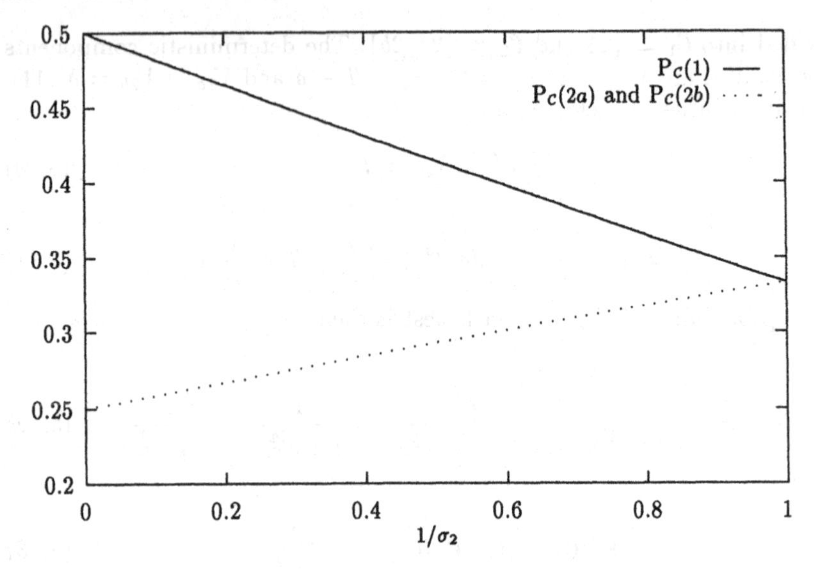

Fig. 5.1. Probability of each alternative as a function of $\frac{1}{\sigma_2}$.

$$P_C(2a) = P_C(2b) = \frac{1}{2}\frac{2^\mu}{1+2^\mu}, \tag{5.17}$$

which is equivalent to (5.14) and (5.15), replacing $1/\sigma_2$ by μ.

A model where the scale parameter μ is arbitrarily constrained to 1 is said to be "normalized from the top". A model where one of the parameters σ_k is constrained to 1 is said to be "normalized from the bottom". The latter may produce a simpler formulation of the model. We illustrate it using the example of Figure 5.2.

In the context of a mode choice with $\mathcal{C}$ = {bus, metro, car, bike}, we consider a model with two nests: $\mathcal{C}_1$ = {bus,metro} contains the public transportation modes and $\mathcal{C}_2$ = {car,bike} contains the private transportation modes. For the sake of the example, we consider the following deterministic terms of the utility functions: $V_{\text{bus}} = \beta_u t_{\text{bus}}$, $V_{\text{metro}} = \beta_u t_{\text{metro}}$, $V_{\text{car}} = \beta_r t_{\text{car}}$ and $V_{\text{bike}} = \beta_r t_{\text{bike}}$, where t_i is the travel time using mode i, β_r and β_u are parameters to be estimated. Note that we have one parameter for private and one for public transportation, and we have not included the alternative specific constants in order to keep the example simple.

Fig. 5.2. A mode choice example

We have

$$P_C(\text{bus}) = \frac{e^{\sigma_1 \beta_u t_{\text{bus}}}}{e^{\sigma_1 \beta_u t_{\text{bus}}} + e^{\sigma_1 \beta_u t_{\text{metro}}}} \times \frac{e^{\frac{\mu}{\sigma_1} \ln(e^{\sigma_1 \beta_u t_{\text{bus}}} + e^{\sigma_1 \beta_u t_{\text{metro}}})}}{e^{\frac{\mu}{\sigma_1} \ln(e^{\sigma_1 \beta_u t_{\text{bus}}} + e^{\sigma_1 \beta_u t_{\text{metro}}})} + e^{\frac{\mu}{\sigma_2} \ln(e^{\sigma_2 \beta_r t_{\text{car}}} + e^{\sigma_2 \beta_r t_{\text{bike}}})}}$$

and

$$P_C(\text{car}) = \frac{e^{\sigma_2 \beta_r t_{\text{car}}}}{e^{\sigma_2 \beta_r t_{\text{car}}} + e^{\sigma_2 \beta_r t_{\text{bike}}}} \times \frac{e^{\frac{\mu}{\sigma_2} \ln(e^{\sigma_2 \beta_r t_{\text{car}}} + e^{\sigma_2 \beta_r t_{\text{bike}}})}}{e^{\frac{\mu}{\sigma_1} \ln(e^{\sigma_1 \beta_u t_{\text{bus}}} + e^{\sigma_1 \beta_u t_{\text{metro}}})} + e^{\frac{\mu}{\sigma_2} \ln(e^{\sigma_2 \beta_r t_{\text{car}}} + e^{\sigma_2 \beta_r t_{\text{bike}}})}}.$$

If we impose $\sigma_1 = 1$, we can define $\theta_1 = \mu$, $\theta_2 = \mu/\sigma_2$, $\tilde{\beta}_u = \beta_u$ and $\tilde{\beta}_r = \sigma_2 \beta_r$ to obtain the following expressions.

$$P_C(\text{bus}) = \frac{e^{\tilde{\beta}_u t_{\text{bus}}}}{e^{\tilde{\beta}_u t_{\text{bus}}} + e^{\tilde{\beta}_u t_{\text{metro}}}} \cdot \frac{e^{\theta_1 \ln(e^{\tilde{\beta}_u t_{\text{bus}}} + e^{\tilde{\beta}_u t_{\text{metro}}})}}{e^{\theta_1 \ln(e^{\tilde{\beta}_u t_{\text{bus}}} + e^{\tilde{\beta}_u t_{\text{metro}}})} + e^{\theta_2 \ln(e^{\tilde{\beta}_r t_{\text{car}}} + e^{\tilde{\beta}_r t_{\text{bike}}})}}$$

and

$$P_C(\text{car}) = \frac{e^{\tilde{\beta}_r t_{\text{car}}}}{e^{\tilde{\beta}_r t_{\text{car}}} + e^{\tilde{\beta}_r t_{\text{bike}}}} \cdot \frac{e^{\theta_2 \ln(e^{\tilde{\beta}_r t_{\text{car}}} + e^{\tilde{\beta}_r t_{\text{bike}}})}}{e^{\theta_1 \ln(e^{\tilde{\beta}_u t_{\text{bus}}} + e^{\tilde{\beta}_u t_{\text{metro}}})} + e^{\theta_2 \ln(e^{\tilde{\beta}_r t_{\text{car}}} + e^{\tilde{\beta}_r t_{\text{bike}}})}}$$

This formulation, proposed by Daly [12], simplifies the estimation process. For this reason, it has been adopted in estimation packages like ALOGIT (Daly [12]) or HieLoW (Bierlaire [5], Bierlaire and Vandevyvere [7]).

ALOGIT (Daly [12]) and HieLoW (Bierlaire [5], Bierlaire and Vandevyvere [7]) estimate nested logit models normalized from the bottom.

We emphasize here that this formulation should be used with caution when the same parameters are present in more than one nest. In this case, specific techniques, inspired from artificial trees proposed by Bradley and Daly [9] must be used to obtain a correct specification of the model. The description of these techniques is out of the scope of this paper.

The scale parameters have to be normalized in nested logit models. Indeed, only their ratio is meaningful. All possible normalizations produce an equivalent model, but the estimation process depends on the particular specification.

A direct extension of the nested logit model consists in partionning some or all nests into sub-nests, which can, in turn, be divided into sub-nests. Because of the complexity of these models, their structure is usually represented as a tree, as suggested by Daly [12]. Clearly, the number of potential structures, reflecting the correlation among alternatives, can be very large.

No technique has been proposed thus far to identify the most appropriate correlation structure directly from the data.

We conclude our introduction of nested logit models by mentioning their limitations. These models are designed to capture choice problems where alternatives *within each nest* are correlated. No correlation *across* nests can be captured by the Nested Logit Model. When alternatives cannot be partitioned into well separated nests to reflect their correlation, Nested Logit Models are not applicable. This is the case for most route choice problems. Several models within the "logit family" have been designed to capture specific correlation structures. For example, Cascetta [10] captures overlapping paths in a route choice context using commonality factors, Koppelman and Wen [15] capture correlation between pair of alternatives, and Vovsha [23] proposes a cross-nested model allowing alternatives to belong to more than one nest. The two last models are derived from the Generalized Extreme Value model, presented in the next section.

6. Generalized extreme value model

The Generalized Extreme Value (GEV) model has been introduced by McFadden [20] in the context of residential location. This general model actually consists in a large family of models that are consistent with random utility theory. The probability of choosing alternative i within $\mathcal{C} = \{1,\ldots,n\}$ is given by

$$\mathrm{P}_{\mathcal{C}}(i) = \frac{e^{V_i}\frac{\partial G}{\partial x_i}(e^{V_1},\ldots,e^{V_n})}{\mu G(e^{V_1},\ldots,e^{V_n})}, \tag{6.1}$$

where $G : \boldsymbol{R}^n_+ \rightarrow \boldsymbol{R}$ is a differentiable function with the following properties.

1. $G(x) \geq 0$ for all $x \in \boldsymbol{R}^n_+$,
2. G is homogeneous of degree[3] $\mu > 0$, that is $G(\alpha x) = \alpha^\mu G(x)$, for all $x \in \boldsymbol{R}^n_+$,
3. $\lim_{x_i \rightarrow +\infty} G(x_1,\ldots,x_i,\ldots,x_n) = +\infty$ for all i such that $1 \leq i \leq n$, and
4. the kth partial derivative with respect to k distinct x_i is non-negative if k is odd, and non-positive if k is even, that is, $\forall i_1,\ldots,i_k$ such that $1 \leq i_j \leq n$ if $1 \leq j \leq k$ and $i_j \neq i_l$ if $1 \leq j,l \leq k$ and $j \neq l$, we have

$$\forall x \in \boldsymbol{R}^n_+, \quad \frac{\partial^k G}{\partial x_{i_1}\ldots\partial x_{i_k}}(x) \begin{cases} \geq 0 & \text{if } k \text{ is odd,} \\ \leq 0 & \text{if } k \text{ is even.} \end{cases}$$

As an example, we consider

$$G(x) = \sum_{i=1}^{n} x_i^\mu, \tag{6.2}$$

[3] McFadden [20] assumed $\mu = 1$. Ben-Akiva and François [4] generalized the result to any $\mu > 0$.

which has the required properties, as it can be easily verified. Then,

$$\mathrm{P}_{\mathcal{C}}(i) = \frac{e^{V_i}\frac{\partial G}{\partial x_i}(e^{V_1},\ldots,e^{V_n})}{\mu G(e^{V_1},\ldots,e^{V_n})} = \frac{e^{V_i}\mu e^{(\mu-1)V_i}}{\mu\sum_{i=1}^{n} e^{\mu}V_i} = \frac{e^{\mu V_i}}{\sum_{i=1}^{n} e^{\mu}V_i}, \tag{6.3}$$

which is the multinomial logit model. Similarly, the nested logit model can be derived with

$$G(x) = \sum_{k=1}^{n}\left(\sum_{i\in C_k} e^{\sigma_k x_i}\right)^{\frac{\mu}{\sigma_k}}. \tag{6.4}$$

It can be shown that property 4 holds if $0 \leq \frac{\mu}{\sigma_k} \leq 1$, which is consistent with condition (5.9).

The multinomial logit model and the nested logit model are both specific Generalized Extreme Value models.

The Generalized Extreme Value model provides a nice theoretical framework for the development of new discrete choice models, like Koppelman and Wen [15] and Vovsha [23].

7. Conclusion

We have covered in this paper the main theoretical aspects of discrete choice models in general, and random utility models in particular. A good awareness of underlying assumptions is necessary for an efficient use of these models for practical applications. In particular, we have focused on the location parameters and the scale parameters in multinomial and nested logit models. Despite its importance, the role of these parameters tend to be underestimated by practitioners. This may lead to incorrect specifications of the models, or incorrect interpretation of the results.

8. Acknowledgments

Comments from the students and other lecturers of the ASI have been very useful to write this paper. Moreover, I am very grateful to Moshe Ben-Akiva and John Bowman for their valuable discussions and comments.

References

1. S. P. Anderson, A. de Palma, and J.-F. Thisse. *Discrete Choice Theory of Product Differentiation.* MIT Press, Cambridge, Ma, 1992.

2. M. E. Ben-Akiva. *Structure of passenger travel demand models.* PhD thesis, Department of Civil Engineering, MIT, Cambridge, Ma, 1973.
3. M. E. Ben-Akiva and S. R. Lerman. *Discrete Choice Analysis: Theory and Application to Travel Demand.* MIT Press, Cambridge, Ma., 1985.
4. M. E. Ben-Akiva and B. François. μ homogeneous generalized extreme value model. Working paper, Department of Civil Engineering, MIT, Cambridge, Ma, 1983.
5. M. Bierlaire. A robust algorithm for the simultaneous estimation of hierarchical logit models. GRT Report 95/3, Department of Mathematics, FUNDP, 1995.
6. M. Bierlaire, T. Lotan, and Ph. L. Toint. On the overspecification of multinomial and nested logit models due to alternative specific constants. *Transportation Science*, 1997. (forthcoming).
7. M. Bierlaire and Y. Vandevyvere. *HieLoW: the interactive user's guide.* Transportation Research Group - FUNDP, Namur, 1995.
8. D. Bolduc, B. Fortin, and M.-A. Fournier. The effect of incentive policies on the practice location of doctors: A multinomial probit analysis. *Journal of labor economics*, 14(4):703, 1996.
9. M. A. Bradley and A. J. Daly. Estimation of logit choice models using mixed stated preferences and revealed preferences information. *Methods for understanding travel behaviour in the 1990's*, pages 116–133, Québec, mai 1991. International Association for Travel Behaviour. 6th international conference on travel behaviour.
10. E. Cascetta. A modified logit route choice model overcoming path overlapping problems. Specification and some calibration results for interurban networks. In *Proceedings of the 13th International Symposium on the Theory of Road Traffic Flow (Lyon, France)*, 1996.
11. C. F. Daganzo. *Multinomial Probit: The theory and its application to demand forecasting.* Academic Press, New York, 1979.
12. A. Daly. Estimating "tree" logit models. *Transportation Research B*, 21(4):251–268, 1987.
13. D. A. Hensher and L. W. Johnson. *Applied discrete choice modelling.* Croom Helm, London, 1981.
14. J. L. Horowitz, F. S. Koppelman, and S. R. Lerman. *A self-instructing course in disaggregate mode choice modeling.* Technology Sharing Program, US Department of Transportation, Washington, D.C. 20590, 1986.
15. F. S. Koppelman and Chieh-Hua Wen. The paired combinatorial logit model: properties, estimation and application. Transportation Research Board, 76th Annual Meeting, Washington DC, January 1997. Paper #970953.
16. R. Luce. *Individual choice behavior: a theoretical analysis.* J. Wiley and Sons, New York, 1959.
17. R. D. Luce and P. Suppes. Preference, utility and subjective probabiblity. In R. D. Luce, R. R. Bush, and E. Galanter, editors, *Handbook of Mathematical Psychology*, New York, 1965. J. Wiley and Sons.
18. C. Manski. The structure of random utility models. *Theory and Decision*, 8:229–254, 1977.
19. A. A. Markov. *Calculation of probabilities.* Tip. Imperatorskoi Akademii Nauk, Sint Petersburg, 1900. (in Russian).
20. D. McFadden. Modelling the choice of residential location. In A. Karlquist et al., editors, *Spatial interaction theory and residential location*, pages 75–96, Amsterdam, 1978. North-Holland.
21. J. Swait. *Probabilistic choice set formation in transportation demand models.* PhD thesis, Department of Civil and Environmental Engineering, Massachussetts Institute of Technology, Cambridge, Ma, 1984.

22. A. Tversky. Elimination by aspects: a theory of choice. *Psychological Review*, 79:281–299, 1972.
23. P. Vovsha. Cross-nested logit model: an application to mode choice in the Tel-Aviv metropolitan area. Transportation Research Board, 76th Annual Meeting, Washington DC, January 1997. Paper #970387.
24. D. K. Whynes, G. Reedand, and P. Newbold. General practitioners' choice of referral destination: A probit analysis. *Managerial and Decision Economics*, 17(6):587, 1996.
25. T. Yai, S. Iwakura, and S. Morichi. Multinomial probit with structured covariance for route choice behavior. *Transportation Research B*, 31(3):195–208, June 1997.

Crew Scheduling and Rostering Problems in Railway Applications

Alberto Caprara[1], Matteo Fischetti[2],
Pier Luigi Guida[3], Paolo Toth[1], and Daniele Vigo[1]

[1] DEIS – University of Bologna – Italy

[2] DEI – University of Padova – Italy

[3] Ferrovie dello Stato SpA – Italy

Summary. Crew planning is a typical problem arising in the management of large transit systems (such as railway and airline companies). Given a set of trips to be covered every day in a given period, the problem is to build a daily assignment of each trip to a crew so as to guarantee that all the trips are covered in the period at minimum cost. In practice, the overall crew management problem is decomposed into two subproblems, called crew scheduling and crew rostering. Crew scheduling deals with the short-term schedule of the crews: a convenient set of pairings is constructed, each representing a subset of trips to be covered by a single crew. Generally this subproblem is solved by generating a very large number of potential pairings, each with a given cost, and by selecting a minimum cost set of pairings covering all the trips. Crew rostering deals with the construction of a set of working rosters which determine the sequence of pairings that each single crew has to perform over the given time period, to cover every day all the pairings selected in the first phase. In this paper, we give an outline of different ways of modeling the two subproblems and possible solution methods. Two main solution approaches are illustrated for real-world applications. In particular we present the solution techniques currently adopted at the Italian railway company, Ferrovie dello Stato SpA, for solving the crew planning problem.

1. Introduction

A typical problem arising in the crew management of large transit systems (such as railway and airline companies) is the following. We are given a planned timetable for the *train services* (i.e., both the actual journeys with passengers or freight, and the transfers of empty trains or equipment between different stations) to be performed every day of a certain time period. Each train service has first been split into a sequence of *trips*, defined as segments of train journeys which must be serviced by the same crew without rest. Each trip is characterized by a departure time, a departure station, an arrival time, an arrival station, and possibly by additional attributes. During the given time period each crew performs a *roster*, defined as a cyclic sequence of trips whose operational cost and feasibility depend on several rules laid down by union contracts and company regulations. To ensure that each daily occurrence of a trip is performed by one crew, for each roster whose length is, say, k days, k crews are needed. In fact, in a given calendar day each of these crews performs the activities of a different day of the roster. Moreover, in consecutive calendar days each crew performs the activities of (cyclically) consecutive days of the roster. The crew planning problem then consists of finding a set of rosters covering once every trip, so as to satisfy all the operational constraints with minimum cost.

Several papers on crew (and vehicle) management appeared in the literature. We refer the interested reader to Wren [36], Bodin, Golden, Assad and Ball [10], Carraresi and Gallo [15], Daduna and Wren [18], Rousseau [31], Desrochers and Rousseau [19], Barnhart, Johnson, Nemhauser, Savelsbergh and Vance [4], Desrosiers, Dumas, Solomon and Soumis [20], and Wise [35]. Crew planning models and methods for railway applications have been recently proposed in Caprara, Fischetti, Toth, Vigo and Guida [13], [14].

Railway crew management represents a very complex and challenging problem due to both the size of the instances to be solved and the type and number of operational constraints. In practice, the overall crew management problem is approached in two phases, according to the following scheme:

1. *Crew scheduling*: the short-term schedule of the crews is considered, and a convenient set of *pairings* (or *duties*) covering all the trips is constructed. Each pairing represents a sequence of trips to be covered by a single crew within a given time period overlapping few consecutive days (typically 1 or 2 days).
2. *Crew rostering*: the pairings selected in phase 1 are sequenced to obtain the final rosters. In this step, trips are no longer taken into account explicitly, but determine the *attributes* of the pairings which are relevant for the roster feasibility and cost.

Decomposition is motivated by several reasons. First of all, each crew is located in a given *depot*, which represents the starting and ending point of his work segments. A natural constraint imposes that each crew must return to his home depot within few days, which leads to the concept of *pairing* as a short-term work segment starting and ending at the home depot and overlapping very few consecutive days. Secondly, constraints affecting the short-term work segments are different in nature from those related to the overall crew rosters. Finally, a global approach to the overall crew management problem is unlikely to be implemented, because of both its intrinsic difficulty and the planners' unwillingness to change their actual practice.

The main objective of crew management is the minimization of the global number of crews needed to perform all the daily occurrences of the trips in the given period. In railway applications, considerable savings can be obtained through a clever sequencing of the pairings obtained in the first phase. Therefore, the objective of the crew scheduling phase has to take into account the characteristics of the pairings selected and their implication in the subsequent rostering phase.

Both crew scheduling and crew rostering problems require finding min-cost *sequences* through a given set of *items*. Items correspond to trips for crew scheduling, and to pairings for crew rostering. Sequences correspond to pairings for crew scheduling, and to rosters for crew rostering.

A natural formulation of both problems in terms of graphs, associates a node with each item, and a directed arc with each possible item transition. More specifically, one can define a directed graph $G = (V, A)$ having one node $j \in V$ for each item, and an arc $(i, j) \in A$ if and only if item j can appear right after item i in a feasible sequence. With this representation, both problems can be formulated as finding a min-cost collection of *circuits* (or *paths*) of G covering each node once.

There are two basic ways of modeling as an integer linear program the problem of covering the nodes of a directed graph through a suitable set of circuits.

The first model associates a cost c_{ij} and a binary variable x_{ij} with each arc $(i, j) \in A$, where c_{ij} is the cost of sequencing item j right after item i in a feasible solution, and $x_{ij} = 1$ if arc (i, j) is used in the optimal solution and $x_{ij} = 0$ otherwise. The model minimizes the sum of the costs of the used arcs, and imposes that each node is covered exactly once. Additional inequalities forbid the choice of the (inclusion-minimal) arc subsets which cannot be part of any feasible solution. Notice that these arc subsets correspond to the arc sequences which cannot be covered by a single crew because of operational constraints. In addition, they may correspond to subsets of arcs which cannot all be selected because of constraints related to the infeasibility of a group of circuits (typically called *crew base constraints*).

This model has a number of drawbacks. First, it can only be applied when the cost of the solution can be expressed as the sum of the costs associated with the arcs. Hence it cannot be used when the cost of a circuit depends on the overall node sequence, or on the "type" of the crew, e.g., on the home location. Second, the number of inequalities forbidding infeasible arc subsets may grow exponentially with $|V|$. Third, the linear programming relaxation of the model can be very weak when tight operational constraints are imposed. This drawback can be partially overcome by introducing additional inequalities taking into account explicitly some specific kinds of infeasibility. On the other hand, the model is particularly suitable for cases in which the most relevant constraints concern the direct transition of the nodes within the sequence, since they can be effectively modeled through an appropriate definition of the arc set A and the arc costs c_{ij}.

The second model has a possibly exponential number of binary variables, each associated with a feasible circuit of G. Each circuit C_j, corresponding to a feasible pairing/roster for a crew, has an associated cost c_j. The binary variable associated with circuit C_j takes value 1 if C_j is part of the optimal solution, and 0 otherwise. We then have to solve a *set partitioning problem* with side constraints. In other words, we have to select a minimum-cost subset of circuits such that each node is covered by exactly one circuit, with additional inequalities forbidding the choice of circuit subsets leading to infeasible solutions (crew base constraints).

A main advantage of the set partitioning model is that it allows for circuit costs depending on the whole sequence of arcs, and possibly on the crew type. In addition, its linear programming relaxation is typically much tighter than in the previous model. Note however that the model often requires dealing with a very large number of variables. In some cases, the explicit generation of all feasible circuits is impractical, and one has to resort to a *column generation* approach, provided that an effective pricing procedure is available to find feasible circuits whose corresponding variable has a negative linear programming reduced cost.

In practice, the choice of the appropriate model and solution algorithm strongly depends on the particular structure of the problem in hand. According to our experience, the second model is particularly suitable for the cases in which feasible circuits cover a small number of nodes, and the constraints on the circuit feasibility are cumbersome and depend on the overall node sequence. This is the situation arising in railway crew scheduling, as described in Section 2. On the contrary, as already mentioned, the first model appears attractive for those cases where the main feasibility constraints concern the direct sequencin g of two nodes, since they can be dealt with implicitly by an appropriate definition of the arc costs. This is the case of railway crew rostering, as described in Section 3.

In the following sections, we discuss in some detail the solution techniques currently adopted at *FS - Ferrovie dello Stato*, the Italian railway company, for solving the crew scheduling and crew rostering phases.

2. Crew scheduling phase

Due to the nature of the services to be carried out, in railway applications a typical crew pairing covers only few trips. Moreover, heavy operational constraints affect pairing feasibility. This makes it practical to effect the explicit generation of all feasible pairings, which are computed and stored in a preprocessing phase calle d *pairing generation*. In addition, operational rules allow a crew to be transported with no extra cost as a passenger on a trip, hence the overall solution can cover a trip more than once. In this situation, in the *pairing optimization* phase, the set partitioning formulation can profitably be replaced by its *set covering problem* relaxation, obtained by imposing that each node is covered by at least one circuit. As a result, only inclusion-maximal feasible pairings, among those with the same cost, need be considered in the pairing generation. This considerably reduces the number of pairings to be generated, and hence the number of variables.

2.1 The pairing generation phase

The pairing generation phase calls for the determination of a set of feasible pairings from the given timetabled trips. Each pairing is a trip sequence starting and ending at the same depot, and satisfying the constraints described in the following.

First of all, *sequencing rules* impose that each pair of consecutive trips i and j in a pairing is *compatible*, namely:

- The arrival station of i coincides with the departure station of j.
- The time interval between the arrival of i and the departure of j is greater than a *technical time*, depending on i and j. This includes the times possibly required to change train, to perform maneuvers and other technical operations in the station.

An *external rest* of a pairing is a rest interval between two compatible consecutive trips, ending and starting at a station different from the depot, which exceeds the technical time by a given number of hours. A pairing can contain at most one external rest. A pairing is defined to be *overnight* if it requires some working during the night. Moreover, additional characteristics are associated with each pairing: *spread time* (defined as the time between the departure of its first trip and the arrival of its last trip), *driving time* (defined as the sum of trip durations plus all short rest periods), *working time* (defined as the driving time plus all the remaining rest periods, but the external rest), *paid time* (defined as the spread time plus possible additional transfer times within the depot). A pairing is defined to be *long* if it does not include an external rest and its working time is longer than a given value. Heavy *operational constraints*, concerning the characteristics of each pairing, are imposed. In addition, each pairing has an associated cost which takes into account its characteristics.

When the number of generated pairings is too big, the best pairings are selected as input for the pairing optimization phase, according to a score taking into account the pairing cost and the trips covered.

The pairing generation is performed through an enumerative algorithm which generates all the feasible pairings according to a depth-first scheme, and backtracks as soon as infeasibilities are detected. Simple bounds on the additional spread, working and driving time required to complete a partial pairing are used to speed-up the enumeration.

2.2 The pairing optimization phase

The pairing optimization phase requires the determination of a min-cost subset of the generated pairings covering all the trips and satisfying additional constraints. These constraints are known as *base constraints* and impose, for each depot: i) a lower and an upper bound on the number of selected pairings associated with the depot; ii) a maximum percentage of selected overnight pairings over all those associated with the depot; and iii) a maximum percentage of selected pairings with external rest over all those associated with the depot. Similar constraints are imposed with respect to the overall set of selected pairings.

Even without the above base constraints, set covering problems arising in the pairing optimization phase appear rather difficult, mainly because of their size. Indeed, the largest instances at the Italian railways involve up to 5,000 trips and more than 1,000,000 pairings.

The pure *Set Covering Problem* (SCP) can formally be defined as follows. Let $I_1, \ldots, I_n$ be the given collection of pairings associated with the trip set $M = \{1, \ldots, m\}$. Each pairing I_j has an associated cost $c_j > 0$. For notational convenience, we define $N = \{1, \ldots, n\}$ and $J_i = \{j \in N : i \in I_j\}$ for each trip $i \in M$. SCP calls for

$$v(\text{SCP}) = \min \sum_{j \in N} c_j x_j \tag{2.1}$$

subject to

$$\sum_{j \in J_i} x_j \geq 1, \qquad i \in M \tag{2.2}$$

$$x_j \in \{0, 1\}, \qquad j \in N \tag{2.3}$$

where $x_j = 1$ if pairing j is selected in the optimal solution, $x_j = 0$ otherwise.

The exact SCP algorithms proposed in the literature can solve instances with up to few hundred trips and few thousand pairings, see Beasley [5], Beasley and Jörnsten [8], and Balas and Carrera [2]. When larger instances are tackled, one has to resort to heuristic algorithms. Classical *greedy* algorithms are very fast in practice, but typically do not provide high quality solutions, as reported in Balas and Ho [3] and Balas and Carrera [2]. The most effective heuristic approaches to SCP are those based on *Lagrangian relaxation*. These approaches follow the seminal work by Balas and Ho [3], and then the improvements by Beasley [6], Fisher and Kedia [22], Balas and Carrera [2], Ceria, Nobili and Sassano [17], Caprara, Fischetti and Toth [11], and Wedelin [34]. Lorena and Lopes [30] propose an analogous approach based on *surrogate relaxation*. Recently, Beasley and Chu [7] and Jacobs and Brusco [28] proposed a genetic and a local search algorithm, respectively.

We briefly outline the heuristic method proposed by Caprara, Fischetti and Toth [11]. The algorithm is based on dual information associated with the widely-used Lagrangian relaxation of model (2.1)-(2.3). We assume the reader is familiar with Lagrangian relaxation theory (see, e.g., Fisher [21] for an introduction). For every vector $u \in R^m_+$ of Lagrangian multipliers associated with the constraints (2.2), the Lagrangian subproblem reads:

$$L(u) = \min \sum_{j \in N} c_j(u) x_j + \sum_{i \in M} u_i \tag{2.4}$$

subject to

$$x_j \in \{0,1\} \quad j \in N \tag{2.5}$$

where $c_j(u) = c_j - \sum_{i \in I_j} u_i$ is the *Lagrangian cost* associated with pairing (column) $j \in N$. Clearly, an optimal solution to (2.4)-(2.5) is given by $x_j(u) = 1$ if $c_j(u) < 0$, $x_j(u) = 0$ if $c_j(u) > 0$, and $x_j(u) \in \{0,1\}$ when $c_j(u) = 0$. The Lagrangian dual problem associated with (2.4)-(2.5) consists of finding a Lagrangian multiplier vector $u^* \in R^m_+$ which maximizes the lower bound $L(u)$. As (2.4)-(2.5) has the *integrality property*, any optimal solution u^* to the dual of the *Linear Programming* (LP) relaxation of SCP, namely problem $\max\left\{\sum_{i \in M} u_i : \sum_{i \in I_j} u_i \leq c_j \ (j \in N), u_i \geq 0 \ (i \in M)\right\}$, is also an optimal solution to the Lagrangian problem (see Fisher [21]). On the other hand, computing an optimal multiplier vector by solving an LP is typically time-consuming for very large scale instances. A commonly used approach for finding near-optimal multiplier vectors within a short computing time, uses the *subgradient vector* $s(u) \in R^m$, associated with a given u, defined by:

$$s_i(u) = 1 - \sum_{j \in J_i} x_j(u), \qquad i \in M. \tag{2.6}$$

The approach generates a sequence $u^0, u^1, \ldots$ of nonnegative Lagrangian multiplier vectors, where u^0 is defined arbitrarily. As to the definition of u^k, $k \geq 1$, a possible choice (Held and Karp [25]) consists of using the following simple updating formula:

$$u_i^{k+1} = \max\left\{u_i^k + \lambda \frac{UB - L(u^k)}{||s(u^k)||^2} s_i(u^k), 0\right\} \qquad \text{for } i \in M, \tag{2.7}$$

where UB is an upper bound on v(SCP), and $\lambda > 0$ is a given *step size parameter*.

For near-optimal Lagrangian multipliers u_i, the Lagrangian cost $c_j(u)$ gives reliable information on the overall utility of selecting pairing j. Based on this property, the Lagrangian (rather than original) costs are used to compute, for each $j \in N$, a *score* σ_j ranking the pairings according to their likelihood to be selected in an optimal solution. These scores are given on input to a simple heuristic procedure, that finds in a greedy way a hopefully good SCP solution. Computational experience shows that almost equivalent near-optimal Lagrangian multipliers can produce SCP solutions of substantially different quality. In addition, no strict correlation exists between the lower bound value $L(u)$ and the quality of the SCP solution found. Therefore it is worthwhile applying the heuristic procedure for several near-optimal Lagrangian multiplier vectors.

The approach consists of three main phases, described in detail in the following. The first one is referred to as the *subgradient phase*. It is aimed at quickly finding a near-optimal Lagrangian multiplier vector, by means of an aggressive policy. The second one is the *heuristic phase*, in which a sequence of near-optimal Lagrangian vectors is determined and, for each vector, the associated scores are given on input to the heuristic procedure to possibly update the incumbent best SCP solution. In the third phase, called *pairing fixing*, a subset of pairings having an estimated high probability of being in an optimal solution is selected, and the corresponding variables are fixed to 1. This leads to an SCP instance with a reduced number of pairings and trips (rows), on which the three-phase procedure is iterated. Computational experience showed that pairing fixing is of fundamental importance to obtain high quality SCP solutions. The above three phases are iterated until either all the trips are covered by the fixed pairings, or, as almost always occurs, the sum of the costs of the fixed pairings plus a lower bound on the cost of the residual problem is not less than the value of the best solution found so far.

In the following, M will denote the set of the trips that are not covered by the currently fixed pairings, and N the set of the pairings covering at least one trip in M.

As already mentioned, the **subgradient phase** is intended to quickly produce a near-optimal Lagrangian multiplier vector. The updating formula (2.7) is used.

The starting vector u^0 is defined in one of two different ways. In the first application of the three-phase procedure, u^0 is defined in a greedy way as follows:

$$u_i^0 = \min_{j \in J_i} \frac{c_j}{|I_j|}, \quad i \in M. \tag{2.8}$$

The other applications of the three-phase procedure start from the best multiplier vector (i.e., the one producing the best lower bound for the subproblem defined by the trips in M), say u^*, computed before the last pairing fixing, and obtain the starting multiplier vector u^0 through random perturbation of u^*. The perturbation lets the subgradient phase converge to a different multiplier vector, hence it allows the subsequent heuristic phase to produce different, and hopefully better, SCP solutions.

The upper bound UB is set to the value of the best SCP solution found so far. As for its initial value, it is computed by applying the greedy heuristic described in the following, by considering $u = 0$ (i.e., the original costs instead of the Lagrangian costs).

Parameter λ controls the step-size along the subgradient direction $s(u^k)$. The classical Held-Karp approach halves parameter λ if for p consecutive iterations no lower bound improvement occurs. In order to obtain a faster convergence, the following alternative strategy have been implemented. Every p subgradient iterations the best and worst lower bounds computed on the last p iterations are compared. If these two values differ by more than a given threshold, the current value of λ is halved. If, on the other hand, the two values are within a too small gap from each other, the current value of λ is increased. This last choice is motivated by the observation that either the current u^k is almost optimal (in which case one is not interested, in this phase, in obtaining a slightly better multiplier vector), or the small lower bound difference is due to an excessively small step-size (that is contrasted by increasing λ).

The computational experience shows that in many cases a large number of pairings happen to have a Lagrangian cost $c_j(u)$ very close to zero. In particular, this occurs for large scale instances with costs c_j belonging to a small range, after a few subgradient iterations. In this situation, the Lagrangian problem has a huge number of almost optimal solutions, each obtained by choosing a different subset of the almost zero Lagrangian cost pairings. As a result, a huge number of subgradients $s(u^k)$ to be used in (2.7) exist. It is known that the steepest ascent direction is given by the minimum-norm convex combination of the above active subgradients. However, the exact determination of this combination is very time consuming, as it requires the solution of a quadratic problem. On the other hand, a random choice of the subgradient direction may produce very slow convergence due to zig-zagging phenomena. This drawback is overcome by heuristically selecting a small-norm subgradient direction, computed by iteratively removing from the set of the currently selected pairings those which are *redundant*, i.e. the pairings whose removal does not change the set of trips covered by the solution of the Lagrangian subproblem.

The subgradient phase ends as soon as the convergence to a near-optimal Lagrangian vector is obtained. This occurs when the lower bou nd improvement obtained in the last subgradient iterations is smaller than a given threshold.

The **heuristic phase** generates a sequence of feasible solutions by using near-optimal Lagrangian vectors. Let u^* be the best Lagrangian vector found during the subgradient phase. Starting with u^*, a sequence of Lagrangian vectors u^k is generated, in an attempt to "explore" a neighborhood of near-optimal multipliers. To this end, the multipliers are updated as in the subgradient phase, but the subgradient norm is not reduced, so as to allow for a change in a larger number of components of u^k. The heuristic phase ends after a given number of iterations.

For each u^k, a greedy heuristic procedure is applied, to produce a "good" SCP solution. At each iteration of the greedy procedure, a *score* σ_j is computed for each not yet selected pairing j, and the pairing having the minimum value of the score is selected. The key step of the procedure is the definition of the pairing scores σ_j. Several rules have been proposed in the literature (see Balas and Ho [3] and Balas and Carrera [2]) which define σ_j as a function of c_j and $\mu_j = |I_j \cap M^*|$ (e.g., $\sigma_j = c_j$, or $\sigma_j = c_j/\mu_j$), where M^* represents the set of the currently uncovered trips. According to the computational experience reported in [11], these rules produce good results when c_j is replaced by

$$\gamma_j = c_j - \sum_{i \in I_j \cap M^*} u_i^k, \tag{2.9}$$

since this term takes into account the dual information associated with the still uncovered trips M^*. The use of γ_j instead of c_j in a greedy-type heuristic was first proposed by Fisher and Kedia [22].

The best solutions were produced by using a new rule giving priority to pairings having low cost γ_j and covering a large number μ_j of uncovered trips: $\sigma_j = \gamma_j / \mu_j$ if $\gamma_j > 0$, $\sigma_j = \gamma_j \mu_j$ if $\gamma_j < 0$.

The solution returned by the greedy procedure may contain redundant pairing. This happens because the pairings selected in a certain iteration to cover some uncovered trips, can lead a previously-selected pairing to become redundant. Removing the redundant pairings in an optimal way leads to an SCP, defined by the redundant pairing set, and by the trip set containing the trips covered only by the redundant pairings. This problem can be solved either exactly through an enumerative algorithm for small cardinalities of the redundant pairing set, or heuristically otherwise.

In the **fixing phase**, the heuristic solution available at the end of the heuristic phase is generally improved by fixing in the solution a convenient set of pairings, and re-applying the whole procedure to the resulting subproblem.

Clearly, the choice of the pairings to be fixed is of crucial importance. After extensive computational testing, the following simple criteria were implemented. Let u^* be the best multiplier vector found during the subgradient and heuristic phases, and define a pairing subset Q containing all the pairings j having Lagrangian cost $c_j(u^*)$ smaller than a given threshold. Each pairing $j \in Q$ for which there is a trip i covered only by j among the pairings in Q is first fixed. Then, the heuristic procedure previously described is applied to the subproblem defined by the trips not covered by the fixed pairings. The heuristic procedure is stopped when a prefixed number of pairings are chosen, and these pairings are fixed.

When very large instances are tackled, the computing time spent on the first two phases becomes very large. It is possible to overcome this difficulty by working in all the phases on a *core problem*, obtained from the original problem by keeping only a subset of the variables (pairings), the remaining ones being fixed to 0. The choice of the pairings in the core problem is often very critical, since an optimal solution typically contains some pairings that, although individually worse than others, must be selected in order to produce an overall good solution. Hence it is better not to "freeze" the core problem. Instead, a **variable pricing** scheme is used to update the core problem iteratively, in a vein similar to that used for solving large size LP's.

In order to dynamically update the core problem, the dual information associated with the current Lagrangian multiplier vector u^k is used. To be specific, at the very beginning of the overall algorithm a "tentative" core is defined, by taking a fixed number of pairings covering each trip. Afterwards, the current core is considered for, say, T consecutive subgradien t iterations, after which the core problem is re-defined as follows. The procedure computes the Lagrangian cost $c_j(u^k)$, $j \in N$, associated with the current u^k, and defines the pairing set of the new core by selecting both the pairings having a Lagrangian cost smaller than a given threshold, and, for each trip, a fixed number of pairings having the smallest Lagrangian costs. Notice that a valid lower bound for the overall problem is only available after the pricing step.

According to the computational experience reported in [11], for large scale instances the use of pricing cuts the overall computing time by more than one order of magnitude.

The solution available at the end of the three-phase procedure is typically close to the optimum, but in some cases it can be improved. For this purpose, a simple **refining procedure** is applied to a given SCP solution. The procedure assigns a score to the selected pairings, fixes to 1 the variables associated with the best scored pairings, and re-optimizes the resulting subproblem.

The algorithm proposed by Caprara, Fischetti and Toth [11], hereafter called CFT, was tested on the real-world instances provided by the Italian railway company within the competition FASTER, aimed at developing effective heuristics for very-large scale SCP instances. Table 2.1 reports the corresponding results. For each instance the table gives the instance name, the number of trips (m) and pairings (n), the density $\sum_{j \in N} |I_j|/(m \cdot n)$, the value of the lower bound LB computed by the subgradient procedure, the value of the heuristic solution found by algorithm CFT, and the best solution obtained by other methods. The reported solutions were obtained within time limits of 3,000 CPU seconds on a PC 486/33 for the first three instances, and 10,000 CPU seconds on a HP 9000 735/125 for the remaining instances.

The table shows that algorithm CFT is capable of providing near-optimal solutions within limited computing time even for very-large size instances. The average percentage gap between the lower bound and the heuristic solution value is 0.9%.

Table 2.1. Results on crew scheduling instances from Ferrovie dello Stato SpA.

Name	$m \times n$	Density	LB	CFT Sol.	Others' Sol.
FASTER507	$507 \times 63,009$	1.2%	173	174	174
FASTER516	$516 \times 47,311$	1.3%	182	182	182
FASTER582	$582 \times 55,515$	1.2%	210	211	211
FASTER2536	$2,536 \times 1,081,841$	0.4%	685	691	692
FASTER2586	$2,586 \times 920,683$	0.4%	937	947	951
FASTER4284	$4,284 \times 1,092,610$	0.2%	1051	1065	1070
FASTER4872	$4,872 \times 968,672$	0.2%	1509	1534	1534

3. The rostering optimization phase

In the rostering optimization phase the pairings selected in the crew scheduling phase are sequenced into cyclic rosters.

Most of the published works on crew rostering refer to urban mass-transit systems, where the minimum number of crews required to perform the pairings can easily be determined, and the objective is to evenly distribute the workload among the crews: see Jachnik [27], Bodin, Golden, Assad and Ball [10], Carraresi and Gallo [16], Hagberg [26], and Bianco, Bielli, Mingozzi, Ricciardelli and Spadoni [9]. Set partitioning approaches for airline crew rostering are described in Ryan [32], Gamache and Soumis [23], Gamache, Soumis, Marquis and Desrosiers [24], and Jarrah and Diamond [29]. Finally, related cyclic scheduling problems are dealt with in Tien and Kamiyama [33], and Balakrishnan and Wong [1].

We next give a description of the real-world crew rostering problem arising at the Italian railways. We are given a set of n pairings to be covered by a set of crew rosters. As previously mentioned, each pairing is characterized by a *start time*, an *end time*, a *spread time*, a *driving time*, a *working time* and a *paid time*. Moreover, each pairing can have additional characteristics: *pairing with external rest*, *long pairing*, *overnight pairing*. A roster contains a subset of pairings and spans a cyclic sequence of groups of 6 consecutive days, conventionally called *weeks*. Hence the number of days in a roster is an integer multiple of 6. The length of a roster is typically 30 days (5 weeks).

The crew rostering problem consists of finding a feasible set of rosters covering all the pairings and spanning a minimum number of weeks. As already discussed, the global number of crews required to cover every day all the pairings is equal to 6 times the total number of weeks in the solution. Thus, the minimization of the number of weeks implies the minimization of the global number of crews required. We next list the main constraints of the crew rostering phase.

Each week can include a limited number of pairings having particular characteristics: pairings with external rest, long pairings, overnight pairings. Furthermore, each week must be separated from the next one in the roster by a continuous rest, called *weekly rest*, which always spans the complete sixth day of the week. There are two types of weekly rests, conventionally called *simple* and *double* weekly rests, each having a minimum length (the minimum length of a double weekly rest being longer).

For each roster, the number of double weekly rests must be at least a given percentage of the total number of weekly rests, and the average weekly rest time must be greater than a given value. Moreover, for each (cyclic) group of 30 consecutive days within a roster, there are bounds on the number of pairings with external rest which can be included, and on the total paid time of the included pairings. Finally, for each (cyclic) group of 7 consecutive days within a roster, the total working time of the included pairings cannot exceed a given length.

Two consecutive pairings of a roster, say i and j, can be sequenced either *directly* in the same week, or with a simple or double rest between them. The break between the end of a pairing and the start of the subsequent pairing lasts at least a given number of hours, depending on the characteristics of both i and j, and on the break type (direct, or with an intermediate simple or double weekly rest).

The solution approach implemented within the Ferrovie dello Stato system is based on a heuristic algorithm proposed by Caprara, Fischetti, Toth and Vigo [12], which is driven by the information obtained through the computation of lower bounds.

Simple lower bounds can easily be obtained by considering each of the operational constraints imposing a limit either on the total number of pairings with a given characteristic, or on the total working and paid time in (cyclic) groups of consecutive days in a roster. A more sophisticated relaxation is proposed in Caprara, Fischetti, Toth and Vigo [12] in order to take into account all the rules for sequencing two consecutive pairings within a roster. The relaxation also imposes that the total number of weekly rests is equal to the total number of weeks making up the rosters, and that the total number of double weekly rests is at least the imposed percentage of the total number of weekly rests. The resulting relaxed problem calls for the determination of a minimum-cost set of disjoint circuits of a suitably defined directed multigraph G, whose nodes represent the pairings and the arcs between each pair of nodes represent the different possible ways of sequencing the corresponding pairings, i.e., directly or with a simple or double weekly rest in between. Each of these circuits corresponds to a, possibly infeasible, roster. Moreover, the set of selected circuits must satisfy the following constraints:

(i) each node of G is covered by exactly one circuit;
(ii) the total number of simple- or double-rest arcs in the circuits has to be at least the total cost of the circuits, expressed in weeks;
(iii) the total number of double-rest arcs in the circuits has to be at least the imposed percentage of the total number of simple- or double-rest arcs.

The relaxed problem is further relaxed in a Lagrangian way by dualizing constraints (ii) and (iii). The corresponding Lagrangian problem is solved as an Assignment Problem (AP) on a suitably defined cost matrix. Computational experience on railway instances showed that a tight lower bound may be obtained by simply considering two specific pairs of Lagrangian multipliers, i.e., by solving only two APs.

We finally describe the constructive heuristic of [12], which extensively uses the information obtained from the solution of the relaxed problem mentioned above. The heuristic constructs one feasible roster at a time, choosing in turn the pairings to be sequenced consecutively in the roster. Once a roster has been completed, all the pairings it contains are removed from the problem. The process is iterated until all pairings have been sequenced.

The procedure used to build each single roster, as it applies to the construction of the first roster, works as follows. One first computes the Lagrangian lower bound, and then starts building the roster by selecting its "initial pairing" which will be performed at the beginning of a week, i.e., preceded by a weekly rest. (The term "initial" is conventional as rosters are cyclic.) Once the initial pairing has been selected, a sequence of iterations is performed where:

a) the best pairing j to be sequenced after the current pairing i is chosen, according to an appropriate *score* taking into account the characteristics of pairing j and the possible lower bound increase occurring when j is sequenced after i;
b) the Lagrangian lower bound is parametrically updated;
c) the possibility of "closing" the roster is considered, possibly updating the best roster found.

The procedure is iterated until no better roster can be constructed, stopping anyway if the current roster spans more than 10 weeks.

When a complete solution to the problem is found, it is tried to improve it by applying a *refining procedure*, which removes the last rosters constructed (which are typically worse than the others) from the solution, and re-applies the heuristic algorithm to the corresponding pairings. To this end, some parameters of the roster construction procedure are either changed with a random perturbation or tuned so as to take into account the constraints that made the construction of the last rosters difficult.

The previously described lower and upper bounding procedures were tested on real-world instances provided by the Italian railway company within the competition FARO, aimed at developing effective heuristics for crew rostering. The results obtained are illustrated in Table 3.1. For each instance we report the instance name, the number of pairings, the best simple lower bound, the Lagrangian lower bound, the heuristic solution value, and the corresponding computing time, expressed in PC Pentium 90 CPU seconds.

The table clearly shows the effectiveness of the approach, since 6 out of 7 instances have been solved to proven optimality within no more than 20 minutes.

References

1. N. Balakrishnan and R.T. Wong. A Network Model for the Rotating Workforce Scheduling Problem. *Networks*, 20:25–42, 1990.
2. E. Balas and M.C. Carrera. A Dynamic Subgradient-Based Branch-and-Bound Procedure for Set Covering. *Operations Research*, 44:875–890, 1996.
3. E. Balas and A. Ho. Set Covering Algorithms Using Cutting Planes, Heuristics and Subgradient Optimization: A Computational Study. *Mathematical Programming Study*, 12:37–60, 1980.

Table 3.1. Results on crew rostering instances from Ferrovie dello Stato SpA.

Name	n	Lower Bounds		Heuristic Solution	
		Simple	*Lagrangian*	weeks	time
FARO021	21	6	7	7	8
FARO033	33	9	11	11	17
FARO069	69	18	19	19	650
FARO134	134	34	39	39	365
FARO164	164	43	48	48	106
FARO360	360	108	108	111	342
FARO386	386	110	118	118	443
FARO525	525	154	164	164	1185

4. C. Barnhart, E.L. Johnson, G.L. Nemhauser, M.W.P. Savelsbergh and P.H. Vance. Branch–and–Price: Column Generation for Solving Huge Integer Programs. In J.R. Birge and K.G. Murty (eds.), *Mathematical Programming: State of the Art 1994*, The University of Michigan, pages 186–207, 1994.
5. J.E. Beasley. An Algorithm for Set Covering Problems. *European Journal of Operational Research*, 31:85–93, 1987.
6. J.E. Beasley. A Lagrangian Heuristic for Set Covering Problems. *Naval Research Logistics*, 37:151–164, 1990.
7. J.E. Beasley and P.C. Chu. A Genetic Algorithm for the Set Covering Problem. *European Journal of Operational Research*, 94:392–404, 1996.
8. J.E. Beasley and K. Jörnsten. Enhancing an Algorithm for Set Covering Problems. *European Journal of Operational Research*, 58:293–300, 1992.
9. L. Bianco, M. Bielli, A. Mingozzi, S. Ricciardelli and M. Spadoni. A Heuristic Procedure for the Crew Rostering Problem. *European Journal of Operational Research*, 58:272–283, 1992.
10. L. Bodin, B. Golden, A. Assad and M. Ball. Routing and Scheduling of Vehicles and Crews: the State of the Art. *Computer and Operations Research*, 10:63–211, 1983.
11. A. Caprara, M. Fischetti and P. Toth. A Heuristic Method for the Set Covering Problem. Technical Report OR-95-8, DEIS University of Bologna, 1995, extended abstract published in W.H. Cunningham, S.T. McCormick and M. Queyranne (eds.) *Proceedings of the Fifth IPCO Conference*, Lecture Notes in Computer Science 1084, Springer, pages 72–84, 1996, to appear in *Operations Research.*
12. A. Caprara, M. Fischetti, P. Toth and D. Vigo. Modeling and Solving the Crew Rostering Problem. Technical Report OR-95-6, DEIS University of Bologna, 1995, to appear in *Operations Research.*
13. A. Caprara, M. Fischetti, P. Toth, D. Vigo and P.G. Guida. Algorithms for Railway Crew Management. *Mathematical Programming*, 79:125–141, 1997.
14. A. Caprara, M. Fischetti, P. Toth, D. Vigo and P.G. Guida. Solution of Large-Scale Railway Crew Planning Problems: the Italian Experience. Technical Report OR-97-4, DEIS University of Bologna, 1997, presented at the 7th International Workshop on Computer-Aided Scheduling of Public Transport, MIT, Cambridge, Massachusetts, August 1997.
15. P. Carraresi and G. Gallo. Network Models for Vehicle and Crew Scheduling. *European Journal of Operational Research*, 16:139–151, 1984.
16. P. Carraresi and G. Gallo. A Multilevel Bottleneck Assignment Approach to the Bus Drivers' Rostering Problem. *European Journal of Operational Research*, 16:163–173, 1984.

17. S. Ceria, P. Nobili and A. Sassano. A Lagrangian-Based Heuristic for Large-Scale Set Covering Problems. Technical Report R.406, IASI-CNR, Rome, 1995, to appear in *Mathematical Programming.*
18. J.R. Daduna and A. Wren. *Computer-Aided Transit Scheduling.* Lecture Notes in Economic and Mathematical Systems 308, Springer Verlag, 1988.
19. M. Desrochers and J.-M. Rousseau. *Computer-Aided Transit Scheduling.* Lecture Notes in Economic and Mathematical Systems 386, Springer Verlag, 1992.
20. J. Desrosiers, Y. Dumas, M.M. Solomon and F. Soumis. Time Constrained Routing and Scheduling. In M.O. Ball et al. (eds), *Handbooks in OR & MS*, Elsevier Science, 8:35–139, 1995.
21. M.L. Fisher. An Applications Oriented Guide to Lagrangian Optimization. *Interfaces*, 15:10–21, 1985.
22. M.L. Fisher and P. Kedia. Optimal Solutions of Set Covering/Partitioning Problems Using Dual Heuristics. *Management Science*, 36:674–688, 1990.
23. M. Gamache and F. Soumis. A Method for Optimally Solving the Rostering Problem. Les Cahiers du GERAD G-93-40, Montréal, 1993.
24. M. Gamache, F. Soumis, G. Marquis and J. Desrosiers, A Column Generation Approach for Large Scale Aircrew Rostering Problems. Cahiers du GERAD G-94-20, Montréal, 1994.
25. M. Held and R.M. Karp. The Traveling Salesman Problem and Minimum Spanning Trees: Part II. *Mathematical Programming*, 1:6–25, 1971.
26. B. Hagberg. An Assignment Approach to the Rostering Problem. In J.-M. Rousseau (ed.). *Computer Scheduling of Public Transport 2*, North Holland, 1985.
27. J.K. Jachnik. Attendance and Rostering Systems. In A. Wren (ed.). *Computer Scheduling of Public Transport*, North Holland, 1981.
28. L.W. Jacobs and M.J. Brusco. A Local Search Heuristic for Large Set-Covering Problems. *Naval Research Logistics*, 52:1129–1140, 1995.
29. A.I.Z. Jarrah and J.T. Diamond. The Crew Bidline Generation Problem. Technical Report, SABRE Decision Technologies, 1995.
30. L.A.N. Lorena and F.B. Lopes. A Surrogate Heuristic for Set Covering Problems. *European Journal of Operational Research*, 79:138–150, 1994.
31. J.-M. Rousseau. *Computer Scheduling of Public Transport 2*, North Holland, 1985.
32. D.M. Ryan. The Solution of Massive Generalized Set Partitioning Problems in Aircrew Rostering. *Journal of the Operational Research Society*, 43:459–467, 1992.
33. J.M. Tien and A. Kamiyama. On Manpower Scheduling Algorithms. *SIAM Review*, 24:275–287, 1982.
34. D. Wedelin. An Algorithm for Large Scale 0-1 Integer Programming with Application to Airline Crew Scheduling. *Annals of Operational Research* 57:283–301, 1995.
35. T.H. Wise. Column Generation and Polyhedral Combinatorics for Airline Crew Scheduling. Ph.D. thesis, Cornell University, 1995.
36. A. Wren. *Computer Scheduling of Public Transport*, North Holland, 1981.

An Introduction to Stochastic Transportation Models

François Louveaux

Département de méthodes quantitatives,
Facultés Universitaires N.-D. de la Paix,
B-5000 Namur, Belgium

Summary. In this paper, we consider routing and location models involving some uncertain data. We motivate through examples in Section 1 why stochastic models are relevant and useful. We indicate in Section 2 how to model uncertainty and give some algorithmic procedures in Section 3.

Introduction

There are many examples of problems in transportation where some elements are uncertain. Demands typically occur in a random fashion, in the distribution of goods as well as in systems responding to calls for emergency. Transportation systems have to be created in face of uncertainty about future level of demands, making strategic decisions difficult to take. Travel routes have to be designed in face of uncertainty about traffic conditions, hence about effective travel times. Models that take uncertainty into account are known as stochastic models, to differentiate them from deterministic models which assume all data are known with certainty.

Even if many decision makers realize the existence of uncertainty, most used decision models are deterministic. The reason for such a choice is that stochastic models are typically more difficult to solve than the deterministic ones. The common tendency is to try to solde very detailed deterministic models. As perfect forecasting does not exist, real data are very often different from the data used in the models. This results in fact in poor decisions being taken.

To avoid such pitfalls, we advocate the use of stochastic models when uncertainty may play a significant role. In particular, it could be wise to solve a more simplified version of a model, but at the same time, allow for some of the data to be random. In this paper, this approach is motivated by examples given in Section 1. These examples illustrate that the use of stochastic models leads to more robust decisions. Section 2 discusses the formulation of stochastic models while Section 3 describes algorithmic solution procedures to some stochastic routing problems.

A recent methodological survey on stochastic routing, that also includes the use of metaheuristics, can be found in Gendreau et al [10]. There are many other problems in transportation for which solutions to stochastic models have been proposed. As a few examples, we may cite the airline yield management (Brumelle and McGill [5]), dynamic vehicle allocation problems (Frantzeskakis and Powell [9]), the location of hazardous materials (Boffey and Karkazis [4]) or the location of emergency units using a queuing theory description of waiting times (Berman et al [1]).

Thus, the purpose of this paper is to provide a number of illustrative examples and certainly not to be an exhaustive review of existing research on stochastic models in transportation.

1. Importance of uncertainty

In this section, we illustrate the importance of using a stochastic model in one simplified example in routing and in one example in location of emergency units. These two examples serve as a motivation for the stochastic modelling approach in transportation and at the same time also illustrate some basic concepts of stochastic programming.

1.1 Routing with stochastic demand

Consider a classical vehicle routing problem where four clients have to be visited in a route starting and ending at the depot. Let $V = \{0, 1, \cdots, 4\}$ consists of the depot $\{0\}$ and the four clients. Assume one single vehicle of capacity $D = 10$ is available. There is no limit on the travel time, so that the vehicle can make several routes if needed and any set of routes is feasible provided the capacity limit of 10 is not violated. We assume here that the vehicle makes collections or deliveries, but not both. We are given in Table 1 a distance matrix $C = (c_{ij})$ where c_{ij} is the symmetrical distance between i and j.

	0	1	2	3	4
0	-	2	4	4	1
1	2	-	3	4	2
2	4	3	-	1	3
3	4	4	1	-	3
4	1	2	3	3	-

Data in Table 1 are illustrated in Figure 1. This distance matrix satisfies the triangle inequality although not necessarily strictly. The demand of clients 1, 2 and 4 is equal to 2 and is supposed to be known. Demand of client 3 is uncertain. It is either 1 or 7, with equal probability 1/2 each. All demands must be served, i.e. collected or distributed, whichever applies.

The treatment of the uncertainty on the level of demand of client 3 depends on the moment when information becomes available to the route planner. A first case is when the client can adequately forecast its demand. This

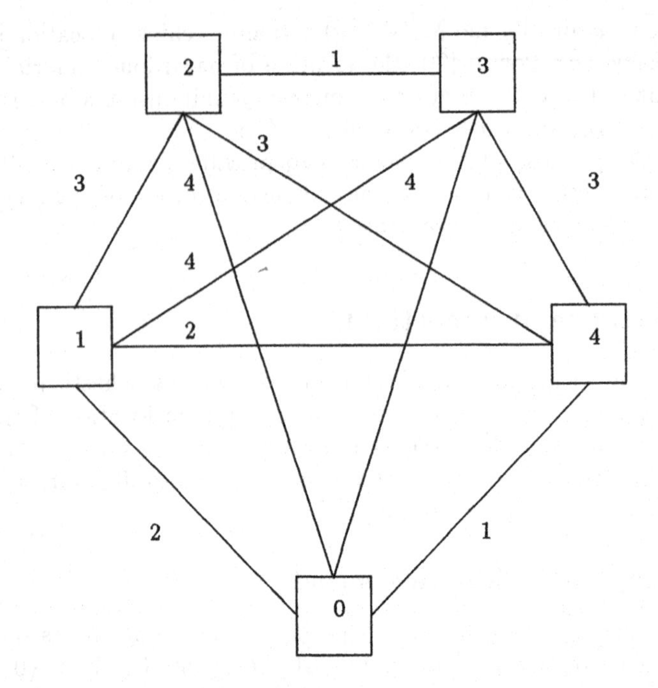

Fig. 1.1. Graph of Example 1

could be the case, for instance, if the products to be delivered are inputs in a fabrication process that works in batches and such that, the product is needed either at the level 1 or at the level 7, depending on the batch.
Similarly, in case of collection, the products may be wastes that are generated during specific periods of production. Then, if an agreement can be reached with the client, it would be possible to have knowledge of the demand before starting the tour. In the stochastic programming litterature, this is known as a situation of "A priori Information". The decision process is then relatively easy, and corresponds to the so-called "Wait and See" approach. It consists of making the route dependant on the received information. We assume in the sequel that demand is collected. All results can easily be adapted for the case of deliveries. Whenever client 3 requires a single unit of the product to be collected the vehicle's capacity is large enough to accomodate the demand of all clients. One single route can be travelled, which is the TSP route of shortest distance, i.e. $\{0, 1, 2, 3, 4, 0\}$, of distance 10. Whenever client 3 requires the collection of seven units of the product, the total demand 13 exceeds the vehicle's capacity, and the vehicle must travel two successive routes, each of which having a total demand less than 10. The combination of two routes of smallest distance is $\{0, 1, 4, 0, 2, 3, 0\}$ with total distance 14. As we assume client 3 has an equal probability 1/2 to call for a level of 1 and a level of 7, optimal routes of 10 and 14 will be travelled half the time each. It follows

that the mean distance travelled, or expected distance travelled, known as the Wait-and-see value, is equal to

$$WS = \frac{1}{2}(10 + 14) = 12 .$$

In many situations however, the information over client's 3 demand may not be available before the route is planned. Thus, the decision on the route must be taken before knowing the level of demand. One first approach then is to plan in view of the expected demand. This approach is known as the "expected value problem" or "mean value problem". Planning in view of the expected demand corresponds to assuming client 3's demand is certainly 4. Under this assumption, the total demand on the route is 10, and can be accomodated by one single route. The optimal tour if demand in 3 is 4 is again $\{0,1,2,3,4,0\}$.
Unfortunately, forgetting uncertainty while chosing the route does not mean uncertainty is absent. In fact, as mentioned before, demand turns out to be 1 half of the time and 7 the other half of the time, but in a random fashion. When demand is 1, the vehicle travels the planned tour, of distance 10. When demand is 7, the vehicle is unable to collect the total demand. Arriving at client 3, it loads 6 units (if the goods are divisible), returns to the depot to unload, then resumes its trip at 3 for the remaining units. Thus, the vehicle travels $\{0,1,2,3,0,3,4,0\}$. In the stochastic routing litterature, the situation when the vehicle is unable to load a client's demand is known as a failure (Dror and Trudeau [8]). The extra distance travelled due to this failure is a return trip to the depot. The total distance travelled is 18, the sum of the shortest length path (10) and the return trip (8). In conclusion, if the information over client 3's demand is only available when arriving at 3 and the planned route corresponds to the solution of the expected value problem, then the effective length travelled by the vehicle is 10 or 18, each with probability 1/2. The expected effective length is 14. It corresponds to the expectation of the expected value problem, denoted by EEV.

At this level, we may also observe that the route $\{0,4,3,2,1,0\}$ has exactly the same length as $\{0,1,2,3,4,0\}$. It corresponds to travelling the route in the opposite direction. In a deterministic setting, due to the assumption of symmetric distances, these two routes are totally equivalent. In the stochastic context, they are no longer equivalent. Indeed, for the case when demand of client 3 is 7, the total demand while traveling in the direction $\{0,4,3,2,1,0\}$ is already 9. It is thus known that the vehicle will be unable to load client 2's demand. The best solution is then to load client 3's demand, return to the depot to unload, then resume the trip at 2, to travel $\{0,4,3,0,2,1,0\}$. Returning to the depot from 3, and resuming the tour at 2, is known as making a preventive return at 3.
Due to the triangle inequality being satisfied, a preventive return is shorter than a return trip. ($c_{30} + c_{02} \leq 2c_{30} + c_{23}$ iff $c_{02} \leq c_{03} + c_{23}$, which is the triangle inequality). The effective length travelled if the planned tour is

$\{0,4,3,2,1,0\}$ and client 3's demand is 7 is now 17 (instead of 18) and the corresponding EEV reduces to 13.5.

Now, a better solution is to plan in view of the uncertainty. Such a solution consists of an a priori route plus either a return trip in case of failure or a preventive return to avoid failure. In the stochastic programming terminology, the a priori route corresponds to a first-stage solution as it is defined prior to the knowledge of the uncertainty and the returntrip or preventive return corresponds to a recourse policy or second-stage solution which is taken to mitigate the effect of uncertainty. An optimal solution is a combination of first-stage plus second stage solution that minimizes total expected cost (here total expected distance). In our example, the optimal solution is the a priori route $\{0,3,2,1,0,4,0\}$, plus a preventive return after 2 when demand at 3 is 7. The effective distance is either 11 if demand at 3 is 1 or 14 if it is 7. The expected effective length is 12.5. It is known as the solution of the recourse problem, denoted by RP.

From this example, we may already draw a number of conclusions. First, we may observe when drawing the optimal routes $\{0,3,2,1,0,4,0\}$ or the optimal route with preventive return $\{0,3,2,0,1,4,0\}$ that optimal routes cross themselves, a situation which never occurs in a deterministic setting. Second, in the optimal solution, a slightly longer route is taken to avoid any return trip. Thus, a slight increase in distance is accepted when demand turns out to be small in order to avoid a larger increase in distance when demand turns out to be high. In that sense, solutions are more robust.
Third, the three solutions concepts are $WS = 12, RP = 12.5$ and $EEV = 14$ (or 13.5 if the tour is travelled in reverse direction). It is relatively easy to show that the following relation always holds

$$WS \leq RP \leq EEV\ .$$

Now, the difference $EVPI = RP - WS$ is known as the "expected value of perfect information". It corresponds to how much is lost by not having the information in advance. In our example, $EVPI = 0.5$. It measures the cost of not having a perfect forecast. The difference $VSS = EEV - RP$ is known as the "value of the stochastic solution". It corresponds to the difference in expected cost between the optimal solution and the solution of the mean value problem. In our example, $VSS = 1.5$ (or 1). It measures the advantage obtained by solving a stochastic programming problem.

Unfortunately, those two values can only be computed a posteriori. Thus, in practice, it is difficult to know whether EVPI and VSS are high or low.

It is therefore advisable to solve the stochastic program whenever it is computationnally tractable. It should be clear from the next sections that such a solution is not always easy to find and is in general more difficult when the number of random elements increases.

1.2 Stochastic location problems

In a covering model, a customer or demand area is said to be covered if the distance (or travel time) to the customer or demand area is below some given threshold, which may depend on some regulations or on the customer's characteristics. As an example, in emergency units location, the threshold may be eight minutes in urban area, under normal traffic condition. The aim of the covering model is to find a set of locations which minimize the total set-up costs (or fixed costs) while covering all customers. Variable costs would typically not be taken into consideration as they are rather unsensitive to the chosen set of locations. Let $i = 1, \cdots, m$ be the index of customers or demand areas and $j = 1, \cdots, n$ be the index of potential site locations. Define x_j to be a binary decision variable equal to 1 if site j is open and 0 otherwise. Let f_j be the corresponding set-up cost.
Also define $N_i = \{j | t_{ij} \leq t_i\}$ to be the set of sites j within travel time t_i of i. This set is sometimes called the set of eligible sites for i, see Toregas et al (1971).
The problem then consists of

$$(CP) \min \quad Z(X) = \sum_{j=1}^{n} f_j x_j \tag{1.1}$$

$$s.t. \quad \sum_{j \in N_i} x_j \geq 1 \; , \; i = 1, \cdots, m \tag{1.2}$$

$$x_j \in \{0, 1\} \quad j = 1, \cdots, n \, . \tag{1.3}$$

Observe that if $f_j = f$ for all j, then the objective (1.1) is simply to minimize the number of open sets. Constraint (1.2) states that each demand area should be covered.

Now, typically in an emergency unit covering system, demands (i.e. calls for the emergency unit) occur in a random fashion. Thus, even if a demand area is covered, as it is close enough to an open site, demand may not always be satisfied. Indeed, when a call is placed, it may turn out that the unit which should serve it is currently busy serving another call. More robust models are now designed to ensure not only coverage but also a sufficiently high level of demand satisfaction. We now use the terminology of queueing theory where demands corresponding to calls and sites correspond to servers (or emergency units). For simplicity in the presentation, we assume each site has one server. Denote by q_j the probability (or fraction of time) that j is busy. Then $q_j^{x_j}$ is the probability that j is not available. Instead of simply requiring coverage as in (1.2), we now require that the probability of demand satisfaction should be langer than $1 - \alpha$, with α small. Such a constraint is known as a probabilistic constraint and is studied in the field of chance-constraint programming (Prékopa, [18]).

Equivalently, the constraint requires that the probability that no server is available to a given demand area i should be less than α, i.e.

$$\prod_{j \in N_i} q_j^{x_j} \leq \alpha \qquad i = 1, \cdots, m \tag{1.4}$$

Taking log's on both sides, one obtains

$$\sum_{j \in N_i} (\log q_j).x_j \leq \log \alpha \tag{1.5}$$

Two remarks are in order. First, one should observe that a major difficulty appears as the calculation of the values of the busy fractions q_j's in fact depends on the set of open sites x_j, see ReVelle and Hogan [19]. Thus, if we replace (1.2) by (1.5), the coefficients $(\log q_j)$ of x_j in the covering problems depend on the solution. Thus, the solution can only be obtained through trial and error. Second, we may better compare (1.5) to (1.2) if we assume identical busy fractions $q_j = q$, for all j. (Even if we only make this assumption in order to compare (1.2) and (1.5), one could also argue that it should be reasonable in a homogeneous region to see the various units having similar loads). When $q_j = q$ for all j, (1.5) becomes

$$\sum_{j \in N_i} x_j \geq \lceil \frac{\log \alpha}{\log q} \rceil \qquad i = 1, \cdots, m \tag{1.6}$$

where $\lceil \; \rceil$ is the rounding up operation, i.e. $\lceil a \rceil$ is the smallest integer greater than or equal to a. The right hand side of (6) is sometimes called the coverage factor. In (2), we have a coverage factor of 1. As an example, consider $\alpha = 0.05$. Then, the coverage factor is 1 if $q \leq 0.05$, 2 if $0.05 < q \leq 0.224$, 3 if $0.224 < q \leq 0.368$, etc $\cdots$ Thus, larger q's imply the need for a larger coverage factor. Here again, the use of a stochastic model through the definition of probabilistic constraints induce more robust solutions.

2. Modelling uncertainty: Recourse models

To model stochastic recourse problems, we define x to be the vector of first-stage decisions, ξ the vector of random variables and y the vector of recourse (or corrective or second-stage) decisions.

To give some examples, in a routing context, x may be the vector $x = (x_{ij})$ that describes the a priori route. If the problem is concerned with stochastic demands and restricted vehicle capacity (as in 1.1), ξ would be the vector $\xi = (d_1, \cdots, d_n)$ of all demands or of these demands that are random, and y would be the vector that describes the return trips or preventive returns. If the problem is concerned with stochastic travel times, $x = (x_{ij})$ would again describe a priori routes, $\xi = (t_{ij})$ would be the vector of uncertain travel times and y may describe the amount of overtimes on various routes.

A general stochastic linear program with recourse is written as

$$\begin{aligned} (SP) \qquad & \min \quad c^T x + E_\xi Q(x,\xi) \\ & s.t. \quad Ax = b \\ & \qquad\quad x \in X \end{aligned} \tag{2.1}$$

where $Q(x,\xi) = \min \{q^T y | Wy = h - Tx, y \in Y\}$ is the recourse or second-stage value function for a given ξ, E_ξ denotes the mathematical expectation w.r.t. ξ and ξ, the random vector, is formed of the random components of q, W, h and T.
x is a $(n_1 \times 1)$ vector, y is a $(n_2 \times 1)$ vector for one given ξ, A is $(m_1 \times n_1)$, W is a $(m_2 \times n_2)$ matrix for given ξ and all other matrices and vectors have conformable dimensions.

Defining $Q(x) = E_\xi Q(x,\xi)$ as the expected recourse function, the stochastic program with recourse is equivalent to

$$\begin{aligned} (SP) \min \quad & c^T x + Q(x) \\ s.t. \quad & Ax = b \\ & x \in X \end{aligned} \tag{2.2}$$

Clearly, all the difficulty of solving a stochastic program relates to the available properties of $Q(x,\xi)$ and $Q(x)$, and the easiness to obtain $Q(x)$ from $Q(x,\xi)$.

Example 2.1

Let (x_{ij}^k) be 1 if vehicle k travels arc (i,j) in a solution and o otherwise. Let $x = (x_{ij}^k)$ and $\xi = (t_{ij})$. Let t_{ij} be the stochastic travel time of arc (i,j). Define y_k= overtime of vehicle k with unit cost q_k. Assume x defines a set of m feasible routes. Then

$$Q(x,\xi) = \min \{\sum_{k=1}^{m} q_k y_k | y_k \geq \sum_{i,j} t_{ij} x_{ij}^k, y_k \geq 0\}$$

is the total cost of overtime over the m routes.

Example 2.2 (Williams 1963)

Consider $i = 1, \cdots, m$ sources and $j = 1, \cdots, n$ destinations. Let x_{ij} be the quantity transported from source i to destination j, with unit cost c_{ij}. Let demand at destination j be the random quantity d_j, with $\xi = (d_j)$. Assume a penalty q_j^- for unmet demand at destination j and a unit penalty q_j^+ for excess delivery at j.
The recourse problem takes the form

$$\begin{aligned} Q(x,\xi) &= \min \{\sum_j q_j^+ y_j^+ + \sum_j q_j^- y_j^- \ s.t. \\ & \quad y_j^+ - y_j^- = \sum_{i=1}^m x_{ij} - d_j \ , \ y_j^+ \geq 0, y_j^- \geq 0\} \, . \end{aligned}$$

Examples 2.1 and 2.2 are cases where the recourse function enjoys nice analytical properties, as it is the solution of a linear program. Such programs can thus routinely be solved by classical stochastic linear programing techniques such as the *L*shaped method of Van Slyke and Wets [22] and its many extensions (see chapter 5 in Birge and Louveaux [3]). Note however that example 2.1 requires in the first-stage the use of a three-index formulation, which might restrict the size of the tractable applications.

Example 2.3

Consider potential sites $j = 1, \cdots, n$. Let $x_j = 1$ if site j is open and o otherwise with fixed cost f_j. Let also z_j be the size of plant j, with variable cost q_j. Let $i = 1 \cdots m$ be the customers with demand d_j. Assume a unit profit of p_i per unit sold and a variable transportation and production cost c_{ij} for serving one unit of customer i's demand.

We than have first-stage decisions $x = (x_j)$ and $z = (z_j)$ and second-stage program

$$\begin{aligned} Q(x,z,\xi) = \max \quad & \sum_{i=1}^m \sum_{j=1}^n (p_i - c_{ij}) y_{ij} \\ & \sum_{j=1}^n y_{ij} \leq d_i \\ & \sum_{i=1}^m y_{ij} \leq z_j \end{aligned}$$

where y_{ij} is the quantity transported from j to i, and $\xi = (d_i)$.

This example is the stochastic extension of the classical simple plant location problem (Daskin [7]). One can observe here that, in this stochastic version, additional decisions are needed. Indeed, it is not sufficient to decide which plants are open to know their sizes, as the quantities plants are needed to serve are random. Thus, in addition to the binary decision variables $x = (x_j)$ for opening plants, the model must also contain variables $z = (z_j)$ on the sizes of these plants. Thus, the recourse model would read

$$\begin{aligned} \max \quad & -\sum_j f_j x_j - \sum_j q_j z_j + \mathcal{Q}(x, z) \\ s.t. \quad & 0 \leq z_j \leq M_j x_j \\ & x_j \in \{0, 1\} \end{aligned}$$

where $\mathcal{Q}(x, z) = E_\xi Q(x, z, \xi)$.
In the second-stage, optimal distribution variables y_{ij} are decided, on the basis of the particular value of the demands and on the previously decided sizes (or capacities) of the plants.
Clearly the recourse problem can be solved only for reasonable sizes of ξ. Thus, problems where the uncertainty corresponds to potential scenarios about future vectors of demands are tractable if the number of such scenarios is limited. On the other hand, problems where all demands d_i's are random independently of each other are untractable at present.

Finally, in some cases, the description of the recourse problem is extremely difficult to obtain in terms of an analytical representation.

Example 2.4

Consider the vehicle routing problem on a graph $G(V, E)$, with $V = \{v_0, \cdots, v_n\}$ the set of vertices, including depot v_0 and E the edge set, $E = \{(v_i, v_j), i < j\}$. Consider $x_{ij} = 1$ if arc (v_i, v_j) is in the solution and 0 otherwise, with travel cost c_{ij}. Assume $C = (c_{ij})$ is symmetrical and satisfies the triangle inequality. Consider a given number m of identical vehicles of given capacity D. Assume demand at vertex $d_i, i \geq 1$ is a random quantity. Define the a priori optimization problem as the one of finding a set of m a priori hamiltonians routes, each starting and ending at the depot, such that each client is visited exactly once. This corresponds to the first-stage decision. Now assume that in the second stage, each route is followed as planned, until possibly the vehicle capacity is exceeded. Assuming demands are collected, the vehicle must return to the depot, unload, then resume its trip where the failure occurs. This is the situation illustrated in the example of section 1. Thus, in the second stage, the recourse value function measures the expected penalty for return trips in case of failures. Even for one given set of demands, i.e. for one given ξ, the formulation of the recourse problem as a mathematical decision problem is extremely difficult to do. It requires in fact a three-index representation as the one given in example 2.1 but with additional binary variables in the second stage to represent the cost of the return trip for failure. Thus, stochastic routing models will typically have a second stage with binary decision variables. As such they belong to the class of stochastic integer programs, which are known to have very few useful analytical properties, see chapter 8 in Birge and Louveaux [3] or Stougie and Vander Vlerk [20]. See also Bertsimas *et al.* [2] for a survey on asymptotic results on a priori optimization.

On the other hand, computing the value of the recourse function for a given first-stage decision is relatively easy. This fact is used in the integer L-shaped method presented in section 3.

3. Solving stochastic routing problems

In this section, we give some examples of solution techniques applicable to stochastic routing problems. Consider a graph $G = (V, E)$, where $V = \{v_0, v_1, \cdots, v_n\}$ is a vertex set, $E = \{(v_i, v_j) : v_i, v_j \in V, i < j\}$ is an edge set. Vertex v_0 represents a depot at which are based m identical vehicles of capacity D. Remaining vertices are customers. A travel cost matrix $C = (c_{ij})$ and a travel time matrix $T = (t_{ij})$ are defined on E. C and T are assumed to be symmetrical. We assume the number of vehicles is given. The VRP consists of designing a set of m least cost vehicle routes starting and ending at the depot, such that each customer is visited exactly once, and possibly satisfying some side constraints.

Let x_{ij} be an integer variable equal to the number of times edge (v_i, v_j) appears in the first-stage solution. If $i, j > 0$, then x_{ij} can only take the values 0 or 1. If $i = 0, x_{ij}$ can also be equal to 2 if a vehicle makes a return trip between the depot and v_j. The problem is then

$$\min \sum_{i<j} c_{ij} x_{ij} \tag{3.1}$$

$$\sum_{j=1}^{n} x_{0j} = 2m \tag{3.2}$$

$$\sum_{i<k} x_{ik} + \sum_{j>k} x_{kj} = 2 \qquad k = 1, \cdots, n \tag{3.3}$$

$$\sum_{v_i, v_j \in S} x_{ij} \leq |S| - 1 \qquad (S \subset V \backslash \{v_0\}, 2 \leq |S| \leq n-2) \tag{3.4}$$

$$0 \leq x_{ij} \leq 1 \qquad (1 \leq i < j \leq n) \tag{3.5}$$

$$0 \leq x_{0j} \leq 2 \qquad (j = 1, \cdots, n) \tag{3.6}$$

$$x = (x_{ij}) \text{ integer.} \tag{3.7}$$

This problem is the m-Traveling Salesman Problem (m - TSP) which generalizes the classical 1-TSP formulation (Dantzig *et al.* [6]). Constraints (3.2) and (3.3) are called degree constraints. Subtour elimination constraints (3.4) prevent the formation of subtours disconnected from the depot.

3.1 Chance constraints

Consider a vehicle route $R = (v_{i_0} = v_0, v_{i_1}, v_{i_2}, \cdots, v_{i_t}, v_{i_{t+1}} = v_0)$. One example of chance constraint applies to the case of stochastic travel times T. Assume a limit B is placed on the duration of any route. Then, under stochastic travel time, a chance constraint model considers a route to be illegal if

$$P\left(\sum_{k=0}^{t} t_{i_k, i_{k+1}} > B\right) > \alpha \tag{3.8}$$

i.e. if the probability that the travel time on the route exceeds the limit B is larger than a given value α, typically a small number such as 5 %.
A second example of chance constraint applies to the case of stochastic demands $d = (d_j)$. There, a chance constraint model considers a route to be illegal if

$$P\left(\sum_{k=1}^{t} d_{i_k} > D\right) > \alpha \tag{3.9}$$

i.e. if the probability of the total demand on the route exceeding the vehicle capacity is larger than α.

Observe that (3.8) and (3.9) are expressed in terms of a given route, not in terms of the decision variables x_{ij} of (3.1)–(3.7).
Finding analytical expressions representing (3.8) or (3.9) is difficult. It can however be done when the route is given. Indeed, (3.8) and (3.9) can take the form

$$P\left(\sum_{k=0}^{t} t_{i_k i_{k+1}} x_{i_k i_{k+1}} > B\right) > \alpha$$

and

$$P\left(\sum_{k=0}^{t} \frac{d_{i_k} + d_{i_{k+1}}}{2} x_{i_k i_{k+1}} > D\right) > \alpha$$

where $x_{i_k i_{k+1}}$ is to be replaced by $x_{i_{k+1} i_k}$ when $i_k > i_{k+1}$.

First, the interested reader may check that these formulations have in general a non linear deterministic equivalent. (Take the case of travel times or demands following independent normal distribution. The variance term will induce non linearity.) Second, observe that these expressions do not generalize to the case of several routes, as the knowledge of the values of the x_{ij}'s variables does not suffice to describe the routes.

Fortunately, illegal routes can nevertheless be eliminated by linear constraints. For the case of stochastic travel times, a route for which (3.8) holds, can be eliminated by imposing

$$\sum_{k=0}^{t} x_{i_k i_{k+1}} \leq t \tag{3.10}$$

(where again $x_{i_k i_{k+1}}$ is to be replaced by $x_{i_{k+1} i_k}$ when $i_k > i_{k+1}$). This constraint only eliminates route R. Under some assumptions, all the permutations of route R can also be eliminated (Laporte *et al.* [16]).

For stochastic demands, define $S = \{v_{i_1}, v_{i_2}, \cdots, v_{i_t}\}$ to be the set of customers visited in route R. Define $V_\alpha(S)$ to be the smallest number of vehicles required to serve S, so that the probability that the total demand over S strictly exceeds D is smaller than or equal to α, i.e. $V_\alpha(S)$ is the smallest integer s.t.

$$P\left(\sum_{j \in S} d_{ij} > D.V_\alpha(S)\right) \leq \alpha\,. \tag{3.11}$$

Then a route for which (3.9) holds can be eliminated by the linear constraint

$$\sum_{\substack{v_i \in S, v_j \in \bar{S} \\ \text{or } v_i \in \bar{S}, v_j \in S}} x_{ij} \geq 2V_\alpha(S) \tag{3.12}$$

where $\bar{S} = V/S$ is the complement of S. Constraint (3.12) is a standard cut to require that at least $V_\alpha(S)$ vehicles serve a set S of customers (see e.g. Laporte [15]). As usual, subset elimination constraints (3.12) are typically embedded in a Branch and Cut procedure, in which they are added each time a subset S is found for which (3.9) holds. To give more insight on the meaning of $V_\alpha(S)$, consider the case where demands d_j's are independently distributed, according to a normal distribution, with mean μ_j and variance σ_j^2. Then, $\sum_{j \in S} d_j$ also follows a normal distribution, with mean $\sum_{j \in S} \mu_j$ and variance $\sum_{j \in S} \sigma_j^2$.

Now, applying the usual transformation on normal variates, (3.11) is transformed into

$$P\left(\frac{\sum_{j \in S} d_j - \sum_{j \in S} \mu_j}{\left(\sum_{j \in S} \sigma_j^2\right)^{1/2}} > \frac{DV_\alpha(S) - \sum_{j \in S} \mu_j}{\left(\sum_{j \in S} \sigma_j^2\right)^{1/2}}\right) \leq \alpha$$

or

$$\frac{DV_\alpha - \sum_{j\in S}\mu_j}{\left(\sum_{j\in S}\sigma_j^2\right)^{1/2}} \geq z ,$$

where z_α is the α-fractile of the $N(0,1)$ or Gaussian distribution. Hence

$$V_\alpha(S) = \lceil \frac{\sum_{j\in S}\mu_j + z_\alpha\left(\sum_{j\in S}\sigma_j^2\right)^{1/2}}{D} \rceil$$

where, as before, $\lceil . \rceil$ is the operation of rounding up to the next integer. This expression of $V_\alpha(S)$ shows, that under chance constraints, the number of vehicles to serve a set of customers is larger than what it would be in the deterministic case, i.e. $\lceil \frac{\sum_{j\in S}\mu_j}{D} \rceil$. This value increases either when smaller risks are allowed (smaller α's imply larger z_α's) or when variances of demands are higher.

Thus, chance constrained models are attractive and relatively simple to solve. However, they do not take into consideration cost issues related to the situations of failures. In order to take such costs into account, it is necessary to resort to a recourse model.

3.2 Recourse models

In a recourse model, the objective function now contains a recourse term and the model reads

$$\begin{aligned} (RM)\ \min \quad & \sum_{i<j} c_{ij}x_{ij} + Q(x) \qquad (3.13)\\ s.t. \quad & (10) - (15) \end{aligned}$$

where $Q(x)$ is the expected recourse function, as defined in section 2. This function takes into account the cost effect of corrective actions which have to be taken in the second-stage. Example 2.4 has considered the case where demands to be collected are uncertain, failure corresponds to the vehicle being unable to collect some client's demand, while second-stage corrective actions correspond to return trips or possibly preventive returns to unload. The function $Q(x)$ then evaluates the expected cost of those return trips. We may say that a chance constraint model as in 3.1 searches for solutions such that undesirable features (such as failures) only occur with little probability, while a recourse model as in 3.2 provides for an evaluation of the costs of corrective actions that cope with failures. By the way, observe that both models agree that it is usually impossible in a stochastic environment to have planning decisions such that failure would never occur.

The Integer L-shaped method is an extension of the Bender's decomposition approach. It operates on a "current problem" (CP) obtained from (RM) by three relaxations :

(i) integrality constraints (3.7) are relaxed,
(ii) subtour elimination constraints (3.4) are relaxed,
(iii) $Q(x)$ is replaced by a lower bound θ in the objective function.

As is classical in Branch-and-Cut techniques, integrality is recovered through a branching process and constraints (3.4) are introduced when they are found to be violated. In addition, a number of so called "optimality cuts" are gradually placed on θ.
The integer L-shaped proceeds as follows

Step 0 : Set $\nu = 0$ and $\bar{z} = +\infty$. The only pendant node corresponds to the initial current problem.
Step 1 : Select a pendant node from the list. If none exists, stop.
Step 2 : Set $\nu = \nu + 1$. Solve (CP). Let (x^ν, θ^ν) be an optimal solution (If no constraint exists on θ, θ is ignored in the computation).
Step 3 : Check for any subtour elimination constraint violation. If at least one violation of (3.4) is identified, introduce the corresponding constraints into (CP) and return to Step 2. Otherwise, if $cx^\nu + \theta^\nu \geq \bar{z}$, fathom the current node and return to Step 1.
Step 4 : If the solution is not integer, branch on a fractional variable. Append the corresponding subproblems to the list of pendant nodes and return to Step 1.
Step 5 : Compute $Q(x^\nu)$ and $z^\nu = cx^\nu + Q(x^\nu)$. If $z^\nu < \bar{z}$, update $\bar{z} = z^\nu$.
Step 6 : If $\theta^\nu \geq Q(x^\nu)$, fathom the current node and return to Step 1. Otherwise impose an optimality cut into (CP) and return to Step 2.

Finite termination of the Integer L-shaped depends on the existence of a complete set of optimality cuts, i.e. a set of cuts of the form $E_l.x + \theta \geq e_l, l = 1, \cdots, s$. s.t.
$(x, \theta) \in \{(x, \theta) | E_l x + \theta \geq e_l, l = 1, \cdots, s\} \Rightarrow \theta \leq Q(x)$.
Such a set is easily obtained through classical linear programming duality when the second-stage problem is defined by a linear program.
Example : Consider a 1-TSP problem with stochastic travel time matrix $T = (t_{ij})$. Assume traffic conditions can be represented by a limited number of scenarios $k = 1, \cdots, K$, with probability p_k. Thus, for one given k, the travel times (t_{ij}^k) are known for all arcs. Assume an allowed duration B for the tour and a penalty q for overtime. Then

$$Q(x) = q \sum_{k=1}^{K} p_k \left(\sum_{i<j} t_{ij}^k x_{ij} - B \right)^+$$

where $a^+ = \max\{a, 0\}$.

Consider a first-stage solution x^ν. Let $A = \{k | \sum_{i<j} t^k_{ij} x^\nu_{ij} > B\}$ be, the set of scenarios such that overtime is due if tour x^ν is chosen. Then

$$\theta \geq q \sum_{k \in A} p_k \left(\sum_{i<j} t^k_{ij} x_{ij} - B \right)$$

is a valid optimality cut. The set of all such cuts for all possible sets A is a complete set of optimality cuts. To see this, observe that $Q(x)$ is obtained by solving for each $k, k = 1, \cdots, K$, a second-stage program of the form

$$\min\{y^k | y^k \geq \sum_{i<j} x_{ij} - B, y^k \geq 0\}$$

and applying classical duality arguments. The finiteness is obvious, as K is finite and the above program has only two possible bases for each k.

As we already have seen earlier, most routing problems do not have a linear second-stage program. Now, assume that $Q(x)$ has some finite lower bound L and that $Q(x^\nu)$ can be computed when an integer x^ν is given. Then the optimality cut

$$\theta \geq L + (Q(x^\nu) - L) \left(\sum_{\substack{0<i<j \\ x^\nu_{ij}=1}} x_{ij} - \sum_{0<i<j} x^\nu_{ij} + 1 \right) \tag{3.14}$$

is valid, see Laporte and Louveaux [14]. These cuts use the fact that a feasible m−TSP solution with m fixed is fully characterized by the variables x_{ij} for which $i > 0$. Under constraint (3.14), either the current solution x^ν is maintained and (3.14) becomes $\theta \geq Q(x^\nu)$ or a new solution is chosen and (3.14) becomes $\theta \geq L$.

Optimality cuts (3.14) for all x^ν's trivially form a complete set of optimality cuts.

Using (3.14) in an efficient algorithm requires the capability of computing $Q(x^\nu)$ for given integer x^ν, of finding a lower bound L on $Q(x)$ and, when possible, on finding additional lower bounding functionals or lifting of (3.14). We now illustrate these points.

3.2.1 Calculation of $Q(x)$.

Example 1. The PTSP problem (Jaillet [13]) is the 1-TSP problem, where client i has probability p_i to place a call. In the first-stage, the problem consists of defining an a priori Hamiltonian path. In the second-stage, the tour is followed as planned but absent clients are skipped. The problem consists in determining an a priori tour of minimal expected length. Thus, here cx is the length of the a priori tour, $-Q(x)$ represents the expected savings obtained by skipping absent clients, and the expected length is $T(x) = cx + Q(x)$.

Thus the PTSP is $\min_x \; T(x) = cx + Q(x)$
s.t. (10) – (15), $m = 1$.
It is sufficient here to be able to compute $T(x)$, as $Q(x) = T(x) - cx$ immediately follows.

Consider a given a priori tour $x^\nu = \{v_{i_0} = v_0, \cdots, v_{i_1}, \cdots, v_{i_n}, v_{i_{n+1}} = v_0\}$. Define $t(k)$ to be the expected length from v_{ik} to depot if v_{i_k} is present. Now, $t(n+1) = 0, t(n) = c_{0,i_n}$ and

$$t(k) = \sum_{r=0}^{n-k} \prod_{j=1}^{r} \left(1 - p_{i_{k+j}}\right) \qquad p_{i_{k+r+1}} \left(c_{i_k i_{k+r+1}} + t(k+r+1)\right) \;,$$
$$k \;=\; n-1, \cdots, 0$$

Then $T(x^\nu) = t(0)$ is obtained by a dynamic programming recursion, which is $0(n^2)$ as $t(k)$ is $O(n)$ provided the successive calculations of $\prod_{j=1}^{r} \left(1 - p_{i_{k+j}}\right)$ are well organized.

Example 2. The m-VRP with m identical vehicles of capacity D and stochastic demands $d_j, j = 1, \cdots, n$ has been illustrated in Example 2.4. The calculation of $Q(x^\nu)$ can be found in Dror and Trudeau [8] under the assumption

i) that the cumulative load of the vehicle at any point is exactly equal to D with probability 0.
ii) the probability of having two failures on the same vehicle route is negligible.
iii) no preventive returns are allowed.

Assumption (ii) is realistic if it is required that the expected demand on a route does not exceed vehicle capacity, i.e. $\sum_{j \in S} \mu_j \leq D$, where S is the set of clients visited on a given route. To illustrate this, assume again demands are independent normal, $d_j \sim N(\mu_j, \sigma_j^2)$, hence $\sum_{j \in S} d_j \sim N\left(\sum_{j \in S} \mu_j, \sum_{j \in S} \sigma_j^2\right)$.
The probability of two failures on one route is

$$P\left(\sum_{j \in S} d_j > 2D\right) = P\left(z > \frac{2D - \sum_{j \in S} \mu_j}{\sqrt{\sum_{j \in S} \sigma_j^2}}\right)$$

where $z \sim N(0,1)$, is usually negligible if $\sum_{j \in S} \mu_j \leq D$ for any realistic coefficient of variations $\sigma_j / \mu_j, j \in S$.

Hjorring and Holt [12] provide a calculation of $Q(x)$ for the 1-VRP, where assumptions (i) – (ii) are removed and preventive returns are made when the cumulative load of the vehicle becomes exactly equal to D.

3.2.2 Calculation of the lower bound L.

1. A lower bound L on $Q(x)$ for the PTSP problem can be found in Laporte *et al.* [17].
2. For the case of the SVRP with stochastic demand and no preventive return, a lower bound L can be obtained as follows (Laporte and Louveaux [14]).

Relabel customers so that $c_{0,i+1} \geq c_{0,i}, i \geq 1$.
Define $f_k(x)$ to be the probability of having exactly k failures in solution x. By associating most likely failures to customers closest to depot, we may thus write

$$
\begin{aligned}
Q(x) &\geq 2\sum_{k=1}^{m} f_k(x).\left(\sum_{i=1}^{k} c_{0,i}\right) \\
&= 2\sum_{i=1}^{m} c_{0,i}\left(\sum_{k=1}^{m} f_k(x)\right) \\
&\geq 2\sum_{i=1}^{m} c_{0,i}q_i
\end{aligned}
$$

where q_i is any lower bound on the probability of having at least i failures, $i \leq m$. One such lower bound is obtained by aggregating the m vehicles' capacity on different routes, i.e.

$$q_i = P\left(\sum_{j=1}^{n} d_j > (m+i-1).D\right)$$

3.2.3 Lower bounding functionals.

1. A lower bounding functional for the PTSP can be found as follows. For a given arc x_{ij}, four possible cases can occur. If v_i and v_j are present, then the effective length is c_{ij}. This happens with probability p_ip_j. When v_i is absent and v_j is present, then the effective length will be at least $\min_{k\neq i}\{c_{jk}\}$. This situation occurs with probability $(1-p_i).p_j$. The case v_i present, v_j absent is symmetrical. Finally, when both vertices are present, we use a lower bound of 0. All together, for any arc (v_i, v_j), we obtain a lower bound b_{ij} on the expected distance actually traveled as

$$b_{ij} = p_ip_jc_{ij} + \frac{1}{2}p_i(1-p_j)\min_{k\neq j}\{c_{ik}\} + \frac{1}{2}p_j(1-p_i)\min_{k\neq i}\{c_{kj}\}$$

where the coefficients $1/2$ are placed to avoid double counting. It follows that

$$\theta \geq \sum_{i<j} (b_{ij} - c_{ij}) x_{ij}$$

is a valid lower bounding functional on θ, that can be placed at the root of the Branch and Cut tree.

2. A lifting of the cut (3.14) can be obtained for the 1-VRP problem with stochastic demands when a partial solution is available. This is established and implemented in Hjorring and Holt [12].

Conclusion

We have illustrated in this paper examples of problems for which uncertainty may play a significant role, in particular we have shown that decisions can be strongly modified by the presence of uncertainty on some data. We have also shown in some examples how the presence of uncertainty can be modelled and have provided illustrative examples of methods of solutions for some stochastic routing problems. As indicated in the introduction, a wide variety of stochastic models exist and can be found in the litterature.

Acknowledgment

This research was in part supported by the "Fonds National de la Recherche Scientifique" of Belgium. This support is gratefully acknowledged.

References

1. O. Berman, R.C. Larson, S.S. Chiu. Optimal server location on a network operating as a M/G/1 queue. *Operations Research*, 33:746–770, 1985.
2. D.J. Bertsimas, P. Jaillet and A.R. Odoni. A priori optimization. *Operations Research*, 38:1019–1033, 1990.
3. J. Birge, and F. Louveaux. Introduction to stochastic programming. Springer-Verlag, New-York, 1997.
4. B. Boffey and J. Karkazis eds. The location of hazadous materials facilities. Location Science, Vol 3, n^0 3, 1995.
5. S.L. Brumelle, J.I. Mc Gill. Airline seat allocation with multiple nested fare classes. *Operations Research*, 41:127–137, 1993.
6. G.B. Dantzig, D.R. Fulkerson and S.M. Johnson. Solution of a large-scale traveling-salesman problem. *Operations Research*, 2:393–410, 1954.
7. M.S. Daskin. Network and discrete location, models, algorithms and applications. Wiley Interscience, 1995.
8. M. Dror and P. Trudeau. Stochastic vehicle routing with modified savings algorithm. *European Journal of Operational Research*, 23:228–235, 1986.

9. L. Frantzeskakis, W. Powell. A successive linear approximation procedure for stochastic dynamic vehicle allocation problems. *Transportation Science*, 24:40–57, 1990.
10. M. Gendreau, G. Laporte, R. Seguin. Stochastic vehicle routing. *European Journal of Operations Research*, 88:3–12, 1996.
11. J. Goldberg, L. Paz. Locating emergency vehicle bases when service time depends on call location. *Transportation Science*, 264–280, 1991.
12. C. Hjorring and J. Holt. New optimality cuts for a single-vehicle stochastic routing problem. to appear in Annals of Operations Research, Recent Advances in Combinatorial Optimization, Theory and Applications, 1997.
13. P. Jaillet. A priori solution of a traveling salesman problem in which a random subset of the customers are visited. *Operations Research*, 36:929–936, 1988.
14. G. Laporte, F. Louveaux. Solving stochastic routing problems with the integer L-shaped method in fleet management and logistics. G. Laporte and T. Crainic, eds., Kluwer, 1997.
15. G. Laporte. Vehicle routing problems. in this volume, 1997.
16. G. Laporte, F. Louveaux, H. Mercure. The vehicle routing problem with stochastic travel times. *Transportation Science*, 26:161–170, 1992.
17. G. Laporte, F. Louveaux, H. Mercure. A priori optimization of the probabilistic traveling salesman problem. *Operations Research*, 42:534–549, 1994.
18. A. Prékopa. Stochastic programming. Kluwer Academic Publishers, Dordrecht, Netherlands, 1995.
19. C. ReVelle and K. Hogan. A reliability-constrained siting model with local estimates of busy fractions. *Environment and Planning B: Planning and Design*, 15:143–152, 1988.
20. L. Stougie, H. Van der Vlerk. Stochastic integer programming, Chapter 8 in Annotated bibliography in Combinatorial Optimization. M. Dell' Amico, F. Maffioli and S. Martello, eds, Wiley, Chichester, 1997.
21. C. Toregas, R. Swain, C. ReVelle and L. Bergmann. The location of emergency service facilities. *Operations Research*, 19:1363–1373, 1971.
22. R. Van Slyke and R. Wets. L-shaped linear programs with application to optimal control and stochastic programming. *SIAM Journal on Applied Mathematics*, 17:638–663, 1969.
23. A.C. Williams. A stochastic transportation problem. *Operations Research*, 11:759–770, 1963.

Facility Location: Models, Methods and Applications

Martine Labbé

Université Libre de Bruxelles
ISRO and SMG
CP 210/01, Boulevard du Triomphe
1050 Brussels - BELGIUM
mlabbe@smg.ulb.ac.be

Summary. This paper is devoted to some of the most important discrete location models. The Uncapacitated Facility Location Problem is first considered. Its properties, the most efficient exact method and some heuristics are presented. Then, extensions and related models proposed recently are reviewed.

1. Introduction

Facility location analysis is concerned with problem of locating one or several new facilities with regard to existing facilities and clients in order to optimize some economic criterion. Examples of facilities are plants, warehouses, depots, schools, hospitals, administrative buildings, departments stores, etc. The interest for such problems traces back to 1869, when Camille Jordan studied a problem about quadratic forms and incidentally considered the first location problem (on a tree network). It is Alfred Weber who is usually given credit for having introduced the first location model in a book in 1909. However, it was only after 1960 that significant progress occurred and that facility location emerged as a field. A bibliography devoted to facility location analysis, published in 1985 by Domschke and Drexl, contains over 1500 references. A glance at the papers reveals a large variance of problems.
Generally, one distinguishes three classes of location models depending on their location space, i.e. the domain of feasible locations. The first class contains those problems for which the location space is continuous. The most representative problem of this class is the so-called Fermat-Weber problem which consists in finding a point in $\mathbb{R}^m$ minimizing the sum of weighted distances to n given points. Of course multifacility problems and/or other objective functions (e.g. minimizing the largest distance) are also considered. Solution techniques for continuous location problems belong to the field of nonlinear programming. See the books of Francis *et al.* [34] and Love *et al.* [72] for introductory texts and the surveys of Michelot [76] and Plastria [83] for a more in-depth presentation and a detailed bibliography. Also, chapters of Drezner [28] present recent developments and new models.

The second class of location problems involves those on networks. One or several facilities must be located on a network in order to minimize an objective function of the distances to the nodes. The distance between two points is given by the length of a shortest path linking them in the network. Since a facility may be located anywhere on the network, (i.e. at a node or at an interior point of an edge), network location problems could be classified as continuous. However, in many of them, it is possible to identify a finite set of points of the network containing the optimal solution. The first such result was obtained by Hakimi [45]. He proved that the set of nodes contains a median, that is a point minimizing the sum of weighted distances to the nodes. In Hakimi [46], this result was extended to the p-median problem in which p points must be located on the network in order to minimize the sum of weighted distances between each node and the closest facility. The whole book of Handler and Mirchandani [47] and chapters of Daskin [26], Evans and Minieka [32], Francis *et al.* [34] and Mirchandani and Francis [77] discuss the main network location problems. Recent surveys include Brandeau and Chiu [12], Hansen *et al.* [50], Hansen *et al.* [51], Labbé *et al.* [64], Tansel *et al.* [93], and Tansel *et al.* [94].

Location problems for which a finite set containing an optimal solution is known are called discrete location problems and constitute the third class. Hence, many network location problems belong to that class. Furthermore, because of land availability, zoning regulations and facility design considerations, most real world facility location problems are modeled as such discrete optimization problems. Finally, the fact of having a finite number of possible locations makes those models very flexible in the sense that they allow to incorporate geographical and economic features.

Several chapters of Daskin [26] and Mirchandani and Francis [77] analyse the basic discrete location models, Labbé and Louveaux [65] present an annotated bibliography of those problems over the last ten years and Hodgson *et al.* [54] review their application literature.

Finally, there exist survey papers focusing on some particular type of location problems. Friesz *et al.* [35] and Eiselt *et al.* [29] consider competitive models, Ghosh and Harche [40] models in the private sector, Erkut and Neuman [30] models for locating undesirable facilities, and Louveaux [71] stochastic models. Several chapters of Drezner [28] also review competitive models and location–routing problems.

The remaining of this paper is devoted to discrete location problems. The so-called "Uncapacitated Facility Location Problem" which should be considered as the most classical and important one is considered into details in the next section. We present its properties, the most efficient exact method and heuristics as well as extensions and related models proposed recently. Section 3 is devoted to p-facility location problems.

2. The uncapacitated facility location problem

The uncapacitated facility location problem (in short UFLP) considered as the prototype of discrete problems was originally formulated by Balinski [6], Kuehn and Hamburger [63], Manne [73] and Stollsteimer [89].

A set I of clients who must be supplied from a facility where a commodity is made available and a discrete set J of possible facility locations are given. The fixed cost for opening a facility at site $j \in J$ is known and denoted by f_j and the cost of supplying the whole demand of client $i \in I$ from facility $j \in J$ is also known and denoted c_{ij} (this cost may include variable production costs and transportation costs). The problem consists in finding the number and locations of the facilities to be operated, as well as the allocation of clients to facilities, in order to minimize total costs. Formally, the *uncapacitated facility location problem* can be expressed as a mixed-integer linear program:

$$
\begin{array}{lrl}
\text{minimize} & \sum_{i \in I} \sum_{j \in J} c_{ij} x_{ij} + \sum_{j \in J} f_j y_j & (Z) \\
\text{s.t.} & \sum_{j \in J} x_{ij} = 1, \; i \in I & (D) \\
 & x_{ij} \leq y_j, \; i \in I, \; j \in J & (B) \\
 & x_{ij} \geq 0, \; i \in I, \; j \in J & (N_x) \\
 & y_j \in \{0,1\}, \; j \in J & (I_y)
\end{array}
$$

Variables y_j correspond to the decision of opening facility j and is a binary yes-no decision (see constraint (Iy)). Variable x_{ij} corresponds to the fraction of client i's demand satisfied from facility j. Constraint (D) means that the demand of each client must be totally satisfied. Constraints (B) states that demand can only be satisfied from open plants.

Variables y_j are considered as strategic since for a given set of $y_j \in \{0,1\}$, $j \in J$, it is easy to find a set of best values for the x_{ij}'s: set $x_{ij} = 1$ for the smallest j such that $y_j = 1$ and c_{ij} is minimum. This also implies that there exists an optimal solution $(x^\star, y^\star)$ to the UFLP in which $x^\star_{ij} \in \{0,1\}$ for all $i \in I$ and $j \in J$.

The above formulation of the UFLP is usually referred as the *strong formulation*. Another formulation of UFLP can be obtained by replacing constraints (B) by the more compact set of constraints

$$\sum_{i \in I} x_{ij} \leq |I|.y_{ij}, \; j \in J. \qquad (A)$$

To see that (B) can be replaced by (A), remark that if $y_j = 0$ than both (A) and (B) imply that $x_{ij} = 0$ for all $i \in I$. Furthermore, if for some $i \in I$ and $j \in J$, $x_{ij} > 0$ then y_j must take the value 1. However, these two formulations are only equivalent when the integrality constraints (Iy) are imposed. If we consider the LP-relaxation, i.e. we replace (I_y) by simple nonnegativity constraints (Ny) $y_j \geq 0, j \in J$, the second formulation appears to be weaker: its feasible solution set strictly contains the feasible solution set of the first formulation. Indeed, first notice that each constraint of type (A) is the sum of $|I|$ constraints of type (B). Second, the following example shows that this is a strict inclusion. We have n clients and n facilities. All fixed costs f_j, $j \in J$, are equal to 1 and for all $i \in I$, $j \in J$, $c_{ij} = 1$ if $i \neq j$ and $c_{ii} = 0$. Then the solution $y_i = 1/n$, $i \in I$ and $x_{ii} = 1$, $i \in I$ and $x_{ij} = 0$ if $i \neq j$ is feasible to the LP-relaxation of the second formulation but not of the first one.

In the remaining of this section, we briefly present the main properties and algorithms. Th reader is referred to Cornuéjols *et al.* [19] and Krarup and Pruzan [62] for further details and others applications of UFLP.

2.1 Properties of the UFLP

The UFLP is $\mathcal{NP}$-hard since the *Node Cover Problem* reduces to it. However, some polynomial special cases are known. Kolen [60] developped an $\mathcal{O}(|V|^3)$ algorithm when the clients and potential facility locations are given by the nodes of a tree with vertex set V and the c_{ij}'s are the distances on the tree between the corresponding nodes. Hassin and Tamir [53] refined this bound to $\mathcal{O}(|V|)$ when the tree is a path.

Let Π be the polytope of feasible solution to the strong formulation of the UFLP and Π_{LP} the polytope of feasible solutions to the LP-relaxation of this formulation. For $|I| \leq 2$ or $|J| \leq 2$, all extreme points of Π_{LP} are integral, see Krarup and Pruzan [62]. For values as small as $|I| = |J| = 3$, Π_{LP} has fractional extreme points. Those were completely characterized for any value of $|I|$ and $|J|$ by Cornuéjols *et al.* [18]. Regarding Π, Cho *et al.* [15] provides its complete description in terms of inequalities when $|I| \leq 3$ or $|J| \leq 3$. For $|I| \geq 4$ and $|J| \geq 4$, only a partial description of Π can be provided. The most significant results can be found in Cho *et al.* [15], [16] and Cornuéjols and Thizy [21]. Further, the inequalities $x_{ij} \leq y_j$, $i \in I$, $j \in J$, $x_{ij} \geq 0$, $i \in I$, $j \in J$ and $y_j \leq 1$, $j \in J$ are facets of Π. They are usually called the *trivial facets* and are very useful in practice since the LP-relaxation of the strong formulation of UFLP often yields an integer solution. This observation is reinforced by the following result of Ahn *et al.* [4].

Consider an instance of the UFLP with $|I| = |J|$ and obtained by generating n points in the unit square according to the uniform distribution. Let c_{ij} be the Euclidian distance between points i and j and assume that all fixed costs f_j are equal to f. If $n^{-\frac{1}{2}+\epsilon} \leq f \leq n^{\frac{1}{2}+\epsilon}$ for some fixed $\epsilon > 0$, then

$$\frac{(Z^* - Z_{LP})}{Z^*} \sim 0.00189$$

almost surely, where Z^* denotes the optimal value of the SPLP and Z^{LP} the optimal value of the LP-relaxation of the strong formulation.

The above results allow to conclude that the LP-relaxation of the strong formulation of the SPLP is very tight. Hence, if we are able to solve it efficiently, there are good chances to be also able to solve the SPLP itself. The algorithm DUALOC of Erlenkotler for the UFLP is based on this idea. Looking at the formulation of UFLP, one immediately notices that the major difficulty is the number of constraints of the model, $\mathcal{O}(mn)$. So, for large values of m and n it is not possible to solve the LP-relaxation with a commercial LP-solver such as CPLEX and a specially tailored method must be derived. The main loop of DUALOC is such a heuristic: it attempts to solve the dual of this LP-relaxation.

2.2 Solving the UFLP with DUALOC

Consider the LP-relaxation of UFLP given by

$$\begin{array}{lrl} \text{minimize} & \sum_{i\in I}\sum_{j\in J} c_{ij}x_{ij} + \sum_{j\in J} f_j y_j & (Z_p) \\ \text{s.t.} & \sum_{j\in J} x_{ij} = 1, \ i \in I & (D) \\ & x_{ij} \leq y_j, \ i \in I, \ j \in J & (B) \\ & x_{ij} \geq 0, \ y_j \geq 0, \ i \in I, \ j \in J & (N) \end{array}$$

Note that the constraints $y_j \leq 1, j \in J$ are redundant given (D) and (B) and assuming that all fixed costs $f_j \geq 0$.

Let λ_i and μ_{ij} denote the dual variables associated with (D) and (B) respectively. The dual problem can then be written as follows:

$$\begin{array}{lll} \text{maximize} & \sum_{i\in I} \lambda_i & (Z_D) \\ \text{s.t.} & \sum_{i\in I} \mu_{ij} \leq f_j, \ j \in J & (F) \\ & \lambda_i - \mu_{ij} \leq c_{ij}, \ i \in I, \ j \in J & (C) \\ & \mu_{ij} \geq 0, \ i \in I, \ j \in J & (N_\mu) \end{array}$$

From $(C)$and (N_μ), one concludes that for any fixed values of the λ_i,

$$\mu_{ij} = (\lambda_i - c_{ij})^+ = \max\{0, \lambda_i - c_{ij}\}.$$

This allows to eliminate the μ_{ij}'s and to derive a condensed version of the dual:

$$\begin{array}{lll} \text{maximize} & \sum_{i\in I} \lambda_i & \\ \text{s.t.} & \sum_{i\in I} (\lambda_i - c_{ij})^+ \leq f_j, \ j \in J & (F^c) \end{array}$$

Bilde and Krarup [11] and Erlenkotter [31] proposed a greedy-like heuristic called *Dual Ascent* for solving this condensed dual. It starts with values for the $\lambda_i, i \in I$ satisfying (F^c) and then tries to increase them one at a time while maintaining (F^c). This heuristic appears to be extremely efficient but there is no guarantee to obtain the optimal solution to UFLP. So, together with some final improvement procedure, it is embedded in a branch-and-bound procedure yielding Erlenkother's algorithm DUALOC.

More precisely, Dual Ascent starts with $\lambda_i = \min_i c_{ij}$. Then, it cycles through the clients $i \in I$ in order to increase λ_i without violating (F^c). If λ_i can be increased, then it is increased to $\min\{c_{ij} \mid c_{ij} > \lambda_i\}$ if (F^c) holds. Otherwise λ_i is increased to the maximum value allowed by (F^c). If λ_i cannot be increased, then λ_{i+1} is considered. When no λ_i can be increased anymore, the procedure stops.

The complementary slackness conditions allows one to construct a primal solution $(x(\lambda), y(\lambda))$ from a dual solution (λ). Let

$$S(\lambda) = \{j \in J \mid \sum_{i \in I} (\lambda_i - c_{ij})^+ = f_j\}.$$

For each $i \in I$, determine $j \in S(\lambda)$ such that c_{ij} is minimum and set $x_{ij}(\lambda) = 1$ and $y_j(\lambda) = 1$. This amounts to allocate each client j to the cheapest (in terms of c_{ij}) facility j of $S(\lambda)$, i.e. for which (F^c) is tight.

If the values of the primal and dual solutions $(x(\lambda), y(\lambda))$ and λ are not equal, the gap between them can possibly be reduced by decreasing one λ_i to its previous value in Dual Ascent. This creates slacks on some constraints (F^c) and Dual Ascent can be applied again. If an improved dual solution is obtained, a new primal solution is then derived. The procedure is iterated until either an optimal solution has been found or no further improvement is possible. In the later case, a branch-and-bound procedure is initiated. Branching is based on facility j for which the complementary condition $\mu_{ij}(x_{ij} - y_j) = 0$ is violated and the branch-and-bound tree is explored with a depth-first-search strategy.

The whole procedure constitutes DUALOC and has been successful in solving problems with $|I| = |J| = 100$ in time varying between 1 and 3 seconds, see Erlenkotter [31]. Körkel [61] improved DUALOC to solve large scale UFLP. On average, Körkel's algorithm, PDLOC is more than 58 times faster than DUALOC. Körkel also solved optimally problems with $|I| = |J| = 3748$ and corresponding to German cities and demands coming from telephone traffic data.

2.3 Primal heuristics

Despite the good performance of exact algorithms such as DUALOC, there is still a need for heuristic methods. First, heuristics are the only way to tackle very large problems. second, most exact procedures converge more rapidly when good initial solutions are available. Several heuristics have been proposed for the UFLP and only the main ones will be presented here.

The primal heuristics are based on the following combinatorial formulation of the UFLP:

$$\underset{S \subseteq J}{\text{minimize}} \quad Z(S)$$

where

$$Z(S) = \sum_{i \in I} \min_{j \in S} c_{ij} + \sum_{j \in S} f_j.$$

This formulation takes into consideration only the strategic decisions of opening facilities and the allocation of clients is implicitly made in the most economical way.

The first method, known as the GREEDY heuristic, has been proposed by Kuehn and Hamburger [63]. For $S \subseteq J$ and $j \in J \setminus S$, let

$$\begin{aligned} \rho_j(S) &= Z(S) - Z(S \cup \{j\}) \\ &= \sum_{i \in I} \left(\min_{k \in S} c_{ij} - c_{ij} \right)^+ - f_j, \end{aligned}$$

which corresponds to the decrement in the objective when opening facility j in addition to the set S. In the initial step, we select facility j such that

$$Z(\{j\}) = \max_{k \in J \setminus S} \rho_k(S)$$

and the procedure stops when $\rho_j(S) < 0$ for all $j \in J \setminus S$.

The GREEDY heuristic requires at most $|J|$ iterations and each iteration requires $\mathcal{O}(|I||J|)$ calculations. So, the overall running time is $\mathcal{O}(|I||J|^2)$.

It can be shown that the GREEDY solution S^G deviates in value from the optimal one by at most $\sum_{j \in S^G} f_j$. Hence, when the fixed costs f_j are small in comparison to the transportation costs c_{ij}, the GREEDY heuristic will yield a small error. This is an *a posteriori* bound which depends of the particular instance of the UFLP we consider. A general relationship can be derived between the values $Z(S^G)$ of the GREEDY solution, Z_{LP} of the optimal solution to the LP-relaxation of the strong formulation and $Z(J)$ of the whole set of possible locations. Cornuéjols *et al.* [18] prove that

$$Z(S^G) \leq \frac{(e-1)}{e} Z_{LP} + (\frac{1}{e}) Z(J).$$

The STINGY heuristic, proposed by Feldman *et al.* [33] proceeds by starting with an initial situation where all facilities $j \in J$ are open: $S := J$. Then for any $j \in S$,

$$\begin{aligned} \sigma_j(S) &= Z(S \setminus \{j\}) - Z(S) \\ &= \sum_{i \in I} (\min_{k \in S \setminus \{j\}} c_{ik} - c_{ij})^- + f_j, \end{aligned}$$

denotes the increase in the objective function when closing facility j, where $(a)^- = \min\{a, 0\}$.
At each iteration, the facility $j \in S$ such that $\sigma_j(S) = \min_{k \in S} \sigma_k(S)$ is selected and j is removed from S, $S := S \setminus \{j\}$. The procedure stops dropping facilities when $\sigma_j(S) > 0$ for all $j \in S$.

Finally, there is the INTERCHANGE procedure developed by Teitzand and Bart [95]. Here we assume that an arbitrary set S of facilities is given a priori and that facilities are moved iteratively, one by one, to vacant sites with the purpose of reducing total cost. For $j \in S$ and $k \in J \setminus S$, let

$$\begin{aligned} \tau_{jk}(S) &= Z(s) - Z((S \setminus \{j\}) \cup \{k\}) \\ &= \sum_{i \in I} \max\{\min_{l \in S} c_{il} - c_{ik}, \min_{l \in S} c_{il} - \min_{l \in S \setminus \{j\}} c_{il}\} - f_j + f_k, \end{aligned}$$

denote the variation of the objective function of the UFLP when facility $j \in S$ is replaced by facility $k \notin S$. INTERCHANGE selects $j \in S$ and $k \in J \setminus S$, computes $\tau_{jk}(S)$ and sets $S := (S \setminus \{j\}) \bigcup \{k\}$ if this value is positive. The procedure stops when no further change decreases the objective function value.

The outcome of INTERCHANGE depends on the initial set of facilities. Empirically, very good results are often obtained by starting with the solution produced by GREEDY or STINGY; performing two consecutive runs initiated with these two candidate solutions and retaining the better solution gives either the optimal solution or a tight approximation.

Even if good solutions are generally obtained by using the above heuristics, one may be willing to apply some metaheuristic such as simulated annealing or tabu search (cf. Reeves [86]). A natural neighbourhood for a solution $S \subseteq J$ is composed of all solution S' obtained by opening, closing or moving one facility in regard to S.

2.4 Recent extensions and related models

It is sometimes meaningful to consider costs associated to the used capacity of open facilities. Verter and Dincer [97] consider the situation where, in addition to a fixed cost for opening a plant, a concave cost of acquisition of capacity is incurred. Due to concavity, each customer is server by only one facility. The algorithm of Verter and Dincer [97] uses successively refined piecewise linear concave approximations. In the model of Desrochers *et al.* [27] a congestion factor is added to the objective function. It corresponds to an increasing convex cost related to the delay incurred by each customer at facility j. The problem is solved using a column generation technique within a branch-and-bound scheme.

Another generalization consists in introducing stochastic elements. Louveaux and Peeters [70] transform the UFLP into a two-stage stochastic program with recourse. Demand, selling prices, production and transportation costs may be considered as random elements. A dual ascent is presented, and the monotone improvements available in the original model are proved to be achievable in the stochastic case. Laporte *et al.* [66] considers the optimal location and sizing of the facilities given that the future demand is uncertain. When demand is known, the available capacity is allocated to clients, which may result in loss sales. An exact algorithm is presented and tested on medium size problems. Psaraftis *et al.* [85] consider an application of a stochastic UFLP in the oil industry. The authors consider the optimal location and size of equipments to fight oil spills. A detailed description of the cost to recover damages and of the environmental cost of the unrecovered damages is provided. Random elements include the arrival of oil spills (independent and non-concomitant Poisson processes) and the size of oil spills (small, medium, large).

Stochastic models have also been proposed for the location of public facilities. Louveaux [69] presents a stochastic version of the p-median where demands and service costs may be random. The p-median problem consists in locating p facilities in order to minimize the total transportation cost (see Section 3). Louveaux [69] provides alternative ways to impose the budget constraint, usually given by the upper bound p of facilities to open, and shows the links to the stochastic UFLP. Gregg *et al.* [42] consider the optimal siting and closing of public facilities. A two stage model is proposed. The level of service to demand regions are decided at the first stage. When demand realize, overage and underage penalties are paid. A detailed application to the public libraries in Queens Borough is provided.

Among the special cases of the UFLP for which there exists a polynomial algorithm, the Economic Lot Sizing Problem is probably the most famous. Wagelmans *et al.* [98] present an algorithm in $\mathcal{O}(n \log n)$ time for solving this problem for some special cases including the Wagner–Within case. This algorithm relates directly to solving the dual of the LP-relaxation of the strong formulation of the UFLP. Jones *et al.* [56] propose an algorithm for solving some specially structured UFLP in $\mathcal{O}(mn)$. They also show that several other problems are instances of this specially structured UFLP. This is the case for the stochastic UFLP, the Economic Lot Sizing Problem and the Capacity Expansion Problem.

Some attempt has been made to include price effects in the UFLP. Logendran and Terrel [68] assume price–sensitive and stochastic demands. First stage decisions consists of selecting facilities to open as well as the allocation of clients to facilities. Quantities to be transported are optimized for each plant–customer combination in view of the demand. Hanjoul *et al.* [49] consider a model in which the price, number, locations, sizes and market areas of the plants supplying the clients must be determined in order to maximize the profit of the firm. Alternative spatial price policies are considered (uniform mill pricing, uniform delivered pricing, spatial discriminatory pricing). The solution methods consist of branch-and-bound algorithms based on a new upper bounding function.

Some applications involve two types or levels of facilities. For example, the distribution networks of many companies often involve major (central) as well as minor (regional) depots. The central facilities supply the regional ones and such shipment quantities are typically large, whereas each client is served from a regional depot where the transport is usually carried out using smaller vehicles. Another application concerns garbage collection. A truck travels from a depot to the client and then to a disposal plant, so each client is assigned to a truck depot and a disposal plant. In situations where spent products are recycled, each client has to be assigned to a supply facility and a recycling facility. Numerous other applications of two-level models exist within areas such as telecommunications and computer network design. Gao and Robinson [36] consider such a model with two levels in which the products are shipped from one level, echelon 1, to the demand points via an intermediate second level of facilities, echelon 2. A fixed cost is associated to each pair of echelon 1 and echelon 2 facilities which serve at least one client. A dual ascent procedure is proposed and embedded into a branch-and-bound algorithm. Computational experiments are reported for problems with up to 25 possible locations at each level and 35 clients. Barros and Labbé [8] consider a two level uncapacitated facility location model involving fixed costs for opening facilities of both types and for having pairs of facilities of different type operating together. A Lagrangean dual approach is used within a branch-and-bound method for solving the problem. Computational experiments are reported for problems with up to 10 or 15 possible locations for each level and 50 clients.

Primal heuristics such as GREEDY, STINGY or INTERCHANGE can be easily adapted to the two-level uncapacitated facility location problem. However no worst case behaviour is known a priori. Such a result holds for GREEDY when applied to the UFLP because this problem is submodular, which is not the case for the two-level problem, see Barros and Labbé [9].

The polyhedral structure of the convex hull of feasible solutions to the two-level uncapacitated facility location problem is studied in Aardal *et al.* [2]. In particular it is shown that all non trivial facets of UFLP define facets of the two-level problems, and conditions when facets of the two-level problem are also facets for the UFLP are derived. New families of facets and valid inequalities are also introduced.

The UFLP can also be generalized to the case where different products are required by the customers. Then, the problem consists of minimizing the total of fixed costs for opening facilities, fixed costs for handling a particular product at a given location and assignment costs for satisfying demands of various products. Klincewicz and Luss [59] extends Erlenkotter's dual ascent and dual adjustment procedures to generate a good feasible solution to the dual of this multiproduct uncapacitated facility location problem. Primal feasible solutions are constructed from the dual solutions and the procedure is embedded in a branch-and-bound algorithm to reach optimality.

A multicommodity location -allocation problem related to the management of a fleet of vehicles over a medium to long-term planning horizon is considered in Crainic *et al.* [22], Crainic *et al.* [23], Crainic and Delorme [24], Crainic *et al.* [25] and Gendron and Crainic [38]. The first paper presents the modelling issues coming from the fact that several modes and several products are involved and that some balancing requirements are imposed. The other four papers propose various solution techniques for the problem.

Maybe the most important extension of the UFLP consists in introducing capacities for the facilities. The problem, called Capacitated Facility Location Problem (CPLP) can then be formulated as the following mixed-integer linear program:

$$\begin{aligned}
\text{minimize} \quad & \sum_{i\in I}\sum_{j\in J} c_{ij}x_{ij} + \sum_{j\in J} f_j y_j && (Z)\\
\text{s.t.} \quad & \sum_{j\in J} x_{ij} = 1,\ i \in I && (D)\\
& x_{ij} \le y_j,\ i \in I,\ j \in J && (B)\\
& \sum_{i\in I} d_i x_{ij} \le s_j y_j,\ j \in J && (C)\\
& x_{ij} \ge 0,\ i \in I,\ j \in J && (N_x)\\
& y_j \in \{0,1\},\ j \in J. && (I_y)
\end{aligned}$$

Constraint (C) imply that the total demand assigned to each facility is not larger than its capacity. Remark that constraints (B) are redundant if the values of the y_j variables are restricted to 0 or 1. However, those constraints appear to be extremely useful in the LP-relaxation (as in the strong and weak formulations of the UFLP).

The CFLP proves to be much more difficult to solve than the UFLP. The exact procedures are of branch-and-bound and of branch-and-cut types. The first ones use different lower bounds on the objective function that are obtained by different relaxations. As the number of constraints increases, the possible relaxations become numerous: Cornuéjols *et al.* [20] review 41 relaxations and find out that they yield only 7 genuinely different bounds.

The first approach, suggested by Geoffrion and McBride [39] and Nauss [80], consists of a Lagrangean relaxation of the constraints (D) with the multipliers λ_i. The relaxed problem can be decomposed into $|J|$ subproblems.

$$\begin{aligned}
\rho_j = \text{maximize} \quad & \sum_{i\in I} (\lambda_i - c_{ij})x_{ij} - f_j\\
\text{s.t.} \quad & \sum_{i\in I} d_i x_{ij} \le s_j\\
& 0 \le x_{ij} \le 1,\ i \in I,
\end{aligned}$$

which are continuous knapsack problems that can be solved in $\mathcal{O}(|I| log|I|)$ by ranking the clients i by decreasing order of $(\lambda_i - c_{ij})/d_i$ and allocating them to j until the corresponding capacity constraint is tight. This decomposition also yields the master problem

$$\begin{aligned} Z_1^D(\lambda) = \text{minimize} \quad & \sum_{i \in I} \lambda_i - \sum_{j \in J} \rho_j(\lambda) y_j \\ \text{s.t.} \quad & y_j \in \{0,1\},\ j \in J, \end{aligned}$$

whose optimal solution can be found by inspection. The Lagrangean dual problem $Z_1^D = \max_\lambda Z_1^D(\lambda)$ which can be solved by subgradient optimization techniques, does not improve upon the linear programming relaxation investigated by Guignard and Spielberg [43]. Moreover, the solution to the above master problem does not always enable one to construct a primal-feasible solution since the sum of the capacities of the open facilities may be smaller than the total demand. To avoid this difficulty, Nauss [80] has proposed to add a total demand constraint to the CFLP:

$$\sum_{j \in J} s_j y_j \geq \sum_{i \in I} d_i. \qquad (TD)$$

Using the Lagrangean relaxation described above, one gets the same subproblems, but the master problem is now a knapsack problem. This implies that the solution Z_2^D of the Lagrangean dual is a tighter bound: $Z_1^D \leq Z_2^D \leq Z^\star$. Since knapsack problems can be solved efficiently (see, e.g. Martello and Toth [74]), the bound Z_2^D can be obtained by subgradient optimization techniques.

The relaxation in which the variable upper bound constraints (B) are dualized in a Lagrangean way is considered in Baker [5]. The subproblem is then a transportation problem and the associated multipliers are updated according to some heuristic. Then, this lower bounding procedure is embedded into a branch-and-bound method. Baker [5] solves problems with up to 40 potential facilities and 80 clients. Guignard and Opaswongkarn [44] dualize constraints (B) and (C) in a Lagrangean way and strengthen the formulation by adding the total demand constraint (TD) and some simple capacity constraints.

A dual ascent procedure is proposed to update the Lagrangean multipliers. The authors report computational experiments on problems with up to 70 possible locations and clients. The gap between the best lower bound obtained and some primal heuristic solution value is generally not larger than 30%. Barcelo *et al.* [7] apply Lagrangean decomposition to a weak formulation of the CFLP (without constraints (B)) plus the total demand constraint (TD). The so-obtained lower bound as well as an associated upper bound procedure. Problems with up to 20 possible locations and 90 customers are solved with this approach.

Another approach called Cross Decomposition and proposed by Van Roy [96], consists in dualizing the capacity constraints (C). The relaxed problem is now an UFLP. The method developed by Van Roy for solving the Lagrangean dual and getting the bound Z_3^D is based on the observation that, for any subset of open facilities, the resulting subproblem is a transportation problem. The optimal dual solutions of this transportation problem are used as Lagrangean multipliers to generate a new UFLP. The algorithm incorporates these alternative steps. When it ceases to make progress, the solutions to both the UFLP and transportation problem allow one to generate new cuts for a Benders master problem. The solution to this problem gives a new seed for alternating procedure.

The bound Z_3^D is tighter than the one given by the linear programming relaxation. It is not dominated by Z_2^D, but Cornuéjols *et al.* [20] report computational experiments where Z_2^D is consistently better than Z_3^D . To improve upon Z_3^D, Van Roy [96] uses the CFLP augmented with the constraint (TD). Using the same relaxation as above yields a UFLP in which (TD) is in turned dualized. The bound Z_4^D obtained in this way requires a substantial computational effort but it dominates Z_2^D and is extremely tight. To sum-up, we have: $Z_1^D \leq \max\{Z_2^D, Z_3^D\} \leq Z_4^D \leq Z^\star$. Van Roy [96] reports computational experiments on problems with up to 100 potential facility locations and 200 clients. These show that the Decomposition algorithm finds near optimal Lagrangean multipliers within just a few iterations and is about 10 times faster than the alternative algorithms using the same Lagrangean relaxation. The polyhedral structure of the CPLP has been studied by Leung and Magnanti [67], Aardal *et al.* [1] and Aardal [3]. The first paper considers several variants of the CPLP. A first version where all facilities have the same capacity and customer's demand does not need to be fully met is considered. The so-called residual capacity inequalities are proved to be valid for that model and conditions under which they are facets are identified. Then, Leung and Magnanti [67] investigate whether these inequalities remain facets for other variants of the CFLP in which the demand of each client must be fully met, is indivisible, and if the facilities have varying capacities. Aardal *et al.* [1] consider the general version of the CPLP. They introduce two new families of valid inequalities which generalize the flow cover inequalities (see e.g. Nemhauser and Wolsey [81]) and the family of submodular inequalities which in turn generalizes the effective capacity inequalities. Two additional families of inequalities are considered: the class of combinatorial inequalities introduced by Cho *et al.* [15] and [16] for the UFLP and for which sufficient conditions for them to be facet defining are provided and the class of (k, l, S, I) inequalities proposed by Pochet and Wolsey [84] for the lot-sizing problem with constant batch sizes. Aardal [3], as a companion paper of Aardal *et al.* [1], considers the cutting plane approach for solving the CFLP. A subfamily of the submodular inequalities and the combinatorial inequalities are used. Heuristics are proposed for solving the associated separation problems. Computational experiments are reported for problems with up to 100 customers and 75 potential facility locations.

The variant of the CFLP in which each demand must be allocated to one single facility (i.e. is indivisible) has been considered by Klincewicz and Luss [58] and by Pirkul [82]. In this model the non-negativity constraints $x_{ij} \geq 0$ are replaced by $x_{ij} \in \{0,1\}$, $i \in I$, $j \in J$. This model appears to be more difficult than the classical CFLP. Indeed, for a given set of open facilities, the client allocation problem is already $\mathcal{NP}$-hard since the Bin Packing Problem (see Martello and Toth [74]) reduces to it. Klincewicz and Luss [58] propose a heuristic procedure based on a Lagrangean relaxation obtained by dualizing the capacity constraints (C). They reports computational experiments for the 12 test problems of Kuehn and Hamburger [63] with two different capacity levels. For all problems but one, the heuristic finds a feasible solution within 12% of the lower bound. Pirkul [82] considers the Lagrangean relaxation obtained by dualizing the demand constraints (D). The so-obtained lower and upper bounds are used in a branch-and-bound procedure. Pirkul [82] reports computational experiments for problems with up to 20 possible locations and 100 demand points.

A more general version of the CFLP may consist in imposing lower and upper bounds on the used facility capacities. Beasley [10] proposes a lower bound on the optimal value of that problem by considering a Lagrangean relaxation obtained by dualizing the demand constraints (D), the new capacity constraints, and some feasible solution exclusion constraints. By incorporating this lower bound and some problem reduction tests in a branch-and-bound procedure, Beasley [10] solves on a Cray-1S computer problems involving up to 500 potential facility locations and 1000 clients.

Another extension of the CFLP consists in allowing to choose each facility capacity among a given set of possible sizes. This problem, called the Multicapacitated Facility Location problem is considered in Hansen *et al.* [52]. This paper compares exact algorithms based on an integer programming code and using either a weak or a strong formulation to a Tabu Search heuristic. Computational results for problems with up to 100 clients and 40 possible locations are reported.

Finally, Shulman [88] considers the special class of Dynamic Capacitated Facility Location problem in which the facility capacities can only take a finite number of given values and several facilities can be opened at the same location. The problem is modeled as a mixed-integer linear program and a resolution method based on the Lagrangean relaxation in which the demand constraints are dualized is proposed. Shulman [88] develops two versions of the algorithm according to the fact that facilities of different capacities are allowed at the same location or not. Computation results for problems with up to 62 facility locations, 62 demand points and 10 planning periods are reported.

3. *P*–facility location problems

In the UFLP, transportation and set-up costs are assumed commensurable. When these two magnitudes are incommensurable, one may look for the locational configuration which minimizes total transportation costs under a budget constraint. The p-median problem is the most common problem of this type, exactly p facilities must be opened and it can be formulated as the following mixed-integer linear program:

$$\begin{array}{llr} \text{minimize} & \sum_{i\in I}\sum_{j\in J} c_{ij}x_{ij} & (Z) \\ \text{s.t.} & \sum_{j\in J} x_{ij} = 1,\ i \in I & (D) \\ & x_{ij} \leq y_j,\ i\in I,\ j\in J & (B) \\ & \sum_{j\in J} y_j = p & (P) \\ & x_{ij} \geq 0,\ i\in I,\ j \in J & (N_x) \\ & y_j \in \{0,1\},\ j\in J & (I_y) \end{array}$$

in which constraint (P) restricts the number of open facilities to p. As for the UFLP, it is readily verified that the problem has an all-integer solution which corresponds to assigning client i to the open facility j for which the transportation cost c_{ij} is minimum, all other x_{ij}'s are set equal to zero. Also, this formulation is not unique: the $|I|.|J|$ constraints $x_{ij} \leq y_{ij}$ can be replaced by the $|J|$ constraints $\sum_{i\in I} x_{ij} \leq |I|y_j$ but then the LP-relaxation is weaker. Finally the LP-relaxation of this strong formulation provides a tight bound on the optimal value $Z^\star$, see ReVelle and Swain [87].

The p-median problem is $\mathcal{NP}$-hard, see Kariv and Hakimi [57]. For any given value of p, the problem becomes polynomially solvable since it suffices to enumerate all sets S of p facility locations, to assign every client i to the nearest one in S, to compute the corresponding value of the objective function Z, and to keep the best value and solution found. Furthermore, when c_{ij}'s represent distances in a tree network, Matula and Kolde [75] have shown that the p-median problem can be solved in polynomial time. Kariv and Hakimi [57] have proposed an $\mathcal{O}(p^2n^2)$ algorithm, where n denotes the number of vertices of the tree and Tamir [92] an $\mathcal{O}(pn^2)$ algorithm. Hassin and Tamir [53] have refined this bound to $\mathcal{O}(pn)$ when the tree is a path. Gavish and Sridhar [37] propose an $\mathcal{O}(n \log n)$ algorithm for the 2-median on a tree.

Cornuéjols *et al.* [18] and Narula *et al.* [79] have proposed to solve the K-median problem by means of a dual approach based on the Lagrangean relaxation of constraint (D). Denoting λ_i the Lagrangean multiplier associated with the i-th constraint, we obtain the relaxed problem:

$$
\begin{aligned}
Z^D(\lambda) = \text{minimize} \quad & \sum_{i \in I} \sum_{j \in J} c_{ij} x_{ij} + \sum_{i \in I} \lambda_i (1 - \sum_{j \in J} x_{ij}) \\
\text{s.t.} \quad & 0 \leq x_{ij} \leq y_j, \; i \in I, \; j \in J \\
& \sum_{j \in J} y_j = p \\
& y_j \in \{0, 1\}, \; j \in J.
\end{aligned}
$$

The objective function can be rewritten as

$$
Z^D(\lambda) = \sum_{i \in I} \lambda_i - \max \sum_{i \in I} \sum_{j \in J} (\lambda_i - c_{ij} x_{ij})
$$

Clearly, x_{ij} must be set equal to y_j when $\lambda_i - c_{ij} > 0$ and to zero if $\lambda_i - c_{ij} < 0$. Defining $\rho_j = \sum_{i \in I} (\lambda_i - c_{ij})^+$, the relaxed problem becomes

$$
\begin{aligned}
Z^D(\lambda) = \quad & \sum_{i \in I} \lambda_i - \max \sum_{j \in J} \rho_j(\lambda) y_j \\
\text{s.t.} \quad & \sum_{j \in J} y_j = p \\
& y_j \in \{0, 1\}, \; j \in J.
\end{aligned}
$$

For a given vector λ this problem can be solved by inspection: the p facilities are located at the point j corresponding to the largest values of $\rho_j(\lambda)$. Since $F(\lambda)$ is a lower bound on $Z^\star$ for any λ, the best possible bound is given by

$$
Z^D = \max Z^D(\lambda)
$$

which corresponds to the Lagrangean dual problem. Solving this problem is done by means of a subgradient method.

It can be shown that $Z^D = Z_{LP}$; even if the bound Z^D does not improve upon the bound given by the LP-relaxation, having Z^D often gives a very tight bound. In case of a genuine duality gap (the subgradient procedure has failed to reach the optimum), an implicit enumeration completes the algorithm. Its performance can be improved by adding tests. For example, let $\overline{Z}$ be an upper bound of $Z^\star$ and relabel the facility locations such that $\rho_1(\lambda) \geq \rho_2(\lambda) \geq \cdots \geq \rho_n(\lambda)$. By definition of $Z^D(\lambda)$ if

$$
Z^D(\lambda) + \rho_j(\lambda) - \rho_{p+1}(\lambda) \geq \overline{Z}
$$

for $j = 1, \ldots, p$ we must have $y_j = 1$ in any optimal solution. On the other hand, if

$$
Z^D(\lambda) + \rho_p(\lambda) - \rho_j(\lambda) \geq \overline{Z}
$$

for $j = p + 1, \ldots, n$ we can set $y_j = 0$ in any optimal solution. These bounds were proposed by Christofides and Beasley [17]. Hanjoul and Peeters [48] have shown that dramatic reductions in the number of implicit enumerations can be reached by implementing these penalties.

A different relaxation, in which the constraint (P) is dualized, has been studied by Hanjoul and Peeters [48]. The relaxed problem is now given by

$$\begin{aligned} Z^D(\mu) = \text{minimize} \quad & \sum_{i \in I} \sum_{j \in J} c_{ij} x_{ij} + \sum_{j \in J} \mu y_j - \mu p \\ \text{s.t.} \quad & \sum_{j \in J} x_{ij} = 1, \; i \in I \\ & o \leq x_{ij} \leq y_j, \; i \in I, \; j \in J \\ & y_j \in \{0, 1\}, \; j \in J. \end{aligned}$$

which is an uncapacitated facility location problem with equal fixed cost μ. The Lagrangean dual

$$Z'^D = \max_{\mu} Z^D(\mu)$$

yields a bound tighter than the linear programming relaxation. The relaxed problem is $\mathcal{NP}$-hard too, but is well solved in many practical instances. Furthermore, the procedure involves a single multiplier μ and Z'^D can be found by considering only a finite number of values for μ. There is rarely a duality gap, in which case an implicit enumeration is required.

The classical heuristics GREEDY, STINGY and INTERCHANGE can be easily adapted to the p-median problem. Golden and Skiscin [41] present a computational study of a simulated annealing approach. Results concerning instances with up to 100 clients and possible locations and $p = 20$ are reported.

Several variants and extensions of the p-median problem exist, see Mirchandani [78] for a survey. One of the most recent is the so-called p-median problem with mutual communication and consists in finding the location of p facilities in order to minimize the total transportation costs between the client locations and the facilities and between pairs of facilities. Tamir [90] shows that the problem is strongly $\mathcal{NP}$-hard even when there are only three clients and p is variable. When the metric space is a tree with n nodes, Tamir [90] proposes an $\mathcal{O}(p^3 \log n + pn + n \log n)$ algorithm. Chhajed and Lowe [13] provide a polynomial time algorithm for the p-median problem with mutual communication when the dependency graph which represents the interactions between pairs of facilities has a special structure. Chhajed and Lowe [14] propose a generic p-facility location model which generalizes the p-median problem with mutual communication and they present a polynomial time algorithm when the dependency graph is a k-tree.

Finally, Tamir [91] considers another extension in which p facilities must be located while satisfying distance constraints between pairs made of clients and a facility and between pairs of facilities in order to minimize the total cost for setting the new facilities. Tamir [91] presents an $\mathcal{O}(p^2 n^2 \log n)$ algorithm for the special case of linear metric space.

References

1. K. Aardal, Y. Pochet and L. A. Wolsey. Capacitated facility location: valid inequalities and facets. *Mathematics of Operations Research*, 20:562–582, 1995.
2. K. Aardal, M. Labbé, J. Leung and M. Queyranne. On the two-level uncapacitated facility location problem. *INFORMS*, 8:289–301, 1996.
3. K. Aardal. Capacitated facility location: separation algorithms and computational experience. *Mathematical Programming*, (to appear), 1997.
4. S. Ahn, C. Cooper, G. Cornuéjols and A. Frieze. Probabilistic analysis of a relaxation for the k-median problem. *Mathematics of Operations Research*, 13:1-30, 1988.
5. B. M. Baker. A partial dual algorithm for the capacitated warehouse location problem. *European Journal of Operational Research*, 23:48–56, 1986.
6. M. L. Balinski. On finding integer solutions to linear programs. (Proceedings of the IBM Scientific Symposium on Combinatorial Problems), IBM Data Processing Division, White Plains (NY), 1966
7. J. Barcelo, E. Fernandez and K. O. Jörnstern. Computational results from a new Lagrangean relaxation algorithm for the capacitated plant location problem. *European Journal of Operational Research*, 53:38–45, 1991.
8. A. I. Barros and M. Labbé. A general model for the uncapacitated facility and depot location problem. *Location Science*, 3:173–191, 1994.
9. A. I. Barros and M. Labbé. The multi-level uncapacitated facility location problem is not submodular. *European Journal of Operational Research*, 72:607–609, 1994.
10. J. E. Beasley. An algorithm for solving large capacitated warehouse location problems. *European Journal of Operations Research*, 33:314–325, 1988.
11. O. Bilde and J. Krarup. Sharp lower bounds and efficient algorithms for the simple plant location problem. *Annals of Discrete Mathematics*, 1:79–97, 1977.
12. M. L. Brandeau and S. S. Chiu. An overview of representative problems in location research. *Management Science*, 35:645–674, 1989.
13. D. Chhajed and T. J. Lowe. m-median and m-center problems with mutual communication: solvable special cases. *Operations Research* , 40(Supp.1):S56–S66, 1992.
14. D. Chhajed and T. J. Lowe. Solving structured multifacility location problems efficiently. *Transportation Science*, 28:104–115, 1994.
15. D. C. Cho, E. L. Johnson, M. W. Padberg and M. R. Rao. On the uncapacitated plant location problem. I. Valid inequalities and facets. *Mathematics of Operations Research*, 8:590–612, 1983.
16. D. C. Cho, M. W. Padberg and M. R. Rao M.R. On the uncapacitated plant location problem. II. Facets and lifting theorems. *Mathematics of Operations Research*, 8:590–612, 1983.
17. N. Christofides and J. E. Beasley. A tree search algorithm for the p-median problem. *European Journal of Operations Research*, 10:196–204, 1982.
18. G. Cornuéjols, M. L. Fisher and G. L. Nemhauser. Location of bank accounts to optimize float: An analytic study of exact and approximate algorithms. *Management Science*, 23:789–810, 1977.
19. G. Cornuéjols, G. L. Nemhauser and L. A. Wolsey. The uncapacitated facility location problem. Discrete Location Theory, (Mirchandani and R.L. Francis, eds.), Wiley, New York, 119–171, 1990.
20. G. Cornuéjols, R. Sridharan and J. M. Thizy. A comparison of heuristic and relaxations for the capacitated plant location problem. *European Journal of Operations Research*, 50:280–297, 1991.

21. G. Cornuéjols and J. M. Thizy. Some facets of the simple plant location polytope. *Mathematical Programming*, 23:50–74, 1992.
22. T. G. Crainic, P. J. Dejax and L. Delorme. Models for multimode multicommodity location with interdepot balancing requirements. *Annals of Operations Research*, 18:279–302, 1989.
23. T. G. Crainic, M. Gendreau, P. Soriano and M.Toulouse. A Tabu search procedure for multicommodity location/allocation with balancing requirements. *Annals of Operations Research*, 41:359–383, 1992.
24. T. G. Crainic and L. Delorme. Dual-ascent procedures for multicommodity location-allocation problems with balancing requirements. *Transportation Science*, 27:90–101, 1993.
25. T. G. Crainic, L. Delorme and P. J. Dejax. A branch-and-bound method for multicommodity location with balancing requirements. *European Journal of Operations Research*, 65:368–382, 1993.
26. M. S. Daskin. Network and Discrete Location, Models, Algorithms and Applications. Wiley Interscience, New York, 1995.
27. M. Desrochers, P. Marcotte and M. Stan. The congested facility location problem. *Location Science*, 3:9–23, 1995.
28. Z. Drezner. Facility Location, A Survey of Applications and Methods. Springer Series in Operations Research, New York, 1995.
29. H. A. Eiselt, G. Laporte, J. F. Thisse. Competitive location models: a framework and bibliography. *Transportation Science*, 27:44–54, 1993.
30. E. Erkut and S. Neuman. Analytical models for locating undesirable facilities. *European Journal of Operations Research*, 40:275–291, 1989.
31. D. Erlenkotter. A dual-based procedure for uncapacitated facility location. *Operations Research*, 26:378–386, 1978.
32. J. R. Evans and E. Minieka. Optimization Algorithms for Networks and Graphs. 2nd edition, Marcel Dekker (ed.), New York, 1992.
33. E. Feldman, F. A. Lehrer and T. L. Ray. Warehouse locations under continuous economies of scale. *Management Science*, 12:670–684, 1966.
34. R. L. Francis, L. M. McGinnis Jr. and J. A. White. Facility Layout and Location: An Analytical Approach. 2nd edition, Prentice Hall, Englewood Cliffs, 1992.
35. T. L. Friesz, T. Miller and R. L. Tobin. Competitive network facility location models: A survey. *Papers of the Regional Science Association*, 65:47–57, 1988.
36. L. L. Gao and E. P. Robinson Jr. A dual-based optimization procedure for the two-echelon uncapacitated facility location problem. *Naval Research Logistic*, 39:191–212, 1992.
37. B. Gavish and S. Sridhar. Computing the 2-median on tree networks in O(n log n). *Networks*, 26:305–317, 1995.
38. B. Gendron and T. G. Crainic. A branch and bound algorithm for depot location and container fleet management. *Location Science*, 3:39–53,1995.
39. A.M. Geoffrion and R. McBride. Lagrangian relaxation applied to facility location problems. *AIIE Transactions*, 10:40–47, 1978.
40. A. Ghosh and F. Harche. Location-allocation models in the private sector: progress, problems and prospects. *Location Science*, 1:81–106, 1993.
41. B. L. Golden and C. C. Skiscin. Using simulated annealing to solve routing and location problems *Naval Research Logistic Quarterly*, 33:261–279, 1986.
42. S. R. Gregg, J. M. Mulvey and J. Wolpert. A stochastic planning system for siting and closing public service facilities. *Environment and Planning*, A20:83–98, 1988.

43. M. Guignard and K. Spielberg. A direct dual method for the mixed plant location problem with some side constraints. *Mathematical Programming*, 17:198–228, 1979.
44. M. Guignard and K. Opaswongkarn. Lagrangean dual ascent algorithms for computing bounds in capacitated plant location problems. *European Journal of Operations Research*, 46:73–83, 1990.
45. S. L. Hakimi. Optimum locations of switching centers and the absolute centers and medians of a graph. *Operations Research*, 12:450–459, 1964.
46. S. L. Hakimi. Optimum distribution of switching centers in a communication network and some related graph theoretic problems. *Operations Research*, 13:462–475, 1965.
47. G. Y. Handler and P. B. Mirchandani. Location on Networks. Theory and Algorithms. The MIT Press, 1979.
48. P. Hanjoul and D. Peeters. Comparison of two dual-based procedures for solving the p-median problem. *European Journal of Operations Research*, 20:387–396, 1985.
49. P. Hanjoul, P. Hansen, D. Peeters and J. F. Thisse. Uncapacitated plant location under alternative spatial price policies. *Management Science*, 36:41–57, 1990.
50. P. Hansen, M. Labbé, D. Peeters and J. F. Thisse. Single facility location on networks. *Annals of Discrete Mathematics*, 31:113–146, 1987.
51. P. Hansen, M. Labbé, D. Peeters and J. F. Thisse. Facility location analysis *Fundamentals of Pure and Applied Economics*, 22:1–70, 1987.
52. P. Hansen, E. de Luna Pedrosa Filho and C. Carneiro Ribeiro. Location and sizing of offshore platforms for oil exploration. *European Journal of Operations Research*, 58:202–214, 1992.
53. R. Hassin and A. Tamir. Improved complexity bounds for location problems on the real line. *Operations Research Letters*, 10:395–402, 1991.
54. M. J. Hodgson, K. E. Rosing and Shmulevitz. A Review of Location – Allocation Application Litterature. *Studies in Locational Analysis*, 5:3–30, 1993.
55. A. P. Hurter and J. S. Martinich. Facility Location and the Theory of Production. Kluwer Academic Publishers, Boston, 1989.
56. P. C. Jones, J. Lowe, G. Muller, N. Xu, Y. Ye and J. L. Zydiak. Specially structured uncapacitated facility location problems. *Operations Research*, 43:661–669, 1995.
57. O. Kariv and S. L. Hakimi. An algorithmic approach to network location problems, II: the p-medians. *SIAM Journal on Applied Mathematics*, 37:539–560, 1979.
58. J. G. Klincewicz and H. Luss. A Lagrangean relaxation heuristic for capacitated facility location with single source constraints. *Journal of the Operational Research Society*, 37:495–500, 1986.
59. J. G. Klincewicz and H. Luss. A dual-based algorithm for multi product uncapacitated facility location. *Transportation Science*, 21:198–206, 1987.
60. A. Kolen. Solving covering problems and the uncapacitated plant location problem on trees. *European Journal of Operations Research*, 12:36–81, 1983.
61. M. Körkel. On the exact solution of large-scale simple plant location problems. *European Journal of Operations Research*, 39:157–173, 1989.
62. J. Krarup and P. M. Pruzan. The simple plant location problem: Survey and synthesis. *European Journal of Operations Research*, 12:36–81, 1983.
63. A. A. Kuehn and M. J. Hamburger. A heuristic program for locating warehouses. *Management Science*, 9:643–666, 1963.

64. M. Labbé, D. Peeters and J. F. Thisse. Location on networks. Handbooks in Operations Research and Management Science, (M.O. Ball et al.,eds.), Elsevier, Amsterdam, 551–624, 1995.
65. M. Labbé and F. V. Louveaux. Location problems. Annotated Bibliography in Combinatorial Optimization. (M. Dell'Amico, F. Maffioli and S. Martello, eds.), Wiley, New York (to appear), 1997.
66. G. Laporte, F. V. Louveaux and L. Van hamme. Exact solution of a stochastic location problem by an integer L-shaped algorithm. *Transportation Science*, 28:95–103, 1994.
67. J. M. Y. Leung and T. Magnanti. Valid inequalities and facets of the capacitated plant location problems. *Mathematical Programming*, 44:271–291, 1989.
68. R. Logendran and M. P. Terrel. Uncapacitated plant location-allocation with price sensitive stochastic demands. *Computers and Operations Research*, 15:189–198, 1988.
69. F. V. Louveaux. Discrete stochastic location models *Annals of Operations Research*, 6:23–34, 1986.
70. F. V. Louveaux and D. Peeters. A dual-based procedure for stochastic facility location. *Operations Research*, 40:564–573, 1992.
71. F. V. Louveaux. Stochastic location analysis. *Location Science*, 1:127–154, 1993.
72. R.F. Love, J. G. Morris and G. O. Wesolowsky. Facilities Location, North-Holland, New York, 1988.
73. A. S. Manne. Plant location under economies-of-scale – Decentralization and computation. *Management Science*, 11:213–235, 1964.
74. S. Martello and P. Toth. Knapsack Problems: Algorithms and Computer Implementations. Wiley, New York, 1990.
75. D. W. Matula and R. Kolde. Efficient multimedian location in a cyclic network. ORSA/TIMS Science Meeting, Miami, 1976.
76. C. Michelot. The mathematics of continuous location. *Studies in Locational Analysis*, 5:59–85, 1993.
77. P.B. Mirchandani and R. L. Francis. Discrete Location Theory. Wiley, New York, 1990.
78. P. B. Mirchandani. The p-median problem and generalizations. Discrete Location Theory (P.B. Mirchandani and R.L. Francis, eds.), Wiley, New York, 1990.
79. S. C. Narula, U. I. Ogbu and H. M. Samuelsson. An algorithm for the p-median problem. *Operations Research*, 25:709–713, 1977.
80. R. M. Nauss. An improved algorithm for the capacitated facility location problem. *Journal of the Operational Research Society*, 29:1195–1201, 1978.
81. G. L. Nemhauser and L. A. Wolsey. Integer and Combinatorial Optimization. Wiley, New York, 1988.
82. H. Pirkul. Efficient algorithms for the capacitated concentrator location problem. *Computers and Operations Research*, 14:197–208, 1987.
83. F. Plastria. Continuous Location Anno 1992: a Progress Report. *Studies in Locational Analysis*, 5:85–128, 1993.
84. Y. Pochet and L. A. Wolsey. Lot-sizing with constant batches: Formulation and valid inequalities. *Mathematics of Operations Research*, 18:767–785, 1993.
85. H. N. Psaraftis, G. G. Tharakan and A. Ceder. Optimal response to oil spills. The strategic decision case. *Operations Research*, 34:203–217, 1986.
86. C. R. Reeves. Modern Heuristic Search Methods. Modern heuristic techniques, (V. J. Rayward-Smith, I. H. Osman, C. R. Reeves and G. D. Smith, eds.), Wiley, New York, 1–24, 1996.

87. C. S. ReVelle and R.Swain. Central facilities location. *Geographical Analysis*, 2:30–42, 1970.
88. A. Shulman. An algorithm for solving dynamic capacitated plant location problems with discrete expansion sizes. *Operations Research*, 39:423–436, 1991.
89. J. F. Stollsteimer. A working model for plant numbers and locations. *Journal of Farm Economics*, 45:631–645, 1963.
90. A. Tamir. Complexity results for the p-median problem with mutual communication. *Operations Research Letters*, 14:79–84, 1993.
91. A. Tamir. A distance constraint p-facility location problem on the real line. *Mathematical Programming*, 66:201–204, 1994.
92. A. Tamir. An $O(pn^2)$ Algorithm for the p-median and related problems on tree graphs. *Operations Research Letters*, 19:59–64, 1996.
93. B. C. Tansel, R. L. Francis and T. J. Lowe. Location on Networks: A Survey-Part I: The p-Center and p-Median Problems. *Management Science*, 29:482–497, 1983.
94. B. C. Tansel, R. L. Francis and T. J. Lowe. Location on Networks: A Survey-Part II: The p-Center and p-Median Problems. *Management Science*, 29:498–511, 1983.
95. M. B. Teitzand and P. Bart. Heuristic methods for estimating the generalized vertex median of a weighted graph. *Operations Research*, 16:955-961, 1968.
96. T. J. Van Roy. A cross decomposition for capacitated facility location. *Operations Research*, 34:145-163, 1986.
97. V. Verter and M. C. Dincer. Facility location and capacity acquisition: an integrated approach. *Naval Research logistic Quarterly*, 42:1141–1160, 1995.
98. A. Wagelmans, S. van Hoesel and A. Kolen. Economic lot sizing: an $O(nlogn)$ algorithm that runs in linear time in the Wagner-Within case. *Operations Research*, 40:S145–S156, 1992.

Eliminating Bias Due to the Repeated Measurements Problem in Stated Preference Data

Cinzia Cirillo[1], Andrew Daly[2], and Karel Lindveld[2]

[1] Politecnico di Torino
Torino, Italy
[2] Hague Consulting Group
The Hague, The Netherlands.

Summary. Usually in SP surveys a number of choice observations is taken from each individual to reduce the cost of data collection. This raises the so-called "repeated measurements" problem: observations taken from the same individual are not independent, which means that simple analysis method may be biased. In this article we describe the application of the Jackknife and the Bootstrap techniques to attack this problem in the case of a simple logit model, and compare their relative performance. The Jackknife and Bootstrap are used to investigate bias due to serial correlation, the influence of the size of the Jackknife sample, the influence of the error distribution for each respondent individually and across respondents, and the variance estimate of the model coefficients obtained in the course of maximum-likelihood estimation.

1. Introduction

Stated Preference (SP) data has been used to an increasing extent in recent years in transport planning as an adjunct to or replacement for "revealed preference" (RP) data of actual choices. SP data has been found to be particularly advantageous

- in handling new alternatives in the transport system, such as new transport infrastructure, where it is not possible to make observations of actual use of the system; and
- in reducing the costs of collecting behavioural data.

SP data can offer considerable cost reductions because the data structure is under the control of the analyst, so that the information content of each observation can be maximised. Moreover, because the observation can be set up by the analyst, it is possible to collect multiple responses from the same individual. Because a major component of survey cost is in contacting the respondent, the ability to collect multiple responses has become important in designing cost-effective surveys.

However, the use of multiple responses from individual respondents has consequences for the analytical procedures that can be used. All simple methods for analysing choice data require the assumption that the separate observations are independent. Of course as soon as repeated observations are taken from an individual this assumption is no longer valid. While it may be reasonable to expect that coefficients estimated on the basis of this false assumption may not be seriously biased, any assertion concerning the accuracy of estimation of the coefficients must be seriously compromised. In general terms, the treatment of the observations as independent overstates the information content of the data. This is the "repeated measurements" problem. Very many practical studies using SP data apply simple analysis methods to data containing repeated measurements and are therefore liable to the errors studied in this paper.

Of course it is possible to design analytical methods that take account of the correlation between the observations from an individual. Unfortunately such methods are very complicated and slow in application. While it is possible to imagine that analytical methods could be used in the future to solve the problem of repeated measurements, at present it is effectively impossible to use such methods in practical work.

The objective of the work reported in this paper was to determine whether simulation methods could be used to attack the problem. In particular, we have applied the two most common methods of "re-sampling", the so-called Jackknife and Bootstrap methods. These methods can be used in circumstances where an analytical solution for an estimator is not available to make unbiased estimates of a statistic and its variance. They therefore appear suitable for the problem at hand.

Jackknife methods had been used in a small number of previous studies based on SP data by Hague Consulting Group (HCG), and a computer program was available to apply this method to logit estimations[1]. However, we wished to compare the effectiveness of the Bootstrap estimator with the Jackknife and to investigate their properties over a wide range of data. The work described in the current paper carried out those investigations.

Essentially, the problem of repeated measurements is that of a specification error in the model. The key interest in this work is then to test the impact of this specification error on the model estimation. A number of SP data sets were available at HCG and could have been used to test the estimation procedures. However, while it is essential to test the applicability of the methods to real data (i.e. with realistic levels of variance and correlation), it is also very helpful in understanding the impact of this type of error to be able to control the extent to which the error is present and other aspects of the choice process that is being observed (in real data, we do not know what the levels of variance and correlation really are). Also, the analysis of real data can be time-consuming and introduce analytical problems of the sort known to every practical analyst. Therefore, in addition to testing real data, for much of the work we made use of simulated data in which the extent of

[1] The program was written by Mark Bradley, repeatedly exploiting HCG's ALOGIT estimation program to derive the Jackknife estimates of parameters and variances.

the specification error and other aspects of the choice were systematically varied.

In the work described here, we have concentrated exclusively on the use of logit models, because these are the most commonly used models in our own and general practice. While the general trend of the results should apply to other model forms, we have found that results can be influenced significantly by apparently minor differences in the form of generally similar distributions (e.g. logistic and normal).

In the following section of the paper a summary explanation is given of the Jackknife and Bootstrap estimators, followed by a brief theoretical comparison. This is followed by a report of their application to two real data sets taken from recent HCG studies. Section 4 describes the generation of simulated data and the results that were obtained from estimations based on it. Finally, conclusions are drawn from the work, together with indications for the further investigations that are necessary.

2. Re-sampling methods: Jackknife and Bootstrap

The Jackknife and Bootstrap are the most commonly used members of a class of technique known as re-sampling. Essentially, these techniques operate by taking samples within the sample of data available. The basic operation of the Jackknife is to sample by systematically omitting a small fraction of the data, while the Bootstrap works by drawing randomly with replacement within the base sample. Repeated application to different samples allows the required results to be derived. Although these methods and more general re-sampling procedures are explained in the literature (e.g. Efron [3], Efron and Tibshirani [4], Shao and Tu [6]), we will give a brief overview here with our application in mind. In particular, as will become clear when results are presented, it is the variance estimators that are important in SP applications.

The relationship between sampling, re-sampling, and parameter estimation is shown in Table 2.1. below. The full population $\mathcal{P}$ is shown in the top-left corner; the sample population P is shown in the middle column, and the re-sampled population P^* is shown in the rightmost column. Estimation of the choice model on information obtained from a set of people ($\mathcal{P}$, P ,or P^*) gives estimated model parameters $\theta, \hat{\theta}, \hat{\theta}^*$ respectively.

As soon as one decides to use re-sampling, the question arises of which method to use since numerous alternatives exist, the most common being the Bootstrap and the Jackknife. This question can be clarified by taking a closer look at the similarities and dissimilarities of the two methods, which is best done in the context of re-sampling statistics in general (see also Efron [3], and Efron and Tibshirani [4]).

Re-sampling statistics starts with a set X of observed data points $x_1, x_2, \ldots,$ x_n which are regarded as fixed: this is the base sample. This set of data points

	Full Population	→	Sample	→	Resample
	$\mathcal{P}$		P		P^*
Estimation	↓		↓		↓
Parameter	θ		$\hat{\theta}$		$\hat{\theta}^*$

Table 2.1. Relationship between Population, Sample, Re-sample and Model Parameters.

defines a so-called empirical distribution F, which assigns probability $1/n$ to each observed data point. Next we have a statistic (e.g. the mean, variance, or a ML estimate of some parameter), $\hat{\theta}$ which we can calculate when we have a specification of F, so that it can be viewed as a function of the empirical distribution function: i.e. $\hat{\theta} = T(F)$. Note that every sub-sample of the base data points will also define an empirical distribution. A re-sampling vector $P^* = (P_1^*, P_2^*, \ldots, P_n^*)$ is any vector for whose elements are all non-negative and have a sum of one:

$$P_i^* \geq 0, \sum_{i=1}^{n} P_i^* = 1$$

The elements of the re-sampling vector P^* can be interpreted as a relative weight of the original data points in a re-sampling from the observed data, and as such define a new, re-sampled, dataset X^*, which defines a re-sampled empirical distribution $F(P^*)$, which in turn gives a re-sampled value of the statistic $\hat{\theta}^*$: $\hat{\theta}^* = T(F(P^*)) = T(P^*)$. The "re-sampling" vector $P^0 = (\frac{1}{n}, \frac{1}{n}, \ldots, \frac{1}{n})$ returns the original sample.

2.1 The Jackknife

The Jackknife uses re-sampling vectors like:

$$P_{(i)} = \left(\frac{1}{n-1}, \frac{1}{n-1}, \ldots, 0, \frac{1}{n-1}, \ldots, \frac{1}{n-1}\right)$$

where the i-th data point is omitted. With the "delete-k" Jackknife, k datapoints at a time are omitted. The Jackknife variance estimator is:

$$\sigma_{JACK}^2(\hat{\theta}) = \frac{n-1}{n} \sum_{i=1}^{n} \left(\hat{\theta}_{(i)} - \hat{\theta}_{(\cdot)}\right)^2$$

with

$$\hat{\theta}_{(\cdot)} = \sum_{i=1}^{n} \frac{\theta_{(i)}}{n}.$$

An alternative formulation found in Bissel and Ferguson [2] gives:

$$\sigma^2_{JACK} = \frac{n-1}{n} \left[\left(\sum_{j=1}^{n} \hat{\theta}^2_{-j} \right) - \left(\sum_{j=1}^{n} \hat{\theta}_{-j} \right)^2 \right], \tag{2.1}$$

which embodies an implict assumption that the $\hat{\theta}_{-j}$ are independent.

2.2 The Bootstrap

The Bootstrap re-samples n data points from the original n observed data points, and has re-sampling vectors that have a rescaled multinomial distribution:

$$P^* \stackrel{\text{def}}{=} \frac{\text{Mult}_n(n, P^0)}{n}.$$

The Bootstrap variance estimator $\sigma^2_{BOOT}\left(\hat{\theta}\right)$ is simply the variance of the statistic $\hat{\theta}^*$ calculated from the individual Bootstrap samples.

2.3 Relationship between Bootstrap and Jackknife variance estimators

The difference between the Jackknife and the Bootstrap is that the Bootstrap creates a completely new sample of n elements each time, whereas the Jackknife takes samples which are much closer to the original sample, but perturbs it by deleting one or more data points (elements of the population). In this way the Bootstrap tries to replicate the distribution of the original population, whereas the Jackknife tests the sensitivity of the parameter estimate on various (groups of) points of the dataset.

It turns out that with respect to a so-called linear statistic the Bootstrap and Jackknife estimates coincide. A linear statistic is a statistic which can be calculated from the data points by considering them one at the time, without interactions between pairs of data points or interactions of higher order.

A linear statistic can be expressed as

$$\hat{\theta}^* = T(P^*) = T(P^0) + (P^* - P^*)^T U,$$

with U a vector satisfying

$$\sum_{i=1}^{n} U_i = 0$$

An example of a linear statistic is the mean, an example of a nonlinear statistic is the median.

The relationship between the bootstrap and the Jackknife is given by the following result (see Efron and Tibshirani [4]):

Theorem 2.1. *Let $(P_{(i)}, T(P_{(i)}))$, $i = 1, \ldots, n$ be the set of Jackknife re-sampling points with their corresponding value of $\hat{\theta}^*$. Then there exists a unique linear statistic T_{LIN} which passes through the Jackknife points, and*

$$\sigma^2_{BOOT}(T_{LIN}) = \frac{n-1}{n}\sigma^2_{JACK}(\hat{\theta}).$$

In this sense, the Jackknife variance estimate can be regarded as $n/n-1$ times the Bootstrap variance estimate applied to a linear approximation of the true statistic. For non-linear statistics this introduces an error, and can even cause the Jackknife to become inconsistent (as is in the case of the median). This can be remedied by using the delete-k Jackknife. We have used the delete-k Jackknife exclusively.

The likelihood function is only a linear statistic if the error terms are uncorrelated; in our context this is only the case if the Logit model is correct. This sounds worse than it is: in Shao and Tu [6], paragraph 2.2.2, it is proved that the Jackknife estimate of variance is consistent for so-called "M-estimators" (of which the maximum likelihood estimate of parameters, universally used in logit estimation, is a special case).

2.3.1 Jackknife or Bootstrap. Although asymptotically the Jackknife and Bootstrap variance estimates are identical, it is the small-sample properties that determine their relative usefulness in practice.

In this context, we note that the Jackknife has a computational edge on the Bootstrap since one has to prepare a complete estimation dataset for each Bootstrap sample, whereas for the Jackknife we can use the same dataset and simply vary the points which the estimation program is to reject in each Jackknife sample.

In Shao and Tu [6], paragraph 3.7 we find the following statement: ...*the Bootstrap variance estimator is often downward-biased and is not as efficient as the Jackknife variance estimator when they are consistent and require the same amount of computation. Hence the bootstrap is not recommended if one only needs a variance estimator.*

However, in Efron [3] and Shao and Tu [6], we find that the Jackknife variance estimator is upward-biased with respect to the real variance.

Generally speaking, the bootstrap re-sampling vectors samples can be further removed from the original sample P^0 than the Jackknife re-sampling vectors $P_{(i)}$. This has the consequence of making the Jackknife unsuitable for highly nonlinear functionals such as the median. On the other hand, the Jackknife will insert less "noise" into the samples, and might therefore be expected to give useful estimates with fewer re-samples than the Bootstrap. This tends to be confirmed in our numerical experiments.

2.3.2 Summary. For variance estimation of parameters estimated through a Maximum-Likelihood procedure, the (delete-k) Jackknife and the Bootstrap are (asymptotically) equivalent. The Jackknife seems to give somewhat smoother results, especially at lower numbers of subsamples. Moreover, the

Jackknife gives 'conservative' estimates of the variance compared with the Bootstrap, which may be biased downwards.

The theoretical discussion concludes that both methods appear suitable and that the Jackknife may be preferable in practice because of its better small-sample properties.

2.4 Implementation

The Bootstrap and Jackknife procedures have been implemented in a program (called "JACKBOOT" which functions as a simple shell around the logit estimation program (ALOGIT). In the case of the Bootstrap, the program performs the sampling, and creates a new estimation dataset; in the case of the Jackknife, the program modifies the control file of the estimation program to skip certain observations. The program accepts a control file telling it whether to use the Jackknife or the Bootstrap, how many re-samples to perform, and how to undertake the sampling[2].

Using an already available program for calculation of the Jackknife estimates, the numerical implementation of the Jackknife estimate we used is the "alternative formulation" (2.1) which can become negative in some cases, as opposed to the original formulation, which does not have this problem.

3. Real data sets

The real data sets come from two recent studies carried out by Hague Consulting Group. They are labelled RDS1 and RDS2.

3.1 The RDS1 Data

This data is drawn from an SP within-mode experiment with about 1200 individuals, each giving a maximum of eight responses, a few of which are deleted in the data processing, leaving about 9500 responses in total.

Both Jackknife and Bootstrap procedures were applied, using numbers of re-samples equal to 10, 20 and 40 in each case. These results were compared with the "naïve" estimates obtained by ignoring the presence of repeated measurements.

The results showed that the coefficient values obtained by the naïve method were highly accurate, re-sampling always gave values within 95%

[2] In a data set containing repeated measurements, it would be possible to sample either among the records without reference to the individuals or among the individuals. However, the former method does not allow the re-sampling to measure the impact of repeated measurements and we have used the latter method in the work reported here.

of the naïve value, without any suggestion of bias. However, the error estimates given by the naïve method were consistently too low, although there was considerable variation.

Table 3.1. compares the relative standard errors of the two re-sampling methods with those of the naïve method. The values presented are the arithmetic mean values of the s.e. ratios over all the coefficients in the model (no reason has been found to use a more sophisticated averaging procedure). The values are consistently less than 1, meaning that the naïve method consistently under-estimates the error in the data. However, these averages conceal a considerable amount of variation: for example the Jackknife-20 results vary between 0.55 and 1.06, giving the average 0.71 as reported in the Table.

n. re-samples	10	20	40
Jack *s.e.* mean	0.6974	0.7116	0.6661
Boot *s.e.* mean	0.7245	0.6944	0.6902

Table 3.1. Standard Error Ratios for Re-sampling and Naïve Methods, RDS1

These ratios can be compared with a simple correction procedure that has been suggested previously[3] of adjusting the s.e. by the square root of the number of observations per individual, in this case giving a factor of 0.35. We conclude that the situation is worse than suggested by the naïve method, but not nearly as bad as would be suggested by the simple correction procedure.

Comparing the two techniques, there is a suggestion that the Bootstrap method is converging (as it should) to a value around 0.69. For smaller numbers of re-samples the Jackknife is closer to that value, but the Jackknife does not converge for larger numbers of re-samples.

3.2 The RDS2 data

The RDS2 data is taken from a study in which RP data and three separate SP games were taken from each respondent. Here we report results only for Games 2 and 3, for which there were 260-270 respondents, each giving maxima of six responses to Game 2 (within-mode) and eight to Game 3 (between-mode).

In the practical study from which the data was taken, model estimation was conducted using all three SP games and the more extensive RP data simultaneously. However, to gain insight into the specific effects of the correlations in the repeated measurements, the separate data sets were analysed individually. This gave rise to a number of practical problems concerning the

[3] By Mark Bradley. This correction effectively assumes that the 'amount of information' in the records for one individual is the same as would be obtained from one record.

statistical identification of some parameters, particularly when working with sub-samples of the data, and it was necessary to make adjustments to the model specification in order to get satisfactory results.

A further practical problem that arose in this work is that one parameter in the Game 3 data was given a negative value by the Jackknife procedure, despite the fact that all of the Jackknife sub-samples and the naïve estimation gave positive results and that the parameter was theoretically constrained to have a positive value. Intensive investigation showed that this result was obtained by correct application of the correct Jackknife formula. The explanation appears to be that the "alternative" formula was applied in our program (see Section 2) and that this relies on an assumption of independence among the re-samples which is not correct.

Apart from this result, the Jackknife method confirms the naïve coefficient estimates, while the standard errors are found to be substantially larger than those found in the naïve analyses. These results confirm the finding with RDS1. The results for Game 2, Game 3 and combined data from Games 2 and 3 (allowing for correlation between responses from the same individual in the separate Games) are set out in Table 3.2.

n. re-samples	10	20	40
SP2	0.7906	0.8038	0.7328
SP3	0.3542	0.3483	0.3349
SP2+SP3	0.5556	0.5479	0.5371

Table 3.2. Standard Error Ratios for Jackknife and Naïve Methods, RDS2

It is interesting to note the difference in the results for Games 2 and 3. In Game 2, we find (as in RDS1) that the s.e. ratio is much larger than the simple correction (which would be 0.41 for six observations per individual). However, for Game 3 the level is approximately equal to the simple correction value (0.35 for eight observations). It is tempting to think that the difference between Game 2 and Game 3 is connected to the nature of the games, perhaps in Game 3 the dominant effect is the mode preference of the individual, so that additional observations per individual add little.

The results obtained by combining Games 2 and 3 are intermediate, as might be expected.

3.3 Summary for Real Data

It appears that in every case the naïve coefficient estimates are confirmed by the Jackknife and Bootstrap, as has been found in previous studies using these methods.

The standard errors found by re-sampling, however, do not reproduce those of the naïve estimations. Substantially smaller s.e. estimates are given by the naïve runs than by the (more correct) Jackknife. The standard errors of the Jackknife SP estimates, shown in the Tables above, indicate that the naïve method gives 35% to 84% of the "true" value. However, the conservative "square root" rule also fails to give reliable estimates of the standard errors.

These results indicate that there is significant specification error in the naïve interpretation of the data and that this leads to the problems found with the s.e. estimates, although (in these data sets) the coefficient estimates themselves are not noticeably affected. Exactly what this specification error is, however, is not known, although it is tempting to associate it with the repeated measurements issue.

4. Simulated data

When working with simulated data it is necessary to distinguish between two models: the data-generating model which is used to produce the artificial data, and the working model which is estimated on the artificial data. In this way we can introduce a controlled specification error into the data, and observe the effect on the estimated model parameters and their estimated variance. In particular, we can introduce the repeated measurements correlation into the data and observe its impact on model estimates.

4.1 The Data-Generating Model

The data generated is of the following type: two alternatives, and one variable x_1 indicating the 'level' of the alternative, and a function to measure the 'utility' of the alternative. The utility function contains a systematic part, given by a constant plus a constant times the level, and error terms. What is observed of the choice process is the simulated choice Y defined as:

$$Y = \begin{cases} 0 & \text{if alternative 1 is chosen} \quad (U_1 > U_2) \\ 1 & \text{if alternative 2 is chosen} \quad (U_1 < U_2) \end{cases}$$

The initial formulation was a simple data-generating model (Model I). In this model the random components of the utility function of the data-generating model were defined in a way that violates the independence requirement for naïve analysis by having separate normally-distributed between-individual and within-individual components in the error term, each with its own variance. The between-individual and within-individual error terms were weighted with parameters s_1 and s_2 which are varied in a systematic way to cover a range of possible specification errors. On this artificial data set we estimated an ordinary logit model as a working model. In this way we introduced a controlled specification error, which could be used to help understand the effect of the specification error in the real data due to repeated measurements being taken from the same individual.

Data-generating Model I is defined by

$$U = \begin{cases} U_1 = 0 \\ U_2 = k_1 + k_2 x + s_1 \varepsilon_{ind} + s_2 \varepsilon_{obs}, \end{cases}$$

where

$$\begin{pmatrix} x \\ \varepsilon_{ind} \\ \varepsilon_{obs} \end{pmatrix} \stackrel{\text{def}}{=} N\left(\begin{pmatrix} 0 \\ 0 \\ 0 \end{pmatrix}, I \right).$$

The variable x (the observed properties of alternative 2) is also a random number. This need not be a problem as long as we are confident that it merely sweeps out all reasonable values. Different forms of distribution of x have been tested without changing the results significantly. In our artificial dataset x, and the error terms $x_{obs} \ldots x_{ind}, \varepsilon_{obs} \ldots \varepsilon_{ind}$ have a joint normal distribution and are independent. If we fix the value of x, to a value (say c), then only ε_{ind} and ε_{obs} have a stochastic nature.

In this case the expectation of the systematic part of the utility difference is: $U^* = E(U|x = c) = k_1|k_2 c$ and the variance of U^* is

$$\sigma^2(U^*) = s_3 = s_1^2 + s_2^2.$$

The idea is to simulate a situation in which we have N_i individuals, each of whom makes a choice in N_r choice situations. This gives $N = N_i * N_r$ responses. The error term ε_{obs} varies across all observations and is called the within-individual variance; the term ε_{ind} varies per individual and is called the between-individual variance; it is constant for the responses given by any fixed individual.

We found behaviour of the variance in the estimation of one of the two model parameters we did not expect from our earlier experience with real data. This led us to change the structure of the error terms in the artificial data to the form of Model II.

Data-generating Model II is defined by

$$U = \begin{cases} U_1 = 0 \\ U_2 = k_1 + k_2 x + \varepsilon_3 \end{cases}$$

where

$$P(\varepsilon_3 \leq x) = F_{\varepsilon_3}(x) = \frac{1}{1 + \exp\left(-\frac{\pi}{\sigma_3\sqrt{3}} x\right)}.$$

The error term ε_3 has a logistic distribution with variance σ_3^2, so that the error term in the utility difference between U_1 has a logistic distribution, which makes it a Logit model. The choice probability is:

$$\begin{aligned} P_{k_1,k_2}(X = 1 \mid x = c) &= P(k_1 + k_2 x + \varepsilon_3 > 0) \\ &= \frac{1}{1 + \exp\left(-\frac{\pi}{\sigma_3\sqrt{3}}(k_1 + k_2 x)\right)}. \end{aligned}$$

When $\sigma_3 = 0$ there is no uncertainty, and the decision maker has a clear choice. When σ_3 becomes very large, all explanatory elements in the model are swamped by the variance in the error terms, and there is no preference one way or another.

4.2 Maximum Likelihood Estimation

The parameters of the working model were estimated using maximum likelihood (ML) estimation; the software used is ALOGIT. With ML estimation, a lower bound (the Cramer-Rao lower bound) on the variance of the estimators is given by the inverse of the matrix of second partial derivatives of the log-likelihood function, provided that there is no specification error. However, repeated measurements imply that there will be correlation the answers provided by the same individual, and therefore in the error terms in the utility function. This violates the assumption that the error terms are independent and means that we can no longer rely on the variance estimates obtained in this way.

The Bootstrap and Jackknife methods apply the ML estimator on each of the resamplings from the original (artificial) dataset generated by the data-generating model, and estimate their variance from the distribution of the estimates across the resamples.

Had there been no specification error, then the ratio between the variance calculated by the Bootstrap/Jackknife, and that calculated by Alogit:

$$R_{p1} = \frac{\sigma^2(P_1^{resample})}{\sigma^2(P_1^{ML})} \quad \text{and} \quad R_{p2} = \frac{\sigma^2(P_2^{resample})}{\sigma^2(P_2^{ML})}$$

should have been 1. Although we have not found a firm theoretical explanation, the ratios found were usually greater than one. That the ratio is clearly not equal to 1 reflects the fact that there is a specification error (both in the real data and in the artificial data where it was deliberately introduced by us), and it is intuitively appealing that it is normally greater than 1.

4.3 The Estimation Runs

By varying the size of the within-individual variation σ_1 and between-individual variance σ_2, we can carry out estimation runs with varying degrees of correlation between the observations for a simulated individual, and hopefully cover the specification error found in real data with repeated measurements.

We have used $k_1 = k_2 = 0.5$; and five levels each of $\sigma_1\sigma_2[0{-}0.2{-}0.5{-}1{-}2]$. It was not necessary to investigate the relative scale of the σ's and the k's, because the overall scale of utility functions is not relevant (only the sign is used).

4.4 The Results

The simulated datasets used to estimate the working model above contained 10000 observations from 2500 respondents, so that four responses are drawn from each individual.

Many of these runs were needed to build confidence in the simulation; because the results with Model I showed that in many instances the variance ratios R_{p1} and R_{p2} turned out to be less then one, contrary to the results with real data.

When we found this, we investigated the effect of the following measures:

- the number of sub-samples
- the number of observations from each individual
- the relative size of k_1 and k_2
- sampling from an orthogonal design: the attributes presented to individuals are varied independently from one another and the analyst defines the alternatives in terms of combinations of levels of attributes.

None of these had the effect of raising R_{p1} and R_{p2}. Finally we modified the error structure by introducing heteroskedasticity into the error-terms as in Model II. This did the trick: both R_{p1} and R_{p2} became greater than 1.

With Model II we have run the following tests:

- Jackknife with four responses from each individual and 20, 40 and 99 resamplings,
- Jackknife with ten responses from each individual and 20, 40 and 99 resamplings,
- Bootstrap with four responses from each individual and 20, 40 and 99 resamplings.

The overall results with 4 responses per individual for the Jackknife are as follows:

- the parameter estimates coincided accurately with the values we put into the simulation; this confirmed our belief that the ML estimation program is working well, and that the resampling procedure was carried out correctly;
- when the between-individual variance is set to zero ($\sigma_1 = 0$; the specification error disappears) we had both $R_{p1} = 1$ and $R_{p2} = 1$.
  This confirmed our faith in the accuracy of the resampling methods and the variance obtained from the ML estimator;
- the specification errors did not introduce a bias;
- when the value of σ_1 was increased, the model specification error was increased, and the ratio's $R_{p1} = 1$ and $R_{p2} = 1$ increased, signifying that the variance estimate obtained from the mis-specified model was too low;
- more extensive runs with 250 resamplings merely produced smoother surfaces, confirming our belief that the resampling statistics had more or less converged.

The results for the Bootstrap are largely the same, except that in some cases the ratios R_{p1} and R_{p2} were less then 1. This reflects our earlier comment that the Bootstrap "injects more noise" into the resamplings, and may not have "converged" at that number of resamplings.

The results with 10 responses per individual for the Jackknife resulted in R_{p1} and R_{p2} always being greater than 1. This result can be explained by noting that as the numbers of observations per individual increases, the "repeated measurement" problem becomes more serious.

5. Conclusions

Overall results on simulated data can be summarised as in Figures 5.1 and 5.2.

5.1 Acknowledgements

The authors wish to thank Joe Whittaker for his comments on the first draft of the article, and Mark Bradley for applying the Jackknife to Logit model estimation, and writing the first version of the automatic Jackknife application software.

The work of the first author was done as part of an academic placement at Hague Consulting Group under the Leonardo programme of the European Union.

References

1. M. Ben-Akiva, S. Lerman. Discrete choice analysis: theory and application to travel demand. MIT press, 1985.
2. A.F. Bissel, R.A. Ferguson. The Jackknife - Toy, Tool or two-edged weapon. *The Statistician*, 24(2):79–100, 1975.
3. B. Efron. The Jackknife, the Bootstrap and other resampling plans. *CBMS-NSF Regional conference series in applied mathematics*, SIAM, 38, 1982.
4. B. Efron, R.J. Tibshirani. An introduction to the bootstrap. *Monographs on Statistics and Applied Probability*, Chapman & Hall, 57, 1993.
5. J.K. Lindsey. Parametric statistical inference. Oxford science publications, 1996.
6. J. Shao, D. Tu. The Jackknife and Bootstrap. *Springer series in statistics*, Springer verlag, 1995.

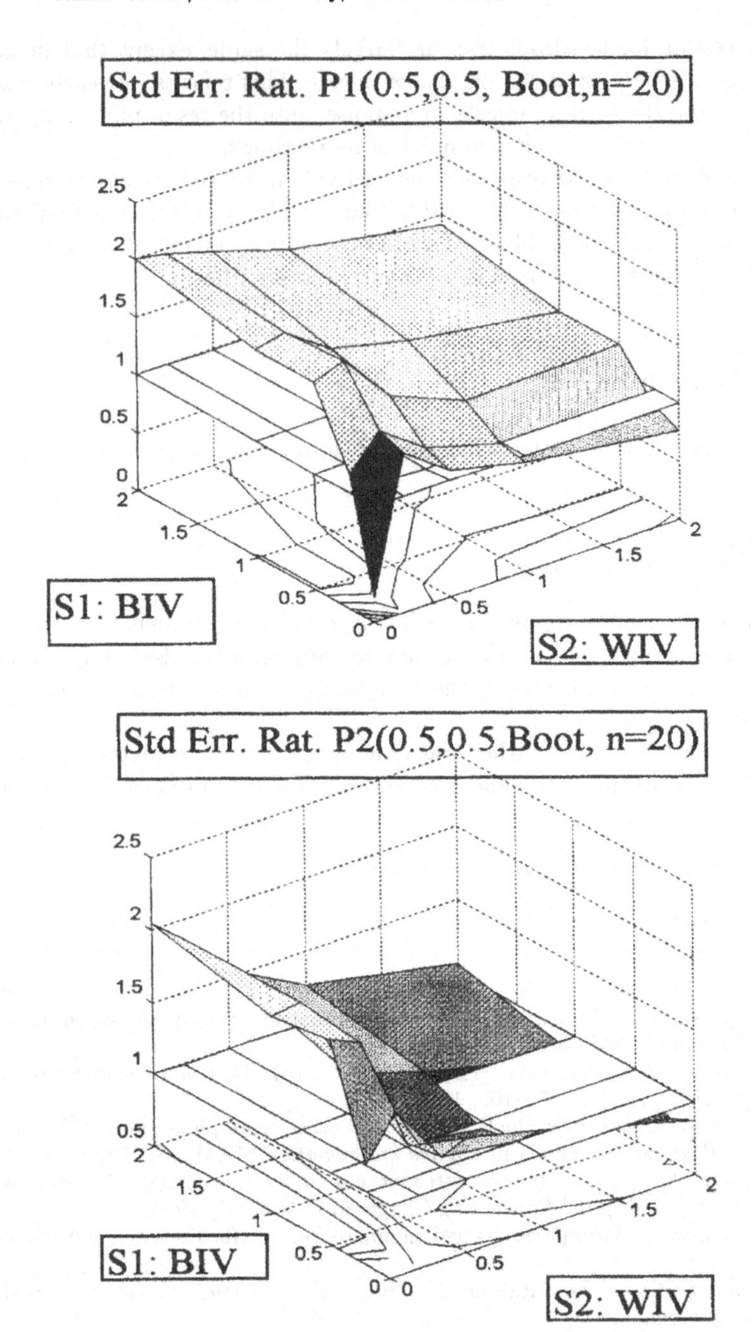

Fig. 5.1. Bootstrap standard error ratio

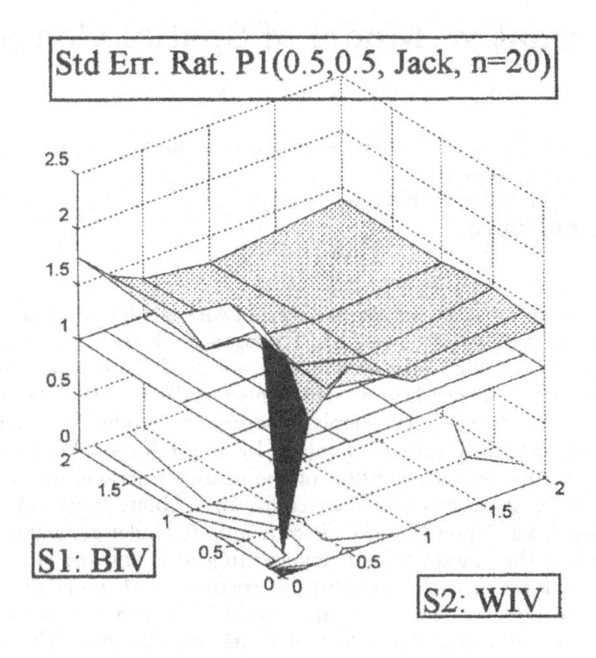

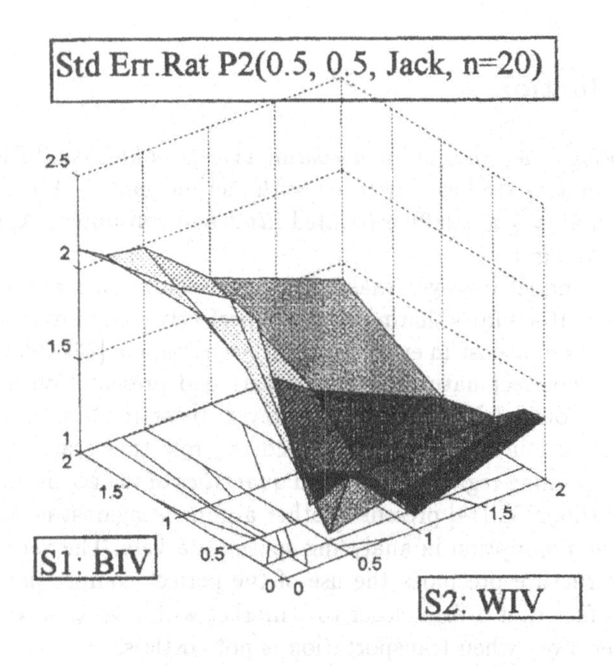

Fig. 5.2. Jackknife standard error ratio

An Alternative Model of Spatial Competition

Vladimir A. Bulavsky[1] and Vyacheslav V. Kalashnikov[2]

[1] Central Economics & Mathematics Institute, Nakhimovsky pr. 47, Moscow 117418, Russia
[2] Sumy University, Rimsky-Korsakov st. 2, Sumy 244007, Ukraine

Summary. In this paper, an alternative network model of oligopolistic markets of homogeneous product is developed. The agents sell their product at several independent markets taking into account the prices of the product unit at different markets, production expenditures, and transportation costs. The unit price at a market depends upon the total supply, whereas the production expenditures may grow along with the total volume of output by all producers. The latter ones choose production volumes and distribution of the output fractions sold at the markets. In order to do that, each agent uses conjectures about the total market supply variations depending upon those of his own supply. Under general enough assumptions concerning the market inverse demand functions and the producers' cost and transportation functions, the equilibrium existence and uniqueness theorems are formulated and proven. Thus, the paper could be considered as a contribution to the analysis of the structuring effect of transportation network on markets and society as a whole.

1. Introduction

Since long, the concept of a *spatial price equilibrium* (SPE) has been the focal point of studies concerned with the movement of goods and with competition among spatially separated firms and consumers. Applications of this model abound.

The SPE model, however, makes a strong assumption concerning the market structure. It assumes that markets are perfectly competitive; that is, price will equal marginal cost in every region. Recent papers [13], [9], [15], [16] recognize the imperfect nature of the markets and present Cournot-Nash and Stackelberg-Cournot-Nash models. However, these models do not recognize the spatial structure of the markets and the role that transportation costs play in determining regional prices and quantities of the goods under study. A series of authors [5], [14] present another argument against using the perfect competition assumption in analyzing spatial markets. The very existence of a spatial structure precludes the use of the perfect competition assumption due to the fact that a firm closer to a market will have an advantage over a firm farther away when transportation is not costless.

Models defined over a continuous plane have been developed that incorporate imperfections in spatial markets. For example, in terms of network models, [7] presents Bertrand-Nash and limit pricing models for electricity markets; [4] develops a model of world cost trade that uses the cooperative

Nash bargaining solution. Then, [12] state a Cournot-Nash model in which producers are spatially distributed, but demand is assumed to be located at one geographical location. Finally, [17] proves existence of spatial Cournot equilibrium.

This paper aims at developing a new approach to modelling and analyzing the economical notion of network oligopoly by mathematical tools. In the standard production-transportation problem, the structure of transportation flows is determined by the way in which the points of production and consumption are located. In the presented model, both production and consumption are influenced, in particular, by transportation conditions; and vice versa, the transportation flows in turn are determined by the demand at the points of consumption. Thus, the transportation flows connected either with production or with consumption experience the mutual influence and must therefore be determined simultaneously.

The authors previously [1],[2] presented a model in which the oligopoly agents presuppose a degree of their influence upon the market price, and they choose their output volumes taking those conjectures into account. However, the market itself and the production costs in those works were treated in somewhat traditional manner. This paper offers an attempt to modify the oligopoly model using both the network approach (cf. [10],[3], [11]) and the idea of production cost depending upon the demand for the raw materials. Such a modification should increase the model's flexibility and its adaptation facilities.

The classical approach assumes no market mechanisms to exist and affect the production costs (or returns to scale). More precisely, each agent's costs depend only upon his own output which means that each subject has a separate production factors market, if any. Of course, such individual markets of the production factors do exists and must be taken into consideration when modelling. They comprise the markets of local resources with limited transportability such as local raw materials, working hands etc. However, there are also some well-transported production factors, and their prices must depend upon not only a single agent's demand for them (and hence, upon his output), but upon the joint demand by all the subjects. The latter quality is proposed to model by the cost functions depending upon two variables: an individual output, and the total production output by all the oligopoly agents.

As for the market sale process, it is modelled with the aid of the classical inverse demand functions. However, instead of a single market, a network of markets with different demand fuctions is considered. Each producer can distribute his output arbitrarily among the sale markets taking into account his conjectures of his influence degree at each market separately.

The paper is organized as follows. Section 2 describes the model with a network of sale markets and individual markets of production factors, and an equilibrium problem for such a model is specified. Section 3 is dedicated to proving the equilibrium existence and uniqueness theorems. Finally, Section 4 deals with the model with the market of production factors in which the producers' returns to scale are dependent upon not only their own outputs but also on the joint output of all the subjects of the model. For that model, the equilibrium existence and uniqueness theorems has been also proven. Notice that the model described in Section 2 is a particular case of that of Section 4. However, restrictions on the functions relevant in the former model are weaker than those in the latter one. Therefore the theorems of Section 3 cannot be formally deduced from those of Section 4.

2. Model With Individual Markets of Production Factors

Consider M firms producing a single homogeneous product with their current outputs denoted by q_i, and their cost functions by $f_i(q_i)$, $i = 1, \dots, M$. The product is supposed to sell at N different markets. Therefore, if s_{ij} is the quantity of the product supplied by i-th producer for sale at market j, then the equality $\sum\limits_j s_{ij} = q_i$ holds. Besides, producer i may suffer supplementary expenses (for instance, transportation costs) at amount of $c_{ij} \geq 0$ per unit of the product.

Denote by G_j the total amount of sale at market j, and by $p_j(G_j)$ the inverse demand function, i.e. price of a unit of product at the market with the total sale volume G_j. For each place j, we also allow some local sources of the product to exist and supply $Q_j \geq 0$ units. Thus, the following equalities must hold $G_j = Q_j + \sum\limits_i s_{ij}$, $j = 1, \dots, N$.

Furthermore, we associate with each pair "producer–consumer" (i, j) a conjectured by producer i quotient $w_{ij} > 0$ of his influence upon the market j as follows. At the current market conjuncture, producer i assumes that his supply variation from s_{ij} to η_{ij} implies the total sale volume at market j to become equal to

$$G_{ij}(\eta_{ij}) = G_j + w_{ij}(\eta_{ij} - s_{ij}). \tag{1}$$

Arrange thus introduced values of sales and local supplies into vectors $G, Q \in R^N$, respectively, and form the product flows from producer i as a vector $S_i = (s_{ij}, j = 1, \dots, N) \in R^N$. Expected profit of each agent depends upon the current situation at the market, his output $\eta_i = (\eta_{ij}, j = 1, \dots, N)$, and conjecture (1):

$$\mu_i(\eta_i) = \sum_{j=1}^{N} \eta_{ij} \left[p_j\left(G_{ij}(\eta_{ij})\right) - c_{ij}\right] - f_i\left(\sum_{j=1}^{N} \eta_{ij}\right). \tag{2}$$

The agent considers his choice to be correct if his expected profit (2) is maximal at $\eta_{ij} = s_{ij}$.

At a balanced market, such states are naturally treated as equilibria. The strict definition of an equilibrium will be given making use of the first order optimality condition for the expected profit function (2). Before doing that in formal terms, we state the requirements for functions f_i, p_j.

A1. Each of the functions $f_i(q_i)$ are defined over $q_i \geq 0$, continuously differentiable, convex and non-decreasing.

A2. Every inverse demand function $p_j(G_j)$ assumes positive values at non-negative G_j, and it is continuously differentiable with $p_j'(G_j) < 0$ for $G_j \geq 0$.

A3. For each i there exists an $H_i > 0$ such that

$$f_i'(H_i) \geq \max_j [p_j(H_i) - c_{ij}] . \tag{3}$$

A4. For each pair i, j the quotients w_{ij} satisfy the constraints

$$0 < w_{ij} \leq 1. \tag{4}$$

A5. For every $j = 1, \dots, N$, the function $p_j(G_j)G_j$ is concave with respect to G_j.

Now differentiate the expected profit function (2) by η_{ij} and obtain

$$\frac{\partial \mu_i}{\partial \eta_{ij}}(\eta_i) = \mu_{ij}'(\eta_i) = \left[p_j\left(G_{ij}(\eta_{ij})\right) + \eta_{ij} w_{ij} p_j'\left(G_{ij}(\eta_{ij})\right) - c_{ij}\right] - \\ -f_i'\left(\sum_{j=1}^{N} \eta_{ij}\right) . \tag{5}$$

If $\eta_{ij} = s_{ij}$ the derivative of function μ_i can be written in the form

$$\mu_{ij}'(S_i) = \left[p_j\left(G_j\right) + s_{ij} w_{ij} p_j'\left(G_j\right) - c_{ij}\right] - f_i'\left(q_i\right) . \tag{6}$$

Remark 1. In the network models with fixed transportation routes it is frequently supposed [10], [3], [11], that the quotients c_{ij} are not constant, but (non-decreasing) functions of the transported volumes s_{ij}. However this assumption does not look natural in our case, since the possibility of transportation by different suppliers along the same routes is not excluded.

Making use of the first order extremum conditions for the expected profit functions of each agent, we specify the equilibrium problem as follows. Given $Q \geq 0$, find a non-negative $(M + 1)$-tuple $(G, S_1, \dots, S_M)$ such that

$$\sum_{i=1}^{M} s_{ij} + Q_j = G_j, \quad j = 1, \ldots, N, \tag{7}$$

$$\sum_{j=1}^{N} s_{ij} = q_i, \quad i = 1, \ldots, M, \tag{8}$$

and for each pair (i,j) the relationships hold:

$$\mu'_{ij}(S_i) = \left[p_j\left(G_j\right) + s_{ij} w_{ij} p'_j\left(G_j\right) - c_{ij}\right] - f'_i\left(q_i\right) \leq 0, \quad s_{ij}\mu'_{ij} = 0. \tag{9}$$

In order to treat a solution to complementarity problem (7) – (9) as an equilibrium for the agents with $q_i > 0$, it suffices to check the concavity of the expected profit function (2) with respect to η_i. In the other words, we demonstrate that the derivative mapping $\mu'_i : R^N \to R^N$ with the components $\mu'_{ij}(\eta_i)$ is antitone over R^N_+. Indeed, consider $\eta_i^1, \eta_i^2 \in R^N_+$ and examine the scalar product

$$\left[\mu'_i(\eta_i^1) - \mu'_i(\eta_i^2)\right]^T \left(\eta_i^1 - \eta_i^2\right) =$$

$$= \sum_{j=1}^{N} \left[p_j\left(G_{ij}\left(\eta_{ij}^1\right)\right) - p_j\left(G_{ij}\left(\eta_{ij}^2\right)\right)\right] \cdot \left(\eta_{ij}^1 - \eta_{ij}^2\right) +$$

$$+ \sum_{j=1}^{N} w_{ij}\left[\eta_{ij}^1 p'_j\left(G_{ij}\left(\eta_{ij}^1\right)\right) - \eta_{ij}^2 p'_j\left(G_{ij}\left(\eta_{ij}^2\right)\right)\right] \cdot \left(\eta_{ij}^1 - \eta_{ij}^2\right) -$$

$$- \left[f'_i\left(\sum_{j=1}^{N} \eta_{ij}^1\right) - f'_i\left(\sum_{j=1}^{N} \eta_{ij}^2\right)\right] \cdot \left(\sum_{j=1}^{N} \eta_{ij}^1 - \sum_{j=1}^{N} \eta_{ij}^2\right). \tag{10}$$

The latter term in (10) is non-negative due to the convexity of f_i. As for the former two sums, we investigate each term in them separately. Fix an arbitrary j and put for certainty $\eta_{ij}^1 \leq \eta_{ij}^2$. Then $G_{ij}\left(\eta_{ij}^1\right) \leq G_{ij}\left(\eta_{ij}^2\right)$, and condition A2 (monotone decreasing of p_j) imply that

$$\left[p_j\left(G_{ij}\left(\eta_{ij}^1\right)\right) - p_j\left(G_{ij}\left(\eta_{ij}^2\right)\right)\right] \cdot \left(\eta_{ij}^1 - \eta_{ij}^2\right) \leq 0. \tag{11}$$

Besides, if $p'_j\left(G_{ij}\left(\eta_{ij}^1\right)\right) \geq p'_j\left(G_{ij}\left(\eta_{ij}^2\right)\right)$, then $\eta_{ij}^1 p'_j\left(G_{ij}\left(\eta_{ij}^1\right)\right) \geq \eta_{ij}^2 p'_j\left(G_{ij}\left(\eta_{ij}^2\right)\right)$ since the derivative is p'_j is negative; therefore,

$$w_{ij}\left[\eta_{ij}^1 p'_j\left(G_{ij}\left(\eta_{ij}^1\right)\right) - \eta_{ij}^2 p'_j\left(G_{ij}\left(\eta_{ij}^2\right)\right)\right] \cdot \left(\eta_{ij}^1 - \eta_{ij}^2\right) \leq 0. \tag{12}$$

Otherwise, if $p'_j\left(G_{ij}\left(\eta_{ij}^1\right)\right) < p'_j\left(G_{ij}\left(\eta_{ij}^2\right)\right)$, we make use of (1) and transform the sum of the first factors in the lefthand sides of (11) and (12) as follows:

$$\left[p_j\left(G_{ij}\left(\eta_{ij}^1\right)\right) - p_j\left(G_{ij}\left(\eta_{ij}^2\right)\right)\right] +$$

$$+\left[G_{ij}\left(\eta_{ij}^1\right)p_j'\left(G_{ij}\left(\eta_{ij}^1\right)\right)-G_{ij}\left(\eta_{ij}^2\right)p_j'\left(G_{ij}\left(\eta_{ij}^2\right)\right)\right]+$$
$$+\left(G_j-w_{ij}s_{ij}\right)\left[p_j'\left(G_{ij}\left(\eta_{ij}^2\right)\right)-p_j'\left(G_{ij}\left(\eta_{ij}^1\right)\right)\right]. \tag{13}$$

The sum of the first two terms in the square brackets in (13) is non-negative due to A5, and the sign of the last term coincides (as (3) and (7) imply) with that of the difference $p_j'\left(G_{ij}\left(\eta_{ij}^2\right)\right)-p_j'\left(G_{ij}\left(\eta_{ij}^1\right)\right)$, which is positive according to our assumption. Thus we have shown that the sum of the lefthand sides of (11) and (12) is non-positive in all cases which implies that term (10) is non-positive, too. But the latter means that the derivative mapping μ_i' is antitone over R_+^N, and consequently, the expected profit function μ_i is concave with respect to $\eta_{i1},\ldots,\eta_{iN}$.

In order to justify our definition of equilibrium from the point of view of agents with $q_i=0$, we remark that the necessary and sufficient maximum condition (9) for the expected profit function can be re-arranged in the following form

$$p_j\left(G_j\right)-c_{ij}-f_i'\left(q_i\right)\le -s_{ij}w_{ij}p_j'\left(G_j\right). \tag{14}$$

The righthand side in (14) can be treated as a threshold that the "price–cost" difference for an extra product unit should overcome in order to make agent i increase his output.

Since the righthand side in (14) vanishes as q_i tends to nothing, it is natural enough to postulate the following behaviour of an inactive agent (with $q_i=0$). Namely, he has an incentive to start producing only if the difference between the price $p_j(G_j)$ at least in one consumption region and his expenditures $f_i'(0)+c_{ij}$ is positive, i.e. $p_j(G_j)-c_{ij}-f_i'(0)>0$. But condition (9) at $q_i=0$ is exactly opposite to that, hence agent i will not produce.

3. Existence and Uniqueness Theorems

Theorem 1. *Let assumptions* A1 – A4 *be valid. Then problem* (7) – (9) *is solvable.*

Proof. Define a mapping $\mathcal{A}: R_+^N\to R_+^N$ as follows: fix an N-tuple $G=(G_1,\ldots,G_N)\ge 0$ and find a (unique) solution $S(G)=(S_1,\ldots,S_M)\in R_+^{MN}$ of a complementarity problem (here $S_i=(s_{i1},s_{i2},\ldots,s_{iN})$):

$$s_{ij}\ge 0,\ \varphi_{ij}=f_i'\left(\sum_{k=1}^N s_{ik}\right)+c_{ij}-w_{ij}s_{ij}p_j'(G_j)-p_j(G_j)\ge 0,$$
$$s_{ij}\varphi_{ij}=0,\ i=1,\ldots,M;\ j=1,\ldots,N. \tag{15}$$

Underline that in this problem, G_j and s_{ij} are not bound by constraint (7).

Problem (15) is solvable uniquely since the mapping $\varphi = (\varphi_{ij})_{i=1,j=1}^{M\ \ N}$ is strongly monotone over R_+^{MN}. Indeed,

$$[\varphi(S^1) - \varphi(S^2)]^T (S^1 - S^2) = \sum_{i=1}^{M}\sum_{j=1}^{N} [\varphi_{ij}(S^1) - \varphi_{ij}(S^2)] \cdot (s_{ij}^1 - s_{ij}^2) =$$

$$= \sum_{i=1}^{M} \left[f_i' \left(\sum_{k=1}^{N} s_{ik}^1 \right) - f_i' \left(\sum_{k=1}^{N} s_{ik}^2 \right) \right] \left(\sum_{j=1}^{N} s_{ij}^1 - \sum_{j=1}^{N} s_{ij}^2 \right) -$$

$$- \sum_{i=1}^{M}\sum_{j=1}^{N} w_{ij} p_j'(G_j) \cdot (s_{ij}^1 - s_{ij}^2)^2 \geq \alpha \|S^1 - S^2\|^2$$

due to assumptions A1, A2 and A4; here $\alpha = \min_{ij} (-w_{ij} p_j'(G_j)) > 0$. Define now entries of the mapping $\mathcal{A}$ by formula

$$\mathcal{A}_j(G) = Q_j + \sum_{i=1}^{M} s_{ij}, \quad j = 1, \ldots, N, \tag{16}$$

where $s_{ij} = s_{ij}(G)$ is the solution to problem (15).

According to the well-known properties of solutions to complementarity problems with strongly monotone operators [6], the mapping $\mathcal{A}$ is continuous over R_+^N. Now construct a compact subset $\Pi \subset R_+^N$ mapped into itself by $\mathcal{A}$. First note that assumptions A1 – A3 imply the following property of the solution $S(G)$ to problem (15). Consider

$$J(G) = \{1 \leq j \leq N \mid G_j > \max_{1 \leq i \leq M} H_i = \hat{H}\}, \tag{17}$$

where H_i are constants from A3. It is clear that for each pair (i, j) with $j \in J(G)$ at least one of the two relationships holds:

$$s_{ij} = 0 \qquad \text{or} \qquad \sum_{k=1}^{N} s_{ik} \leq H_i,$$

whence

$$s_{ij} \leq H_i, \qquad i = 1, \ldots, M, \ j \in J(G). \tag{18}$$

For every subset $J \subseteq \{1, 2, \ldots, N\}$ define a parallelepiped

$$\Pi(J) = \{G \in R_+^N \mid G_j > \hat{H}, \ j \in J; \ Q_k \leq G_k \leq \hat{H}, \ k \notin J, \}$$

and demonstrate the mapping $\mathcal{A}$ to be bounded over each of them. Indeed, for $J = \{1, \ldots, N\}$, the boundedness follows from (18). When $J = \emptyset$, the subset $\Pi(\emptyset)$ is compact, hence $\mathcal{A}(G)$ is bounded because solutions $s_{ij} = s_{ij}(G)$ of problem (15) are continuous by G. Finally, let $J \neq \emptyset$ and $J \neq \{1, \ldots, N\}$, and vectors $G = (G_1, \ldots, G_N)$ run through the subset $\Pi(J)$. From (18), it again follows that elements $s_{ij}(G)$, $j \in J$ of solution to problem (15) are uniformly bounded. The structure of the problem implies all the rest elements of the solution $s_{ik}(G)$, $k \notin J$, to depend (continuously) upon only G_k, $k \notin J$, and upon the elements s_{ij}, $j \in J$, values of which conform a compact subset. Therefore, all the elements of solutions $S = S(G)$ for $G \in \Pi(J)$ are bounded by some $H(J) \geq \hat{H}$. Take $H = \max_J H(J)$. Then from definition (16) we deduce

$$Q_j \leq \mathcal{A}_j(G) \leq Q_j + MH, \qquad \forall\, G \in R_+^N + Q. \tag{19}$$

Now apply the Brouwer theorem to the convex compact subset

$$\Pi = \{G \in R^N \mid Q_j \leq G_j \leq Q_j + MH,\ j = 1, \ldots, N\},$$

with the continuous mapping $\mathcal{A}$. Indeed, from (16) and (19) it follows that $\mathcal{A}(\Pi) \subseteq \Pi$. Hence, there exists a fixed point $G^* \in \Pi$, i.e. the equality holds $\mathcal{A}(G^*) = G^*$ which is tantamont to (7) – (9) for the element $(G^*, S(G^*))$. The theorem is proved completely. ∎

The question of uniqueness of the equilibrium can be reduced to that of the equilibrium total sales G_j. Indeed, when proving the existence theorem, we have established that given G_j, relationships (9) determine the values of variables s_{ij} uniquely.

In order to examine the uniqueness of the total sales in the markets, we keep on assumptions A1 – A5, and add the following one.

A6. Functions f_i, p_j are twice differentiable, and for each $j = 1, \ldots, N$ and all $G_j > 0$ the inequality holds

$$p_j''(G_j)G_j - 2p_j'(G_j) \geq 0. \tag{20}$$

Definition: An equilibrium $Z(Q) = [G(Q), S(Q)]$ is referred to as *non-monopolistic* for market j, if $s_{ij} < G_j, i = 1, \ldots, M$.

Theorem 2. *Let assumptions* A1 – A6 *be valid. If an equilibrium* $Z(Q) = [G(Q), S(Q)]$ *is non-monopolistic for each market* j, *the total sales* $G_j(Q)$ *at them are determined uniquely.*

Proof. Consider an operator $\psi : R^{MN} \to R_+^{MN}$ with components

$$\psi_{ij} = f_i'\left(\sum_{r=1}^{N} s_{ir}\right) + c_{ij} - w_{ij}s_{ij}p_j'\left(Q_j + \sum_{t=1}^{M} s_{tj}\right) - p_j\left(Q_j + \sum_{t=1}^{M} s_{tj}\right), \tag{21}$$

$i = 1, \ldots, M$, $j = 1, \ldots, N$, and calculate entries of its Jacobi matrix $D = (d_{ij}^{k\ell})$, omitting the arguments of functions f_i'', p_j', and p_j'' for the brevity purpose:

$$d_{ij}^{ij} = \frac{\partial \psi_{ij}}{\partial s_{ij}} = f_i'' - (1 + w_{ij})p_j' - w_{ij}s_{ij}p_j''; \tag{22}$$

$$d_{ij}^{i\ell} = \frac{\partial \psi_{ij}}{\partial s_{i\ell}} = f_i'', \quad \text{if} \quad \ell \neq j; \tag{23}$$

$$d_{ij}^{kj} = \frac{\partial \psi_{ij}}{\partial s_{kj}} = -p_j' - w_{ij} s_{ij} p_j'', \quad \text{if} \quad k \neq i; \tag{24}$$

$$d_{ij}^{kl} = \frac{\partial \varphi_{ij}}{\partial s_{k\ell}} = 0, \quad \text{if} \quad k \neq i, \ \ell \neq j. \tag{25}$$

Now investigate whether matrix D is positive definite. For an arbitrary collection $\alpha = (\alpha_{ij})$, write down the product

$$\alpha^T D \alpha = \sum_{i,k=1}^{M} \sum_{j,\ell=1}^{N} \alpha_{k\ell} d_{ij}^{k\ell} \alpha_{ij}. \tag{26}$$

Making use of (22) – (25) and introducing

$$\beta_j = \sum_{i=1}^{M} \alpha_{ij}, \tag{27}$$

re-arrange (26) as follows

$$\alpha^T D \alpha = \sum_{i=1}^{M} \left[f_i'' \cdot \left(\sum_{j=1}^{N} \alpha_{ij} \right)^2 \right] - \sum_{j=1}^{N} B_j, \tag{28}$$

where each B_j, $j = 1, \ldots, N$, can be represented in the form

$$B_j = p_j' \sum_{i=1}^{M} w_{ij} \alpha_{ij}^2 + \beta_j p_j'' \sum_{i=1}^{M} w_{ij} s_{ij} \alpha_{ij} + p_j' \beta_j^2. \tag{29}$$

In order to demonstrate each B_j to be non-positive, solve the extremal problem

$$B_j = B_j(\alpha_{1j}, \ldots, \alpha_{Mj}) \longrightarrow \max, \tag{30}$$

subject to condition (27) by the standard Lagrange techniques. Consider the Lagrange function of problem (27), (30)

$$\mathcal{L}(\alpha_{1j}, \ldots, \alpha_{Mj}) = B_j(\alpha_{1j}, \ldots, \alpha_{Mj}) + \lambda \cdot \left(\sum_{i=1}^{M} \alpha_{ij} - \beta_j \right)$$

and make its partial derivatives equal zero:

$$\frac{\partial \mathcal{L}}{\partial \alpha_{ij}} = 2 p_j' w_{ij} \alpha_{ij} + \beta_j p_j'' w_{ij} s_{ij} + \lambda = 0, \ i = 1, \ldots, M. \tag{31}$$

Solving the linear equations with respect to α_{ij}, find

$$\alpha_{ij} = -\frac{\lambda + \beta_j p_j'' s_{ij} w_{ij}}{2p_j' w_{ij}}, \ i = 1, \ldots, M. \tag{32}$$

Note that B_j are concave with respect to α_{ij} due to the inequality $p_j' < 0$. Therefore, by substituting terms from (32) into (29), we find the maximal value $\hat{B}_j = \max B_j$ of the objective function of problem (30) subject to (27):

$$\hat{B}_j = \sum_{i=1}^{M} \left[\frac{\lambda^2 + 2\lambda\beta_j p_j'' w_{ij} s_{ij} + \beta_j^2 \left(p_j''\right)^2 w_{ij}^2 s_{ij}^2}{4 w_{ij} p_j'} - \right.$$

$$\left. -\beta_j \frac{\lambda s_{ij} p_j''}{2p_j'} - \beta_j^2 \frac{w_{ij} s_{ij}^2 \left(p_j''\right)^2}{2p_j'} \right] + p_j' \beta_j^2 =$$

$$= \frac{\lambda^2}{4p_j'} \sum_{i=1}^{M} \frac{1}{w_{ij}} - \beta_j^2 \frac{\left(p_j''\right)^2}{4p_j'} \sum_{i=1}^{M} w_{ij} s_{ij}^2 + p_j' \beta_j^2. \tag{33}$$

Again make use of negativity of p_j', boundedness of quotients w_{ij} by the unity from above, and assumption that market j is non-monopolistic. Combined with the non-negativity of s_{ij}, the latter assumption implies the strict

$$\sum_{i=1}^{M} s_{ij}^2 < \left(\sum_{i=1}^{M} s_{ij} + Q_j \right)^2 = G_j^2. \tag{34}$$

Omitting the (non-positive and even strictly negative when $\lambda \neq 0$) first term in righthand side of (33) and making use of assumptions A5 and A6, finally come to the estimates

$$\hat{B}_j \leq -\beta_j^2 \frac{\left(p_j''\right)^2}{4p_j'} \sum_{i=1}^{M} s_{ij}^2 + p_j' \beta_j^2 \leq \frac{\beta_j^2}{4p_j'} \left[\left(2p_j'\right)^2 - G_j^2 \left(p_j''\right)^2 \right] =$$

$$= \frac{\beta_j^2}{4p_j'} \left(2p_j' - G_j p_j''\right) \cdot \left(2p_j' + G_j p_j''\right) \leq 0; \tag{35}$$

moreover, (34) implies the strict inequality $\hat{B}_j < 0$, if $\beta_j \neq 0$ or $\lambda \neq 0$. Remark that if $\beta_j = 0$ and $\lambda = 0$, then from (32) it follows that $\alpha_{ij} = 0$, $i = 1, \ldots, M$.

Coming back to equality (28), we deduce that

$$\alpha^T D \alpha = \sum_{i=1}^{M} \left[f_i'' \cdot \left(\sum_{j=1}^{N} \alpha_{ij} \right)^2 \right] - \sum_{j=1}^{N} B_j \geq 0 \tag{36}$$

due to the convexity of functions f_i and inequalities (35). Besides, the inequality in (36) is strict unless all of α_{ij}, $i = 1, \ldots, M$, $j = 1, \ldots, N$ are equal to zero.

Therefore, the matrix D is positive definite, and hence the operator $\Psi = (\psi_{ij})$ is strongly monotone. Thus, we can conclude that the equilibrium exists uniquely. The proof is complete. ∎

Remark 2. If $Q_j > 0, j = 1, \ldots, N$, then every equilibrium is non-monopolistic. Consequently, in this case, the equilibrium sales $G_j(Q)$ are determined uniquely, according to Theorem 2. In general, it is possible that a single network has both monopolistic and non-monopolistic equilibria with different sales at the markets (for instance, it can be the case if the functions f_i and $p_j(G)G$ are piece-wise linear).

Remark 3. If we consider the equilibrium vectors $Z(Q)$ as functions of the influence quotients w_{ij}, then it is evident from the proof of theorem 2 that $Z(Q)$ depend continuously upon w_{ij} due to the fact that $0 < w_{ij} \leq 1$.

Remark 4. Inequality (20) in assumption A6 is valid, for example, if function p_j satisfies assumption A2 and is convex.

4. Model with a Market of Production Factors

Here again we consider the problem of transportation of a certain commodity. However, in contrast to Section 2, here we suppose that the cost function f_i of the i-th producer depends upon not only of his output $q_i = \sum_{j=1}^{N} s_{ij}$ but also upon the total output T by all the producers:

$$T = \sum_{i=1}^{M} \sum_{j=1}^{N} s_{ij}. \tag{37}$$

For example, this dependence may be of the form

$$f_i(q_i, T) = q_i h_i(T), \tag{38}$$

where $h_i(T)$ is the expenditure of producer i for production of a unit of a good under an assumption that the total output by all the producers is equal to T. In what follows however we do not suppose that function f_i must be represented by equality (38). Thus, the performance of the transportation network influences the way in which the commodity may be distributed, and this in turn has an impact on the market structure as well as on the structure of the demand on the production factor.

We keep on the hypotheses of section 2 and add an assumption that each producer is aware of his cost as a function of the total output by all the producers. Hence making use of his influence quotients w_{ij} he can predict variations of his expenditures as a result of varying his output. Namely, agent i predicts the variation of the total output T according to the rule

$$T_i(\eta_i) = T + \sum_{j=1}^{N} w_{ij} (\eta_{ij} - s_{ij}); \tag{39}$$

here, just as above, $\eta_i = (\eta_{ij})_{j=1}^N$. Then the i-th agent expects his profit to constitute

$$\mu_i(\eta_i) = \sum_{j=1}^{N} \eta_{ij} [p_j (G_j(\eta_{ij})) - c_{ij}] - f_i \left(\sum_{j=1}^{N} \eta_{ij}, T_i(\eta_i) \right), \tag{40}$$

and he makes his choice in order to maximize the value of function (40).

Like in Section 2, we define an equilibrium through the first order optimality condition for the expected profit function (40). First of all, list the assumptions concerning functions f_i and p_j.

AA1. Each of the functions $f_i(q_i, T)$ are defined over R_+^2, continuously differentiable, convex by the pair of variables, and non-decreasing by T.

AA3. There exists such scalar $a > 0$ that

$$\frac{\partial f_i}{\partial q_i}(0, T) \geq a \quad \text{at every} \quad T \geq 0. \tag{41}$$

AA5. For each $j = 1, \ldots, N$, function $p_j(G_j)G_j$ is bounded over $G_j \in [0, +\infty)$.

AA2 and **AA4** coincide with assumptions **A2** and **A4**, respectively.

Differentiating the expected profit function (40) with respect to η_{ij}, we get

$$\frac{\partial \mu_i}{\partial \eta_{ij}}(\eta_i) = \mu'_{ij}(\eta_i) = [p_j (G_j(\eta_{ij})) + \eta_{ij} w_{ij} p'_j (G_j(\eta_{ij})) - c_{ij}] -$$

$$-\frac{\partial f_i}{\partial q_i} \left(\sum_{j=1}^{N} \eta_{ij}, T_i(\eta_i) \right) - w_{ij} \frac{\partial f_i}{\partial T} \left(\sum_{j=1}^{N} \eta_{ij}, T_i(\eta_i) \right). \tag{42}$$

At the point $S = (S_i)_{i=1}^M$, the derivative of function μ_i can be represented in the form

$$\mu'_{ij}(S) = [p_j (G_j) + s_{ij} w_{ij} p'_j (G_j) - c_{ij}] -$$

$$-\frac{\partial f_i}{\partial q_i} (q_i, T) - w_{ij} \frac{\partial f_i}{\partial T} (q_i, T). \tag{43}$$

Now making use the first order extremality conditions for the expected profit functions of each agent, we specify the equilibrium problem as follows. Given $Q_j \geq 0$, $j = 1, \ldots, N$, find a non-negative $(M+1)$-tuple $(G, S_1, \ldots, S_M)$ such that

$$\sum_{i=1}^{M} s_{ij} + Q_j = G_j, \quad j = 1, \ldots, N, \tag{44}$$

$$\sum_{j=1}^{N} s_{ij} = q_i, \quad i = 1, \ldots, M, \tag{45}$$

$$\sum_{i=1}^{M} \sum_{j=1}^{N} s_{ij} = T, \tag{46}$$

and for each pair (i, j) the following holds

$$\mu'_{ij}(S) = \left[p_j\left(G_j\right) + s_{ij} w_{ij} p'_j\left(G_j\right) - c_{ij}\right] - \frac{\partial f_i}{\partial q_i}\left(q_i, T\right) -$$

$$-w_{ij} \frac{\partial f_i}{\partial T}\left(q_i, T\right) \leq 0, \quad s_{ij} \mu'_{ij}(S) = 0. \tag{47}$$

In order to treat a solution to complementarity problem (44) – (47) as an equilibrium for agents with $q_i > 0$, it again suffices to verify the concavity of the expected profit function (40) with respect to η_i. But the latter is established immediately since the first sum in (40) coincides exactly with that of Section 2, whereas the second term is a concave function being a superposition of a concave and a linear ones.

Theorem 3. *Let assumptions* AA1 – AA5 *be valid. Then problem* (44) – (47) *is solvable.*

Proof. According to AA5, there exists a constant B such that $\sum_{j=1}^{N} p_j(G_j) G_j \leq B$ for all $G_j \geq 0$. Estimate the sum

$$V = \sum_{i=1}^{M} \sum_{j=1}^{N} \mu'_{ij}(S) s_{ij}. \tag{48}$$

Due to (47) and assumptions AA1 – AA5,

$$V = \sum_{i=1}^{M} \sum_{j=1}^{N} \left\{ \left[p_j(G_j) + s_{ij} w_{ij} p'_j(G_j) - c_{ij}\right] - \right.$$

$$-\frac{\partial f_i}{\partial q_i}(q_i,T) - w_{ij}\frac{\partial f_i}{\partial T}(q_i,T)\Bigg\} s_{ij} \le$$

$$\le \sum_{j=1}^{N} p_j(G_j)G_j - \sum_{i=1}^{M} q_i \frac{\partial f_i}{\partial q_i}(q_i,T) \le B - a\sum_{i=1}^{M}\sum_{j=1}^{N} s_{ij}. \tag{49}$$

Thus, $V < 0$ if the vector $\{s_{ij}\}_{i,j=1}^{M,N}$ belongs to the subset

$$\left\{ s_{ij} \ge 0 \mid \sum_{i=1}^{M}\sum_{j=1}^{N} s_{ij} > \frac{B}{a} \right\}. \tag{50}$$

According to [8, Theorem 4], there exists a solution to the complementarity problem

$$\varphi_{ij}(S) = -\mu'_{ij}(S) \ge 0,\ s_{ij} \ge 0,\ -s_{ij}\mu'_{ij}(S) = 0, \tag{51}$$

which completes the proof. ■

In order to investigate the uniqueness of the amounts of market sales, we preserve assumptions AA1 — AA5 and add the following one.

AA6. All the agents are identical, i.e. they have the same twice continuously differentiable cost function $f(q_i,T)$ with a non-negative mixed derivative as well as the identical influence quotients at all the markets, i.e. $w_{ij} = w$. Besides, as in assumption A6, suppose that functions p_j are twice differentiable.

Remark 5. Assumption of non-negativity of the mixed derivative of the cost function $f(q_i,T)$ looks natural enough. It means that the marginal costs of a producer do not decrease along with the growth of the total production output.

Theorem 4. *Under assumptions* AA1 – AA6, *for an equilibrium that is non-monopolistic at each market* j, *the values of total market sales* $G_j(Q)$ *are determined uniquely.*

Proof. For the case of identical producers, consider an operator $\varphi : R^{MN} \to R_+^{MN}$ of problem (51) with components

$$\varphi_{ij} = \frac{\partial f}{\partial q_i}(q_i,T) + w\frac{\partial f}{\partial T}(q_i,T) + c_{ij} - ws_{ij}p'_j(G_j) - p_j(G_j), \tag{52}$$

$$i = 1,\ldots,M, \qquad j = 1,\ldots,N,$$

and calculate entries of its Jacobi matrix $D = \left(d_{ij}^{k\ell}\right)$ (omitting the arguments of functions f, p'_j E p''_j):

$$d_{ij}^{ij} = \frac{\partial \varphi_{ij}}{\partial s_{ij}} = \frac{\partial^2 f}{\partial q_i^2} + (1+w)\frac{\partial^2 f}{\partial q_i \partial T} + w\frac{\partial^2 f}{\partial T^2} -$$

$$-(1+w)p_j' - w s_{ij} p_j''; \tag{53}$$

$$d_{ij}^{i\ell} = \frac{\partial \varphi_{ij}}{\partial s_{i\ell}} = \frac{\partial^2 f}{\partial q_i^2} + (1+w)\frac{\partial^2 f}{\partial q_i \partial T} + w\frac{\partial^2 f}{\partial T^2},$$

$$\text{if} \quad \ell \neq j; \tag{54}$$

$$d_{ij}^{kj} = \frac{\partial \varphi_{ij}}{\partial s_{kj}} = (1+w)\frac{\partial^2 f}{\partial q_i \partial T} + w\frac{\partial^2 f}{\partial T^2} - p_j' - w s_{ij} p_j'',$$

$$\text{if} \quad k \neq i; \tag{55}$$

$$d_{ij}^{k\ell} = \frac{\partial \varphi_{ij}}{\partial s_{k\ell}} = (1+w)\frac{\partial^2 f}{\partial q_i \partial T} + w\frac{\partial^2 f}{\partial T^2},$$

$$\text{if} \quad k \neq i,\ \ell \neq j. \tag{56}$$

Now examine if the matrix D is positive definite. For an arbitrary vector $\alpha = (\alpha_{ij})$ write down the product

$$\alpha^T D \alpha = \sum_{i,k=1}^{M} \sum_{j,\ell=1}^{N} \alpha_{k\ell} d_{ij}^{k\ell} \alpha_{ij}. \tag{57}$$

Making use of (53) – (56) and introducing

$$\sigma = \frac{\partial^2 f}{\partial q_i^2} + (1+w)\frac{\partial^2 f}{\partial q_i \partial T} + w\frac{\partial^2 f}{\partial T^2}, \tag{58}$$

$$\tau = (1+w)\frac{\partial^2 f}{\partial q_i \partial T} + w\frac{\partial^2 f}{\partial T^2}, \tag{59}$$

one can re-arrange (57) in the form

$$\alpha^T D \alpha = \sum_{i=1}^{M} \sigma \left(\sum_{j=1}^{N} \alpha_{ij}\right)^2 + \sum_{k\neq i} \tau \left(\sum_{j=1}^{N} \alpha_{ij}\right)\left(\sum_{j=1}^{N} \alpha_{kj}\right) - \sum_{j=1}^{N} B_j, \tag{60}$$

where B_j are defined in the same way as when proving Theorem 2. Due to the convexity of function f and non-negativity of its mixed derivative, we obtain the series of inequalities

$$0 \leq \tau \leq \sigma, \tag{61}$$

which clearly implies

$$\sum_{i=1}^{M} \sigma \left(\sum_{j=1}^{N} \alpha_{ij}\right)^2 + \sum_{k\neq i} \tau \left(\sum_{j=1}^{N} \alpha_{ij}\right)\left(\sum_{j=1}^{N} \alpha_{kj}\right) \geq 0 \tag{62}$$

for any vectors α. By an exact repetition of the discourse from the proof of Theorem 2, we can confirm the inequality

$$\sum_{j=1}^{N} B_j \leq 0 \qquad \forall\alpha, \tag{63}$$

to be valid, and strict unless all the α_{ij} equal zero.

Making use of relationships (62), (63) and repeating exactly the discourse that completes the proof of Theorem 2, we establish the uniqueness of the equilibrium total sales G_j at all the (non-monopolistic) markets j. The Theorem is proved. ■

Acknowledgements

This work was supported by the Russian Humanitarian Scientific Foundation (RHSF), project No. 96-02-02251.

References

1. V. A. Bulavsky and V. V. Kalashnikov. One-Parametric Driving Method to Study Equilibrium. *Economics and Mathematical Methods*, 30(4): 129–138, 1994 (*in Russian*).
2. V. A. Bulavsky and V. V. Kalashnikov. Equilibria in Generalized Cournot and Stackelberg Models. *Economics and Mathematical Methods*, 31(3): 164–176, 1995. (*in Russian*).
3. S. Dafermos and A. Nagurney. A network formulation of market equilibrium problems and variational inequalities. *Operations Research Letters*, 3(3): 247–250, 1984.
4. J. Falk and G. P. McCormick. Mathematical Structure of the International Coal Trade Model. Report DOE/NBB-0025, U.S. Department of Energy, Washington, D.C.
5. P. T. Harker. Alternative Models of Spatial Competition. *Operations Research*, 34(3): 410–425, 1986.
6. P. T. Harker and J. S. Pang. Finite-dimensional variational inequalities and nonlinear complementarity problems: a survey of theory, algorithms and applications. *Math. Programming*, 48(2): 161–220, 1990.
7. B. Hobbs. Network Model of Spatial Oligopoly with an Application to Deregulation of Electricity Generation. *Operations Research*, 34(3): 395–409, 1986.
8. V. V. Kalashnikov. Fixed Point Existence Theorems Based upon Topological Degree Theory. Working Paper, Central Economics and Mathematics Institute (MOscow), 1995. (*in Russian*).
9. F. Murphy, H. Sherali and A. Soyster. A Mathematical Programming Approach for Determining Oligopolistic Market Equilibria. *Mathematical Programming*, 24(1): 92–106, 1962.
10. A. Nagurney, Jie Pan and Lan Zhao. Human migration network. *European Journal of Operational Research*, 59(3): 262–274, 1992.

11. A, Nagurney. A network model of migration equilibrium with movement costs. *Mathematical and Computer Modelling*, 13(1): 79–88, 1990.
12. R. Rovinsky, C. Shoemaker and M. Todd. Determining Optimal Use of Resources among Regional Producers under Differing Levels of Cooperation. *Operations Research*, 28(6): 859–866, 1980.
13. S. W. Salant. Imperfect Competition in the International Energy Market: A Computerized Nash-Cournot Model. *Operations Research*, 30(2):252–280, 1982.
14. E. Sheppard and L. Curry. Spatial Price Equilibria. *Geographical Analysis*, 14(3): 279–304, 1982.
15. H. D. Sherali, A. Soyster and F.H. Murphy. Stackelberg-Nash-Cournot Equilibria: Characterizations and Computations. *Operations Research*, 31(2):252–280, 1983.
16. T. E. Smith and T. L. Friesz. Spatial Market Equilibria with Flow-Dependent Supply and Demand: The Single Commodity Case. *Regional Science and Urban Economics*, 15(2): 181–218, 1985.
17. A. Weskamp. Existence of Spatial Cournot Equilibria. *Regional Science and Urban Economics*, 15(2): 219–228, 1985.

An Integrated Approach to Extra-Urban Crew and Vehicle Scheduling*

Maddalena Nonato

Istituto di Elettronica,
Facoltà di Ingegneria,
Università di Perugia, Italy

Summary. The scheduling of vehicles and crews, traditionally performed sequentially by scheduling vehicles prior to crews, has to be carried out simultaneously in particular settings such as the extra-urban mass transit, where crews are tightly dependent on vehicle activity or crew deadheadings are highly constrained. In this paper we propose an integrated approach to vehicle and crew scheduling which exploits the network structure of the problem. A heuristic method based on Lagrangean relaxation is presented, which determines a set of pieces of work suitable for both vehicle duties and crew duties. Crew duties are fixed step by step, while vehicles are scheduled once all the trips have been partitioned into pieces.

Extended use of Bundle methods for polyhedral functions and algorithms for constrained shortest path and assignment is made within a dual greedy heuristic procedure for the Set Partitioning problem.

Computational results are provided for Italian mass transit public companies, showing some improvements with respect to the results of the sequential approach.

1. Introduction

In spite of a variety of very efficient methods for the sequential scheduling of vehicles and crews, few proposals have been made in the literature for the solution of the joint problem. Scheduling vehicles prior to crews does not affect the overall quality of the solution except for a few particular cases, among which extra-urban public transport. Since in most industrialised countries crew related costs largely outweigh vehicle costs, it is reasonable to tackle the joint problem in those cases when a very efficient vehicle schedule may lead to a quite poor crew schedule or even to no feasible solution at all. The size of the problem, which is a critical aspect of this class of combinatorial problems, dramatically increases when the two problems are handled together. In fact, most models describing the joint problem are solved in practice by means of heuristic approaches where the problem is indeed decomposed: either crew constraints are taken into account at vehicle scheduling time [26] or crews are scheduled before vehicles as in the pioneering work of Ball et al. [2]. Other attempts have been made to solve the joint problem for the single depot case alone [21] which does not fit into many extra-urban settings where several depots are scattered over a large geographical area.

* This work has been supported by the "Progetto Finalizzato Trasporti 2", grant 91.02479.PF74.

In order to better explain the drawbacks of the sequential process in an extra-urban setting, let us briefly review the basic terminology. Transit service is given by a set of *lines* and *timetables*. Each traversal of a line is a *trip* which is described by both starting and ending times and locations. A trip is the unit of service for the purpose of vehicle scheduling, since a trip must be operated by a single vehicle. Trips that can be operated in sequence by the same vehicle are said to be *compatible*. Vehicle transfers with no passenger service that connect compatible trips or either come from or go to a depot are called *deadheadings* (hereafter *dh-trips*). A vehicle *block* is a sequence of compatible trips starting and ending at any pair of depots. Between consecutive blocks vehicles wait at the depots, where no crew attendance is needed. A *vehicle duty* is a feasible sequence of blocks such that the last block ends at the starting depot of the first block, namely the *depot of residence* of the vehicle. A *relief point* along a block provides time and place coordinates of possible crew relieving. A *piece of work* is defined as a continuous crew working period on a single vehicle between two relief points on a block. A piece between two consecutive relief points is called a *task*. On a *crew dead-heading*, drivers move from one relief point to another. A *crew duty*, often called a *run*, is a feasible sequence of pieces that starts and ends at the drivers residence depot.

Since vehicle scheduling packages would usually minimize fleet size and operational costs, they are likely to yield pieces of work which are too long to meet union regulations as well as situations in which drivers have no means to return to their depot of residence at the end of their duty. While the first issue, which affects individual blocks, can somehow be dealt with by considering some union constraints at vehicle scheduling time, such as maximum spreadtime and labour time, issues related to the latter question cannot be solved without knowledge of crew schedule. Moreover, while in the urban case drivers can move from a relief point to another regardless of the vehicle schedule, thus allowing for many ways to combine pieces of work within a crew duty, in the extra-urban setting distances are such that drivers are tied to their vehicle for the purpose of reaching relief points, so that pieces and blocks collapse into the same entity. For this reason, it would sometimes be convenient for the crew to split a block into two and force a vehicle to detour in order to pass by a depot or by another relief point.

In conclusion, we feel that, since in the extra-urban setting vehicle schedule heavily affects crew schedule, crews must be scheduled at the same time as vehicles in order to save on the vehicle costs without spoiling the crew schedule.

In our approach we focus on the intermediate stage of the solution process where trips are partitioned into pieces of work out of which both crew duties and vehicles duties are constructed. The Lagrangean relaxation of the constraints enforcing consistency between the pieces of crew duties and the pieces of vehicle duties provides a guideline on how to partition the trips into pieces.

In Section 2 a review of the literature on both vehicle scheduling and crew scheduling is provided, focusing on graph theory models. In the following sections our model is introduced and a solution approach via Lagrangean relaxation is proposed. Finally, computational results are provided for real life cases.

2. Network Models for Crew and Vehicle Scheduling Problems

Both vehicle and crew scheduling problems (VSP and CSP, respectively) consist of covering a set of service units by a set of duties. Since these units are fully characterised in terms of time and place, the compatibility relation holding between units that can be operated in sequence is often exploited through the *compatibility graph*; here duties are modelled as a constrained path, which has inspired several network-based solution approaches [8]. When feasibility constraints are overwhelming set partitioning/covering models are preferred, where duties are explicitly modelled and the network structure is eventually exploited within column generation mechanisms.

2.1 The Vehicle Scheduling Problem

The VSP consists of finding a set of vehicle duties such that: each trip is covered by exactly one vehicle, each vehicle returns to its starting depot at the end of its duty, and the number of duties operated from each depot does not exceed the number of vehicles available. When dealing with a single depot and a homogeneous fleet, the VSP can be formulated and polynomially solved as a network flow problem, but it becomes *NP*-Hard when multiple depots are introduced as proved in [5]: both heuristic and exact approaches have been proposed for the latter case (see [14] for a review), to which most of real life problems in both urban and extra-urban mass transportation systems belong.

Multiple Depot Vehicle Scheduling Problem, hereafter MD-VSP, can be formalised as follows: let $V = \{v\}$ be the set of trips to be operated, $(v = 1,..,n)$, starting from location s_v at time st_v and arriving at location e_v at time et_v; travelling time τ_{uv} from e_u to s_v is provided for each pair $u, v \in V$. Trips u and v are compatible if and only if $et_u + \tau_{uv} \leq st_u$. Let $D = \{d^1, .., d^K\}$ be a set of depots, each d^k being used as a base for at most C^k vehicles, k=1,..,K. Hereafter, vehicles are assumed to be homogeneous but our approach can easily be generalised to deal with different vehicle types.

MD-VSP is usually modelled as a multicommodity flow problem on the compatibility graph extended with depot connections, with one commodity for each depot [14]. A vehicle duty corresponds to a cycle through exactly one depot node, where no more than C^k cycles through each depot k are allowed.

Ribeiro and Soumis [23] describe a very efficient exact method where the column generation subproblem yielded by the Dantzig Wolfe decomposition applied to the linear relaxation of the multicommmodity flow formulation reduces to K shortest path problems on an acyclic graph.

2.2 The Crew Scheduling Problem

In the CSP blocks are first partitioned into pieces, and then pieces are sequenced to yield a set of feasible duties. This approach possess a nice network representation in terms of constrained path covering problems on acyclic graphs [8], [10]. However, the most popular approach is the one based on the set partitioning/covering model (SP/C), where each task must be covered by at least one duty. Several of these approaches exploit the network structure within a column generation phase, which can either be embedded within a primal branch and bound scheme [15] or within a Lagrangean dual context. Unfortunately, the related subproblem has to deal with union regulations which are usually cumbersome and vary locally. This may be the reason why many packages still resort to a once-for-all philosophy, where a subset of duties is first generated according to some qualitative criterion, and then an SP/C problem is optimally solved on this subset [18]. Dual heuristics for the SP/C based on Lagrangean relaxation have recently received some attention [1], [6] partly due to improved algorithms for the Lagrangean dual (see [25] and [9]).

2.3 The Joint Case

One of the most significant works on this subject, from Ball, Bodin & Dial [2], dates back to 1983 and deals with the single-depot urban case, where crews are not bound to vehicles. The structure of the problem is exploited through two different graphs sharing the same set of nodes, which model trips or portions of trips that must be operated by a single driver and a single vehicle. On each graph the arc set models vehicle/crew dead-headings, with different kinds of arc modelling different kinds of dead-headings. For the first time the joint problem is modelled as a feasible path cover on both graphs (here a *path cover* of a graph refers to a set of node disjoint paths covering all the nodes). However, two drawbacks prevent this model from yielding a practical approach: first, network dimensions are prohibitive; second, coherence between the two covers must be guaranteed. The authors resort to a problem decomposition, emphasising the crew scheduling phase when the service is initially partitioned into a set of crew pieces by heuristically solving a constrained path cover by a sequence of matching problems. This approach, suitable for an urban setting, would not straightforwardly apply to an extra-urban setting where the cost of a piece varies depending on the residence depot of the driving crew.

Since then, few proposals follow, such as [22] and [27], that reverse the sequential approach by scheduling crews first disregarding vehicles completely. Since scheduling crews directly on the trip set gives rise to a huge number of runs, a set covering problem is solved on a small subset, following a once-for-all scheme.

The second attempt after [2] to solve the joint problem by exploiting the network structure is described in [21], where vehicles and crews are explicitly considered in the same model. However, the authors tackle the single depot case, exploiting the fact that in such a case the VSP is polynomial within the solution of a Lagrangean dual.

In conclusion, no practical models and approaches have been proposed so far for the *extra-urban setting with multiple depots*, which is where the sequential approach most often fails, and which will be specifically addressed in this paper.

3. A Lagrangean Based Heuristic

Let us introduce our model by describing the joint problem in terms of networks. Consider the standard compatibility graph where nodes (representing trips) are linked by compatibility arcs. Connections with the depots are explicitly represented by introducing depot source and sink nodes connected to all the trip nodes. A block corresponds to a path from a source to a sink depot. The idea is to find a set of blocks covering the service such that this set can be partitioned into feasible vehicle duties as well as into feasible crew duties, all complying with some global constraints. Note that, since in this particular context vehicle blocks correspond to pieces of work, we shall refer to both simply as *pieces*. Moreover, crew duties will be referred to as *runs* while vehicle duties will be more simply referred to as *duties*.

3.1 Definitions and Notations

Two sets of variables are explicitly introduced in order to model runs and duties, while pieces are implicitly derived from variables modelling dh-trips. Let x_j be the Boolean variable associated with run j with cost c_j, $j \in J$, where J is the set of feasible runs. Let z_i be the Boolean variable associated with duty i with cost a_i, $i \in I$, where I is the set of feasible duties. Let $D = \{d^k\}$, $k = 1,..,K$, be the set of depots with capacity C^k, and let $I(k)$ be the set of duties having d^k as the residence depot. Let $V = \{v\}$ be the set of trips to be covered; let $J(v) \subseteq J$ ($I(v) \subseteq I$) be the set of runs (duties) covering trip v, and conversely let $V(j)$ ($V(i)$) be the set of trips covered by run j (duty i). We shall say that run j and run j' intersect each other if they share a common trip, i.e. $V(j) \cap V(j') \neq \emptyset$; likewise, duties i and i' intersect if $V(i) \cap V(i') \neq \emptyset$. Let $T = \{t\}$ be the set of the different types of run, let $J(t)$

be the set of runs of type t, and let R^t be the maximum number of runs of type t allowed; let us assume that $J(t) \cap J(t') = \emptyset \, \forall t \neq t'$. Moreover, let y_{uv} be the variable modelling the dh-trip from u to v which indicates whether trips u and v are operated in sequence by the same vehicle and the same driver, and let b_{uv} be its cost; for $u = d \in D$, $(v = d)$, let y_{dv} (y_{ud}) be the variable modelling the dh-trip from (to) depot d. Finally, let $J(u,v)$ $(I(u,v))$ be the subset of runs (duties) covering the dh-trip from e_u to s_v.

3.2 An Integer Linear Programming Model for the Joint Problem

The following ILP model fully describes our problem:

$$\min \sum_{j \in J} c_j x_j + \sum_{i \in I} a_i z_i + \sum_{(u,v) \in A'} b_{uv} y_{uv} \qquad \textbf{(P1)}$$

subject to :

$$\sum_{u \in V \cup D} y_{vu} = 1 \qquad \forall v \in V \tag{3.1}$$

$$\sum_{u \in V \cup D} y_{uv} = 1 \qquad \forall v \in V \tag{3.2}$$

$$\sum_{j \in J(uv)} x_j = y_u v \qquad \forall (u,v) \tag{3.3}$$

$$\sum_{i \in I(uv)} z_i = y_u v \qquad \forall (u,v) \tag{3.4}$$

$$\sum_{i \in I(k)} z_i \leq C^k \qquad \forall d_k \in D \tag{3.5}$$

$$\sum_{j \in J(t)} x_j \leq R^t \qquad \forall t \in T \tag{3.6}$$

$$x_j, z_i, y_{uv} \in \{0,1\} \qquad \forall j \in J, \forall i \in I, \forall (u,v) \in A'.$$

Constraints (3.1)-(3.2) require that for each trip $v \in V$ a driver drives a vehicle to s_v within st_v and leaves from e_v at time et_v, and correspond to the constraints of an assignment problem. Constraints (3.3)-(3.4) require that exactly one run and one duty (no runs and no duties) among the ones including the dh-trip from u to v belong to the schedule if and only if arc (u,v) does (does not) belong to the matching. Therefore, set partitioning constraints such as:

$$\sum_{j \in J(v)} x_j = 1 \text{ and } \sum_{i \in I(v)} z_i = 1 \qquad \forall v \in V$$

which are usually included in order to guarantee that the service is covered, are redundant here since they are surrogate constraints of (3.1)-(3.4), due to the fact that drivers are bound to vehicles.

Finally, constraints (3.5) and (3.6) model depot capacity and the maximum number of runs of type t allowed. As long as the different types of run in T are disjoint over J, as is the case when *split-shifts, straight-runs, and trippers* are concerned, these constraints can be implicitly handled as described in [CGN95] and will be disregarded hereafter. We shall refer to (**P1**) when disregarding (3.5)-(3.6) as problem (**P**).

This model allows a nice network representation. Consider the usual compatibility graph $G = (N, A)$, where N is isomorphic to V and $A = \{(u,v) : u \text{ is compatible with } v)\}$ where compatibility is now intended for both vehicles and drivers. Depot connections are modelled by adding additional node sets D' and D'', with $d'_k = n + k \in D'$ and $d''_k = n + K + k \in D''$ $\forall d_k \in D$, and additional arcs connecting each d'_k to each trip v, and each trip v to each d''_k, respectively.

Let $A^D = \cup_{k=1..K} \cup_{v \in V} \{(n+k, v), (v, n+K+k)\}$ be the set of additional arcs, and let this new graph be $G' = (N', A')$ with $N' = N \cup D \cup D'$ and $A' = A \cup A^D$. In Figure 3.1, G' is depicted, with arcs in A being dotted. Note that, if interlining is not allowed, as is usually the case, G' is much sparser than the usual compatibility graphs for MD-VSP. Actually, a threshold value on layover time and distance of the dh-trips is introduced in order to further limit the number of possible connections.

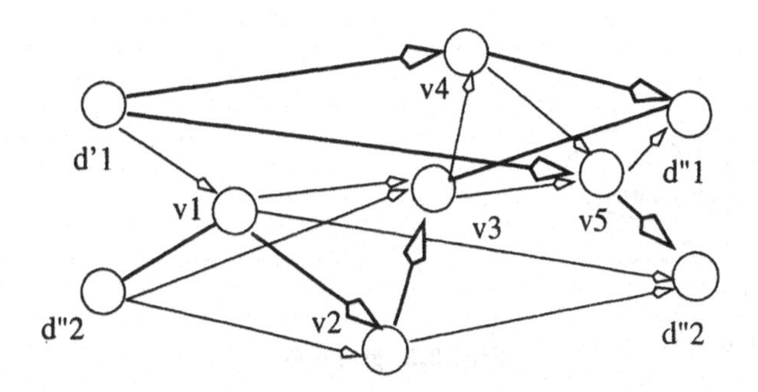

Fig. 3.1. graph G'

Let P be the set of paths on G' from D' to D'' that correspond to feasible pieces. Consider the graphs $G^V = (P, A^V)$ and $G^C = (P, A^C)$, where A^V is the set of arcs connecting $p, q \in P$ such that p ends in d''_k at time t_p and q starts at depot d'_h at time t_q with $h{=}k$ and $t_p < t_q$, that is to say p and q can be operated in sequence by the same vehicle. The set of arcs A^C is a subset of A^V since compatibility among pieces is restricted by union regulations. For example, if $t_p >$ 12:00 am and $t_q <$ 2:00 pm, then p and q are compatible only if the meal- break can take place in between p and q.

Each $(\mathbf{y}, \mathbf{x}, \mathbf{z})$ feasible for (**P**) induces:

- $P(\mathbf{y}) \subseteq P$ a set of paths on G' from D' to D'' corresponding to a partition of the trip set into pieces;
- $G^V(P(\mathbf{y}))$ subgraph of G^V;

- $G^C(P(\mathbf{y}))$ subgraph of G^C;
- a partition of $P(\mathbf{y})$ into a set of feasible paths on $G^V(P(\mathbf{y}))$ according to the value of $\mathbf{z}$;
- a partition of $P(\mathbf{y})$ into a set of feasible paths on $G^C(P(\mathbf{y}))$ according to the value of $\mathbf{x}$.

Note that once $\mathbf{y}$ has been given (**P**) reduces to two set partitioning problems.

In Figure 3.2 subgraph $G^V(P(\mathbf{y}))$ is depicted where $\mathbf{y}$ is the solution shown in bold in Figure 3.1; subgraph $G^C(P(\mathbf{y}))$ is shown in Figure 3.3. Note that because of union restrictions the piece corresponding to (d_1, v_4, d_1) cannot be operated by the same crew staffing piece (d_1, v_5, d_2).

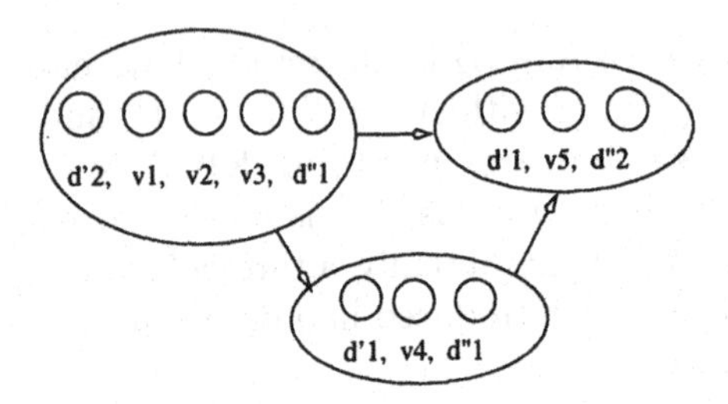

Fig. 3.2. graph G^V

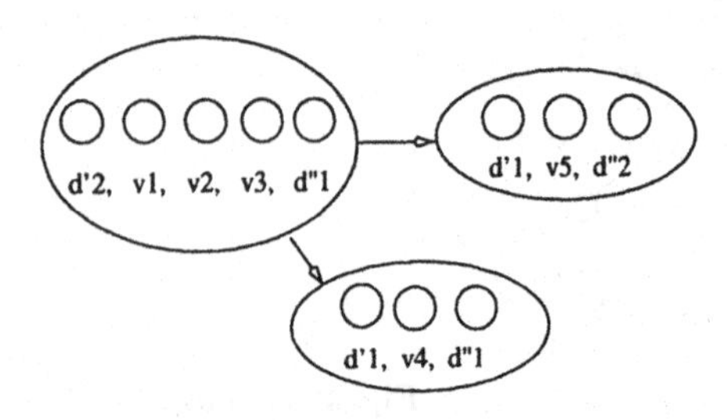

Fig. 3.3. graph G^C

A few remarks on these two graphs: first note that G^V and G^C differ from other *vehicle graphs* and *crew graphs* described in the literature since here nodes model pieces of work rather than trips. On the other hand, there is a tighter link with the *crew duty graph* of [8] and [10], although only subgraphs of G^V and G^C will be explicited in our approach. Moreover, performing column generation as constrained shortest paths is much easier on G^C than directly on the compatibility graph since the complexity dramatically increases with the number of nodes of the paths [10], [17].

The Lagrangean relaxation of constraints (3.5)-(3.6) leads to the following:

(LRP) $\Phi(\lambda, \mu) =$

$$\min \sum_{j \in J} x_j \,(c_j - \sum_{(u,v) \in A'(j)} \lambda_{uv}) + \sum_{i \in I} z_i \,(a_i - \sum_{(u,v) \in A'(i)} \mu_{uv})$$

$$+ \sum_{(u,v)\in A'} y_{uv} \left(b_{uv} + \lambda_{uv} + \mu_{uv}\right)$$

subject to :

$$\sum_{u\in N'} y_{vu} = 1 \quad \sum_{u\in N'} y_{vu} = 1 \qquad \forall v \in V$$

$$x_j, z_i, y_{uv} \in \{0,1\} \qquad \forall j \in J,\ \forall i \in I,\ \forall (u,v) \in A'.$$

In order to simplify the notation let us define the following *reduced costs* as

$$\begin{aligned} \bar{c}_j = &\ \left(c_j - \textstyle\sum_{(u,v)\in A'(j)} \lambda_{uv}\right) \quad \forall j \in J \\ \bar{a}_i = &\ \left(a_i - \textstyle\sum_{(u,v)\in A'(i)} \mu_{uv}\right) \quad \forall i \in I \\ \bar{b}_{uv} = &\ \left(b_{uv} + \lambda_{u,v} + \mu_{u,v}\right) \qquad \forall (u,v) \in A'. \end{aligned}$$

The evaluation of the Lagrangean function $\Phi(\lambda,\mu)$ reduces to the solution of an assignment on G' with respect to reduced costs $\bar{b}_{uv}$ and to the listing of variables x_j and z_i with negative reduced costs. The size of J and I calls for a column generation scheme to be applied within the solution of the Lagrangean dual (**D**) $= \{\max_{\lambda,\mu} \Phi(\lambda,\mu)\}$ but, since in this case the size of set P is critical too, then also P will be dynamically handled. $\Phi(\lambda,\mu)$ is maximized via a bundle algorithm for which we refer to [9].

For each value of λ, μ let $\boldsymbol{y}^*, \boldsymbol{x}^\star, \boldsymbol{z}^*$ be any optimal solution to $\Phi(\lambda,\mu)$: a vector $[g_\lambda, g_\mu] \in \mathbb{R}^{2A'}$ such that

$$g_{\lambda_{uv}} = y^*_{uv} - \sum_{j\in J(uv)} x^*_j \quad \text{and} \quad g_{\mu_{uv}} = y^*_{uv} - \sum_{i\in I(uv)} z^*_i$$

is a subgradient of the function in (λ,μ). Because of the integrality property of the relaxed problem, the optimal multipliers λ^*, μ^* – for which a null subgradient does exist – define a set of pieces (the solution to the assignment problem with respect to λ^*, μ^*) and a set of duties and runs with zero reduced costs that belong to the optimal solution of the continuous relaxation of (**P**).

3.3 A Heuristic Procedure Based on Lagrangean Relaxation

Let us first describe our solution method assuming that we can handle the whole set P. A heuristic procedure computes a list of feasible crew schedules, each obtained by fixing runs in a greedy fashion guided by the values of the Lagrangean multipliers. During this process, vehicle duties are also made explicit and they influence the selection of runs by means of the Lagrangean multipliers. A feasible vehicle schedule is then devised on top of the pieces of the best computed crew schedule. Although we cannot by any means certify the (sub)optimality of such a solution other than by considering the gap with respect to the value of the Lagrangean function which provides a lower bound, it is a widespread opinion that for this class of SP/C problems many almost-equivalent and nearly-optimal solutions exist and tailored dual heuristics pay off because of the large duality gap [6], [1], [19].

3.3.1 Computing a Single Crew Schedule. Let us suppose that we apply a column generation scheme in a dual context, where the master problem at iteration h applies the bundle algorithm to optimize $\Phi(\lambda, \mu)$ over $J_h \subseteq J$ and $I_h \subseteq I$ yielding (λ_h^*, μ_h^*), and the subproblem generates runs and duties with negative reduced costs (if any) with respect to (λ_h^*, μ_h^*) as constrained shortest paths on G^C and G^V, respectively. The sequence $\{(\lambda_h^*, \mu_h^*)\}$ will eventually converge to the actual optimal multipliers (λ^*, μ^*). Let $(\tilde{\lambda}, \tilde{\mu})$ indicate the multipliers at some step of this process after a stabilising phase - again, we are not so interested in (λ^*, μ^*) rather than in *good* multipliers because of the duality gap and only require that $(\tilde{\lambda}, \tilde{\mu}) \simeq (\lambda^*, \mu^*)$. Finally, let $\tilde{J}$ be the set of active variables (i.e. $\bar{c}_j \geq 0 \, \forall j \in \tilde{J}$) at that stage, and let $\tilde{P}^C$ and $\tilde{P}^V$ be the set of nodes in G^C and G^V, respectively.

A feasible crew schedule is iteratively computed in a greedy fashion, which closely relates to a family of Lagrangean based heuristics which were mainly developed for Set Covering problems, although here we combine it with a column generation scheme. At each step of our procedure we solve a subproblem in a subspace of the previous dual space as well as in a subspace of the primal space. This is achieved by setting a subset of the Lagrangean multipliers to their current values and by restricting ourselves to some subsets of the node sets $\tilde{P}^C$ and $\tilde{P}^V$ of the graphs where columns are generated as constrained shortest paths.

At each step of the greedy procedure a set of runs is selected and scheduled in the following way: $\tilde{J}$ is sorted according to non-increasing reduced costs and then scanned in that order up to a threshold value. When j is selected, x_j is fixed to 1 and added to the partial solution, until run j' has been found such that j' intersects a run which has already been selected. Then j' is set to 0 and the search is stopped. Note that for this partial solution, complementary slackness holds with respect to the current multipliers.

Let J^+ be the set of runs that have been scheduled so far, P^+ the set of pieces of the runs in J^+ and $V^+ = \cup_{j \in J^+} V(j)$ the set of trips covered by the runs in J^+. Since each trip must be covered by exactly one run and one duty, it must be guaranteed that within the column generation process no run intersecting with runs in J^+ will ever be generated. This is achieved by deleting from $\tilde{P}^C$ all pieces intersecting with pieces in P^+, P^+ included.

As far as the vehicle schedule is concerned, fixing a partial solution for the crew scheduling problem simply restricts the set of pieces that can appear in the vehicle schedule; this is achieved by deleting from $\tilde{P}^V$ all the pieces that intersect with pieces in P^+ except for P^+ itself. Moreover, graph G' is accordingly modified by deleting all the arcs incident to nodes in V^+ other than the arcs corresponding to the dh-trips included in the partial solution.

As a result of this fixing operation, the value of the Lagrangean multipliers of the arcs involved is fixed to their current value and will not be further modified, thus defining a subproblem both in the primal as well as in the dual space. $\Phi(\lambda, \mu)$ is again approximately optimized in the resulting subspace performing column generation with respect to the updated sets $\tilde{P}^C$ and $\tilde{P}^V$, thus yielding a different point $(\tilde{\lambda}, \tilde{\mu})$. Then the process is repeated until all trips have been covered, yielding a feasible crew schedule $\boldsymbol{x}^F$, a set of dh-trips $\boldsymbol{y}^F$, and a set of multipliers (λ^F, μ^F).

The idea of computing a lower bound for the Set Covering problem by way of a dual heuristic, instead of solving the linear relaxation and of using reduced costs within a greedy primal heuristic, has recently received much attention [11], [6], [1], and it is exploited in depth in [19] where the mixed problem, Covering and Partitioning, is addressed. Nevertheless, in most studies the cardinality of the primal variables set, although considerable, can still be treated, and the computation of a primal feasible solution is carried out with respect to a feasible dual solution whose value does not change during the process. On the other hand, we are embedded in a column generation scheme where only a subset of the primal variables is made explicit. Therefore, once a primal partial solution has been set, the optimal continuous solution of the resulting subproblem might involve variables that have not been generated and that correspond to a different point in the dual space according to which their reduced costs are zero. We suggest that this situation may happen as soon as the currently selected run intersects with runs in the partial solution, which is the point where we update our primal and dual solution by way of solving a related subproblem, as previously mentioned.

Complementary slackness obviously holds for $\boldsymbol{x}^F$, $\boldsymbol{y}^F$, and λ^F since runs in the current solution all have a zero reduced cost with respect to λ^F. However, λ^F may be infeasible, since other runs in J may have negative reduced cost, which in turn suggests that the current integer solution could be improved. If on the one hand the duality gap would partly explain it, on the other hand this is mainly due to the lack of feed-back of the greedy procedure. For this reason the process is repeated a few times in order to collect several solutions, from which the best is eventually selected. Nevertheless, since each step is rather time consuming, only a subset of the current solution is put back into play according to the following criterion. Starting from (λ^F, μ^F), the procedure for maximizing $\Phi(\lambda, \mu)$ is applied by performing column generation on the whole primal space. This yields a new sub-optimal point (λ', μ'), which differs form $(\tilde{\lambda}, \tilde{\mu})$ as will be explained later. Let J^F be the set of runs of the current schedule: their reduced cost is evaluated with respect to (λ', μ') – let us call it $\bar{c}'_j$ – to identify $J' = \{j \in J^F : \bar{c}'_j \geq \epsilon > 0\}$ the set of runs whose reduced cost has increased over a threshold ϵ. The partition on $V(J') = \cup_{j \in J'} V(j)$, where $V(J')$ is the set of trips covered by runs in J', is brought back into play.

3.3.2 A Modified Column Generation Scheme. In the previous section, at certain steps of the algorithm column generation is assumed to be performed on the whole primal space, which entails computing shortest paths on graph G^V and constrained shortest paths on G^C. While the first is straightforward, the second is prohibitive, which suggests a dynamic strategy.

At each step a list of active pieces P', runs J', and duties I' is maintained. Periodically, pieces p not in P' resulting from the solution of the assignment are added to P' and column generation is performed to add runs to J' and duties to I' involving piece p. Each p has a label representing its contribution to the last evaluations of the Lagrangean function. This label is progressively decreased besides being set to the initial value, whenever p belongs either to the assignment or to duties and runs in J' or in I' with negative reduced cost. An overflow control based on reduced cost keeps J' and I' within a fixed size.

This strategy differentiates our procedure from classical column generation schemes since in our case master problem and subproblem intertwine, thus causing the polyhedral description of the Lagrangean function to dynamically vary according to the updating of P', J', and I'.

This procedure is integrated with the algorithm described in the previous section, thus introducing diversification in the outcomes of the different iterations. This fact can be seen as a dual variant of cost perturbation which is quite popular in heuristic approaches to SP/C problems. In fact, when it is almost impossible to compute the actual cost of each single run and the main target is to cover the service with the minimum number of crews, the cost function does not discriminate among feasible runs and the problem is heavily degenerate, which is when cost perturbation techniques are most exploited. Also, note that knowledge of the cost of a previous primal feasible solution can be exploited within Bundle algorithms, since it provides an upper bound to the value of $\Phi(\lambda, \mu)$ which improves the convergence rate of the algorithm.

3.4 A Feasible Solution for the Vehicle Scheduling Problem

Once all the pieces of work induced by the crew schedule solution have been set, vehicle duties must be devised. Note that our problem is slightly different from a usual MD-VSP. Since in our case pieces coincide with blocks, blocks are already defined. Consider the compatibility graph extended with depot connections depicted in Figure 3.1, where now each node corresponds to a block. Note that, for each block, the starting depot and the ending depot are univocally determined. Therefore, compatibility holds between two nodes only if the associated blocks share the same connecting depot.

The block compatibility graph extended with depot connections is depicted in Figure 3.4. Note how compatibility is affected by depot connections. A feasible vehicle duty is a path such that the ending depot of the last node is the same as the starting depot of the first node. The resulting network is much smaller than the trip compatibility graph in the original MD-VSP and very sparse too, potentially allowing for a good performance of exponential solution algorithms. On the other hand the problem might be unfeasible, i.e. a disjoint feasible path cover satisfying depot capacities might not exist – note that the crew schedule itself provides a feasible solution when it meets capacity constraints. According to our experience, a feasible solution

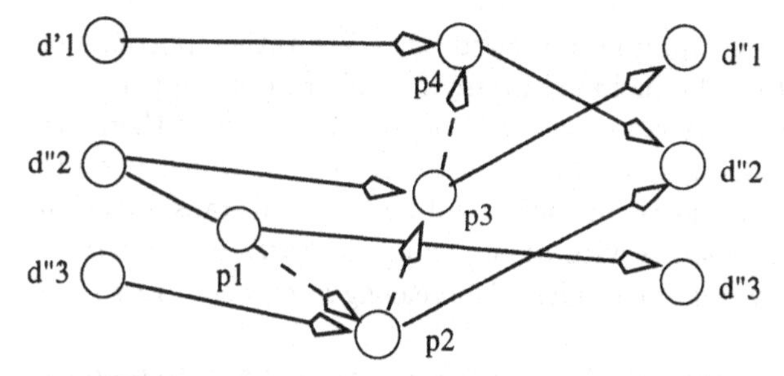

Fig. 3.4. block compatibility graph extended with depot connections

to the MD-VSP can be straightforwardly reached by solving a set partitioning problem with side constraints, with column generation performed on this network.

4. Computational Results

The integrated approach was implemented in *C++*C and run on a *Power PC 604*. In order to compare it with the sequential approach, our prototype was tested against the commercial package MTRAM©, which is currently in use at a few Italian municipal mass transit companies. MTRAM contains two applications specific for scheduling purposes, namely VS-Alg©and BDS-Alg©; the former solves MD-VSP while the latter addresses CSP. VS-Alg and BDS-Alg were used in pipeline in order to implement the sequential approach. The two approaches were tested on two different sets of data, *S1* and *S2*, which are of a different nature. Set *S1* is a collection of samples of typical extra-urban service, while set *S2* concerns mixed urban and extra-urban service. The aim of this testing was to compare the performance of the two approaches in these two different settings, in order to identify under which conditions which of the two is more suitable, if any. In the extra-urban setting as well as in the urban setting, union constraints that define feasible runs mainly concern bounds on spread time and driving time, and involve meal-breaks. However, a new type of run is peculiar to the extra-urban setting, the so-called extended spread-time run, together with a global constraint allowing no more than 20% of extended runs over the set of runs to be assigned to each depot. We found that this new feature remarkably increases the computational burden.

In Tables 4.1 and 4.2, the number of trips is shown for each sample, D-Seq and R-Seq stand for the number of duties and runs obtained by MTRAM, while D-Seq and R-Seq refer to the number of duties and runs obtained by our prototype. When MTRAM failed to provide a feasible solution, some trippers were produced and their number is reported in the relative column. Actually, these do not necessarily correspond to the usual trippers, which are runs much shorter than the minimum allowed; they are rather portions of vehicle duties that could not be included in any feasible run or tripper.

Set *S1* refers to the extra-urban service of Bologna, Italy, where there are 28 depots at an average distance of 25 *km* between depots, which is quite typical for Italy. Here, MTRAM is currently used by the local transportation company to schedule the urban service alone, while the extra-urban service is still scheduled manually. Eight samples of increasing sizes were drawn out of the whole trip set.

Let us comment the results in Table 4.1: MTRAM was not able to produce a feasible crew schedule in five tests out of eight, while our prototype was. For example, in test 1 VS-Alg yielded four duties which could not be staffed: two contained tasks too long to be included in a piece and the other two had tasks covering the meal-break period. Moreover, note that, when MTRAM is able to solve the problem as in tests 3, 5 and 6 our approach solves it as well or even outperforms MTRAM, since it yields the same number of duties and saves few runs.

These results encourage the use of an integrated approach in the extra-urban setting, although we should mention that running time is still critical for the prototype we implemented, since it varies from two to six times more than MTRAM, which in turn ranges from 2 to 6 hours on a *Power PC 604, 180Mh*.

Table 4.1. Samples from set S1

test	Trips	D-Seq	R-Seq	Tripper	D-Int	R-Int
1	60	15	17	6	15	21
2	79	21	31	4	21	33
3	117	41	27		41	27
4	125	18	24	3	20	23
5	139	23	30		23	28
6	180	27	35		27	33
7	200	41	58	4	44	65
8	257	34	49	2	36	45

As mentioned before, the two approaches were also tested on a mixed urban and extra-urban service. Set *S2* refers to a case with four residence depots plus ten accounted relief points and more than one-thousand trips. Six sets of data were drawn out of it, considering not necessarily disjoint subsets of the whole service with trips ranging from 120 to 135. In these tests, the characteristics of the samples are further described in terms of the number of pairs of compatible trips, which corresponds to the cardinality of set A, and the number of allowed depot connections, which corresponds to the size of set A^D. We noticed a positive correlation among the incidence of urban service in the sample and the number of pairs of compatible trips, which is plausible since in the urban setting the service is concentrated in a relatively small geographical area and travel distance between trips is low, on the average.

Table 4.2. Samples from set S2

test	Trips	A	A^D	D-Seq	R-Seq	D-Int	R-Int
1	121	2188	630	23	41	23	42
2	130	608	1475	23	45	24	44
3	110	821	703	21	40	21	39
4	135	681	1320	25	47	27	45
5	127	1147	609	24	42	24	42
6	120	871	806	22	43	23	42

In test 2, 3, 4 and 6 of Table 4.2 our prototype saves a few runs at the cost of introducing additional vehicles, while in tests 1 and 5 it equals the result of the sequential approach or even worsens it. This may be explained by the fact that these two samples include a relevant percentage of urban service, for which the integrated approach is not suitable. In fact, the number of compatible trips has a combinatorial effect on the number of pieces unless blocks are already defined, as it happens in the sequential approach where vehicles are scheduled first, since in that case pieces must be contained inside blocks.

In conclusion, we can state that in the extra-urban setting specific tools are required to solve scheduling problems. The integrated approach here described seems to be suitable for this specific purpose although computational time is still an issue. Moreover, further research should tackle the case of mixed urban and extra-urban service, for which none of the two tested approaches proved fully successful.

References

1. E. Balas, M. C. Carrera. A dynamic subgradient based branch and bound procedure for set covering. *Operations Research*, 44:875–890, 1996.
2. M. Ball, L. Bodin, R. Dial. A matching based heuristic for scheduling mass transit crews and vehicles. *Transportation Science*, 17:4–13, 1983.
3. M. Ball, H. Benoit-Thompson. A lagrangean relaxation based heuristic for the urban transit crew scheduling problem. J. R. Daduna, A. Wren Eds, Springer Verlag, Berlin Heidelberg, 54–67, 1988.
4. J. E. Beasley. A lagrangean heuristic for set-covering problems. *Naval Research Logistics*, 37:151–164, 1990
5. B. Bertossi, P. Carraresi, G. Gallo. On some matching problems arising in vehicle scheduling models. *Networks*, 17:271–281, 1987.

6. A. Caprara, M. Fischetti, P. Toth. A heuristic procedure for the set covering problem. *Operations Research*, (to appear), 1997.
7. G. Carpaneto, M. Dell'Amico, M. Fischetti, P. Toth. A branch and bound algorithm for the multi depot vehicle scheduling problem. *Networks*, 19:531–548, 1989.
8. P. Carraresi, G. Gallo. Network models for vehicle and crew scheduling. *European Journal of Operational Research*, 38:121–150, 1991.
9. P. Carraresi, A. Frangioni, M. Nonato. Applying bundle methods to the optimization of polyhedral functions. An application-oriented development. *Ricerca Operativa*, 25:4–49, 1995.
10. P. Carraresi, L. Girardi, L. Nonato. Network models, lagrangean relaxation and subgradient bundle approach to crew scheduling problems. J. R. Daduna, I. Branco, J. P. Paixão Eds, Springer Verlag, Berlin, Heidelberg, 188–212, 1995.
11. S. Ceria, P. Nobili, A. Sassano. A lagrangean-based heuristic for large scale set covering problems. Report # 406, University of Roma, La Sapienza, Roma, Italy, 1995.
12. J. R. Daduna, A. Wren. Computer aided transit scheduling. Lecture notes in Economics and Mathematical Systems 308, Springer Verlag, Berlin, Heidelberg, 1988.
13. J. R. Daduna, I. Branco, J. P. Paixão. Computer aided transit scheduling. Lecture notes in Economics and Mathematical Systems 430, Springer Verlag, Berlin, Heidelberg, 1995.
14. J. R. Daduna, J. P. Paixão. Vehicle scheduling for public mass transit: an overview. J. R. Daduna, I. Branco, J. P. Paixão Eds. Springer Verlag, Berlin, Heidelberg, 76–90, 1995.
15. M. Desrochers, E. Soumis. A column generation approach to the urban transit crew-scheduling problem. *Transportation Science*, 23:1–13, 1989.
16. M. Desrochers, E. Soumis. Computer aided transit scheduling. Lecture notes in Economics and Mathematical Systems 386, Springer Verlag, Berlin, Heidelberg, 1992.
17. J. Desrosiers, Y. Dumas, M. M. Solomon, E. Soumis. Time constrained routing and scheduling. Les Cahiers du GERAD G-92-42, Ecole des Hautes Etudes Commerciales, Montreal, Canada, 1993.
18. J. C. Faulkner, D. M. Ryan. Express: set partitioning for bus crew scheduling in Christchurch. M. Desrochers, J.-M. Rousseau Eds, Springer Verlag, Berlin, Heidelberg, 358–378, 1992.
19. M. Fisher, P. Kedia. Optimal solution of set covering / partitioning problems using dual heuristics. *Management Science*, 36:674–688, 1990.
20. M. A. Forbes, J. N. Holt, A. M. Watts. An exact algorithm for multiple depot bus scheduling. *European Journal of Operational Research*, 72:115–124, 1994.
21. R. Freling, J. P. Paixão. An integrated approach to vehicle and crew scheduling. Proceedings of TRISTAN II, Capri, 319–333, 1994.
22. I. Patrikalakis, D. Xerocostas. A new decomposition scheme for the urban public transport scheduling problem. M. Desrochers J.-M. Rousseau Eds, Springer Verlag, Berlin, Heidelberg, 407–425, 1992.
23. C. Ribeiro, E. Soumis. A column generation approach to the multiple depot vehicle scheduling problem. *Operations Research*, 42:41–52, 1991.
24. J.-M. Rousseau. *Computer scheduling of public transport 2.* North-Holland, Amsterdam, 1985.
25. H. Schramm, J. Zowe. A version of the bundle idea for minimizing a nonsmooth function: conceptual idea, convergence analysis, numerical results. *SIAM Journal of Optimization*, 2:121–152, 1992.

26. D. Scott. A large scale linear programming approach for the public transport scheduling and costing problem. J.-M. Rousseau Ed., North-Holland, Amsterdam, 473–491, 1985.
27. E. Tosini, C. Vercellis. An interactive system for the extra urban vehicle and crew scheduling problems. J. R. Daduna, A. Wren Eds, Springer Verlag, Berlin, Heidelberg, 41–53, 1988.

Index

NATO ASI Series F

Including Special Programmes on Sensory Systems for Robotic Control (ROB) and on Advanced Educational Technology (AET)

Vol. 122: Simulation-Based Experiential Learning. Edited by D. M. Towne, T. de Jong and H. Spada. XIV, 274 pages. 1993. *(AET)*

Vol. 123: User-Centred Requirements for Software Engineering Environments. Edited by D. J. Gilmore, R. L. Winder and F. Détienne. VII, 377 pages. 1994.

Vol. 124: Fundamentals in Handwriting Recognition. Edited by S. Impedovo. IX, 496 pages. 1994.

Vol. 125: Student Modelling: The Key to Individualized Knowledge-Based Instruction. Edited by J. E. Greer and G. I. McCalla. X, 383 pages. 1994. *(AET)*

Vol. 126: Shape in Picture. Mathematical Description of Shape in Grey-level Images. Edited by Y.-L. O, A. Toet, D. Foster, H. J. A. M. Heijmans and P. Meer. XI, 676 pages. 1994.

Vol. 127: Real Time Computing. Edited by W. A. Halang and A. D. Stoyenko. XXII, 762 pages. 1994.

Vol. 128: Computer Supported Collaborative Learning. Edited by C. O'Malley. X, 303 pages. 1994. *(AET)*

Vol. 129: Human-Machine Communication for Educational Systems Design. Edited by M. D. Brouwer-Janse and T. L. Harrington. X, 342 pages. 1994. *(AET)*

Vol. 130: Advances in Object-Oriented Database Systems. Edited by A. Dogac, M. T. Özsu, A. Biliris and T. Sellis. XI, 515 pages. 1994.

Vol. 131: Constraint Programming. Edited by B. Mayoh, E. Tyugu and J. Penjam. VII, 452 pages. 1994.

Vol. 132: Mathematical Modelling Courses for Engineering Education. Edited by Y. Ersoy and A. O. Moscardini. X, 246 pages. 1994. *(AET)*

Vol. 133: Collaborative Dialogue Technologies in Distance Learning. Edited by M. F. Verdejo and S. A. Cerri. XIV, 296 pages. 1994. *(AET)*

Vol. 134: Computer Integrated Production Systems and Organizations. The Human-Centred Approach. Edited by F. Schmid, S. Evans, A. W. S. Ainger and R. J. Grieve. X, 347 pages. 1994.

Vol. 135: Technology Education in School and Industry. Emerging Didactics for Human Resource Development. Edited by D. Blandow and M. J. Dyrenfurth. XI, 367 pages. 1994. *(AET)*

Vol. 136: From Statistics to Neural Networks. Theory and Pattern Recognition Applications. Edited by V. Cherkassky, J. H. Friedman and H. Wechsler. XII, 394 pages. 1994.

Vol. 137: Technology-Based Learning Environments. Psychological and Educational Foundations. Edited by S. Vosniadou, E. De Corte and H. Mandl. X, 302 pages. 1994. *(AET)*

Vol. 138: Exploiting Mental Imagery with Computers in Mathematics Education. Edited by R. Sutherland and J. Mason. VIII, 326 pages. 1995. *(AET)*

Vol. 139: Proof and Computation. Edited by H. Schwichtenberg. VII, 470 pages. 1995.

Vol. 140: Automating Instructional Design: Computer-Based Development and Delivery Tools. Edited by R. D. Tennyson and A. E. Barron. IX, 618 pages. 1995. *(AET)*

Vol. 141: Organizational Learning and Technological Change. Edited by C. Zucchermaglio, S. Bagnara and S. U. Stucky. X, 368 pages. 1995. *(AET)*

Vol. 142: Dialogue and Instruction. Modeling Interaction in Intelligent Tutoring Systems. Edited by R.-J. Beun, M. Baker and M. Reiner. IX, 368 pages. 1995. *(AET)*

Vol. 143: Batch Processing Systems Engineering. Fundamentals of Chemical Engineering. Edited by G. V. Reklaitis, A. K. Sunol, D. W. T. Rippin, and Ö. Hortaçsu. XIV, 868 pages. 1996.

Vol. 144: The Biology and Technology of Intelligent Autonomous Agents. Edited by Luc Steels. VIII, 517 pages. 1995.

NATO ASI Series F

Including Special Programmes on Sensory Systems for Robotic Control (ROB) and on Advanced Educational Technology (AET)

Vol. 145: Advanced Educational Technology: Research Issues and Future Potential. Edited by T. T. Liao. VIII, 219 pages. 1996. *(AET)*

Vol. 146: Computers and Exploratory Learning. Edited by A. A. diSessa, C. Hoyles and R. Noss. VIII, 482 pages. 1995. *(AET)*

Vol. 147: Speech Recognition and Coding. New Advances and Trends. Edited by A. J. Rubio Ayuso and J. M. López Soler. XI, 505 pages. 1995.

Vol. 148: Knowledge Acquisition, Organization, and Use in Biology. Edited by K. M. Fisher and M. R. Kibby. X, 246 pages. 1996. *(AET)*

Vol. 149: Emergent Computing Methods in Engineering Design. Applications of Genetic Algorithms and Neural Networks. Edited by D.E. Grierson and P. Hajela. VIII, 350 pages. 1996.

Vol. 150: Speechreading by Humans and Machines. Edited by D. G. Stork and M. E. Hennecke. XV, 686 pages. 1996.

Vol. 151: Computational and Conversational Discourse. Burning Issues – An Interdisciplinary Account. Edited by E. H. Hovy and D. R. Scott. XII, 202 pages. 1996.

Vol. 152: Deductive Program Design. Edited by M. Broy. IX, 467 pages. 1996.

Vol. 153: Identification, Adaptation, Learning. Edited by S. Bittanti and G. Picci. XIV, 553 pages. 1996.

Vol. 154: Reliability and Maintenance of Complex Systems. Edited by S. Özekici. XI, 589 pages. 1996.

Vol. 155: Cooperation: Game-Theoretic Approaches. Edited by S. Hart and A. Mas-Colell. VIII, 328 pages. 1997.

Vol. 156: Microcomputer-Based Labs: Educational Research and Standards. Edited by R.F. Tinker. XIV, 398 pages. 1996. *(AET)*

Vol. 157: Logic of Computation. Edited by H. Schwichtenberg. VII, 396 pages. 1997.

Vol. 158: Mathematical Methods in Program Development. Edited by M. Broy and B. Schieder. VIII, 528 pages. 1997.

Vol. 159: Fractal Image Encoding and Analysis. Edited by Y. Fisher. XIV, 368 pages. 1998.

Vol. 160: Discourse, Tools, and Reasoning: Essays on Situated Cognition. Edited by L.B. Resnick, R. Säljö, C. Pontecorvo and B. Bunge. XII, 474 pages. 1997. *(AET)*

Vol. 161: Computational Methods in Mechanical Systems. Edited by J. Angeles and E. Zakhariev. X, 425 pages. 1998.

Vol 162: Computational Intelligence: Soft Computing and Fuzzy-Neuro Integration with Applications. Edited by O. Kaynak, L.A. Zadeh, B. Türkşen, I.J. Rudas. IX, 538 pages. 1998.

Vol. 163: Face Recognition: From Theory to Applications. Edited by H. Wechsler, P.J. Phillips, V. Bruce, F. Fogelman Soulié, T.S. Huang. IX, 626 pages. 1998.

Vol. 164: Workflow Management Systems and Interoperability. Edited by A. Doğaç, L. Kalinichenko, M.T. Özsu, A. Sheth. XVII, 481 pages. 1998.

Vol. 166: Operations Research and Decision Aid Methodologies in Traffic and Transportation Management. Edited by M. Labbé, G. Laporte, K. Tanczos and P. Toint. XIII, 339 pages. 1998.

Vol. 167: Neural Circuits and Networks. Edited by V. Torre and J. Nicholls. VIII, 247 pages, 1998.

Vol. 168: Neural Networks and Machine Learning. Edited by Ch.M. Bishop. XII, 353 pages, 1998.